Specificity
in Plant Diseases

NATO ADVANCED STUDY INSTITUTES SERIES

A series of edited volumes comprising multifaceted studies of contemporary scientific issues by some of the best scientific minds in the world, assembled in cooperation with NATO Scientific Affairs Division.

The series is published by an international board of publishers in conjunction with NATO Scientific Affairs Division

A	Life Sciences	Plenum Publishing Corporation
B	Physics	New York and London
C	Mathematical and Physical Sciences	D. Reidel Publishing Company Dordrecht and Boston
D	Behavioral and Social Sciences	Sijthoff International Publishing Company Leiden
E	Applied Sciences	Noordhoff International Publishing Leiden

Specificity in Plant Diseases

Edited by

R. K. S. Wood, F.R.S.

Imperial College of Science and Technology
London, England

and

A. Graniti

University of Bari
Bari, Italy

SPRINGER SCIENCE+BUSINESS MEDIA, LLC

Library of Congress Cataloging in Publication Data

NATO Advanced Study Institute on Specificity in Plant Diseases, Alghero, Sardinia, 1975.

Specificity in plant diseases.

(NATO advanced study institutes series: Series A, Life sciences; v. 10)
"Lectures presented at the NATO Advanced Study Institute on Specificity in Plant Diseases, held in Porto Conte, nr. Alghero (Sardinia), Italy, May 4-17, 1975."
1. Plant diseases—Congresses. 2. Host—parasite relationships—Congresses. I. Wood, R. K. S. II. Graniti, A. III. Title. IV. Series. [DNLM: 1. Plant diseases—Congresses. SB731 Nlls 1975]

SB731.N28 1975	581.2'32	76-10644

ISBN 978-1-4684-2771-4   ISBN 978-1-4684-2769-1 (eBook)
DOI 10.1007/978-1-4684-2769-1

Lectures presented at the NATO Advanced Study Institute on Specificity in Plant Diseases, held in Porto Conte, nr. Alghero (Sardinia), Italy, May 4-17, 1975

© 1976 Springer Science+Business Media New York
Originally published by Plenum Press New York in 1976
Softcover reprint of the hardcover 1st edition 1976

Preface

A NATO Advanced Study Institute on "Phytotoxins in Plant Diseases" was held in Pugnochiuso (Italy) in June 1970. It was concerned mainly with the chemistry and mode of action of substances toxic to higher plants which are produced by pathogenic bacteria and fungi. The role of such substances in specificity was considered but largely in relation to host-specific toxins.

In 1973, in light of the success of the 1970 Institute, we decided to plan for another in 1975 and after discussion with colleagues during the Second International Congress of Plant Pathology, we selected "Specificity in Plant Diseases" as the theme for the 1975 Institute.

Our chief reasons for the choice were that specificity is undoubtedly the dominant problem in plant pathology and that with the rapid increase during the last decade or so in the diversity and sophistication of biochemical techniques, we should now expect during the next few years much more research on specificity and major advances in our understanding of the mechanisms that control it. It seemed to us that a residential meeting with the advantages and status of a NATO Advanced Study Institute would do much to stimulate interest and research in this most important field. The theme also had the merit of continuity with that of the 1970 Institute. Our application to the Scientific Affairs Division of NATO for a grant to cover most of the costs of the meeting was successful and we decided that it should be held at Porto Conte nr. Alghero (Sardinia), Italy on 4 - 17 May 1975.

At first our programme included viruses as plant pathogens on the same scale as bacteria and fungi but we soon decided that to do so would unduly widen the scope of the meeting. Instead we aimed to include no more than one lecture which would review specificity in plant diseases caused by viruses. Unfortunately, this plan for one lecture had to be abandoned at a late date. We were, however, fortunate in that early in our planning Dr. R.C. Hughes accepted our invitation to lecture on animal membranes as recognition sites, and then Professor H.F. Linskens agreed to survey specific interactions in higher plants. The other lectures which were the bulk of the programme dealt with almost every aspect of specificity of interactions between pathogenic bacteria and fungi and higher plants and were given by distinguished workers in this field. We are particularly indebted to all lecturers for their contributions upon which the many meetings and discussions were based and which comprise the main part of this book. We were also very grateful to those whom we were unable to include as main lecturers but who accepted our invitation to be both Chairmen of meetings and Discussion Leaders, and who later summarized the discussions and short papers which followed each main lecture; these summaries are also in this book.

The organization and administration of an Advanced Study Institute would be impossible without the help of innumerable colleagues and assistants. We thank each of them and we regret that we can name only a few. First and foremost we thank Dr. R.D. Durbin who helped us for many months before the Institute and then during the meetings. We thank Professor A. Ciccarone and Professor M. Shaw who served with Dr. Durbin on our Advisory Committee and gave us much useful advice on the programme and scientific content of the meeting. We are especially indebted to Mrs. June Cheston for her secretarial work before the start of the Institute, for her editorial work with manuscripts and above all for preparing the final typescripts from which this book was copied. We also thank Dr. A. Bottalico and Dr. L. Sparapano, University of Bari, for their help in the organization and running of the Institute, and Professor F. Marras, University of Sassari, for his advice on and help with the local arrangements. We are grateful to Mrs. Teresa Jardella and Miss Maria Palumbo for their excellent secretarial work during the meetings.

The Institute was helped financially or in other ways by the Ministry of Agriculture, Rome, Italy; Consiglio Nazionale delle Richerche, Rome, Italy; Regione Sarda, Cagliari, Italy; University of Sassari, Italy; and the National Science Foundation, Washington, U.S.A.; we thank each for their contributions.

Finally, and on behalf of all those who participated in the Institute we thank most warmly the Scientific Affairs Division of NATO for its sponsorship of the Institute and for the generous financial grant which met a large part of the expenses of participants; and we are particularly grateful to Dr. T. Kester of the Scientific Affairs Division for his interest and his help and advice in organizing the Institute.

                                                    A.G.

December 1975   .   R.K.S.W.

Contents

OPENING ADDRESS

ANTONIO CICCARONE

Istituto di Patologia vegetale dell' Università

Bari, Italy

To open this Institute is not an easy task, if the introduction is not limited to a hearty welcome, but the idea of the organizers, whom I thank most sincerely for the honour conferred upon me, is that my talk has some technical content. They also suggested that I speak generally on specificity, speciation and taxonomy. This is rather hard for a plant pathologist such as I am, so I hope you will listen sympathetically to a somewhat indisciplined discourse in mycological plant pathology, and a somewhat careful approach to the main theme of the Institute.

In my talk I shall refer only briefly to bacteria of which I know all too little. The criteria for identifying and recognizing bacterial species are different from those used for the eucaryotes. As Savile (43) comments, in simple organisms, biochemistry is more important than morphology.

Among specialized fungi there are, as is well known, some which are obligate parasites such as powdery mildews and rusts, some which Van der Plank (52) calls "near obligate" such as *Phytophthora infestans* and non-obligate parasites such as *formae speciales* of *Fusarium oxysporum*. Notwithstanding the acute reasoning of Van der Plank (52), it is prudent to remember that populations of specialized non-obligate parasites such as *formae speciales* of *F. oxysporum* usually show a reduced ability to compete freely in the soil with saprophytic populations of the same species and may perennate better as resting structures in plant residues, in seeds or propagating organs, and also in occasional hosts and carriers. The normalizing selection that should eliminate phenodeviant individuals and counteract directional selection by the resistant hosts is not always easy to interpret.

1

Furthermore, when we speak of individuals we refer to something which is difficult to recognize in mycology. Multinuclearity, heterocaryosis and related phenomena induced Buller (6) to visualize a "social organization" of hyphae in higher fungi and Hansen (16) to refer to multinucleate spores as groups of individuals and to the single nucleus as the basic unit. More recently, Esser (12) has conceived the mycelium not as a multicellular individual, but as a population of numerous nuclei incorporated in a metabolic unit.

In recent years, hypotheses on the macro-evolution of fungi have somewhat altered, with indirect consequences for our ideas on the origin of parasitism. The derivation of fungi from the red algae was proposed in the time of de Bary, accepted for some decades, and then re-discussed some years ago (26, 43) when an autonomous origin of heterotrophs was again favourably considered. Currently, the derivation of fungi from *Rhodophyta* is stressed again (7, 23), with the ascomycetes *Spathulosporales*, or *Laboulbeniales*, or both, as the most probable links between the two classes. There are, however, authorities who, although they accept a Floridean ancestry for the *Ascomycetes*, nevertheless consider that the early typical fungi were parasitic and not derived from saprotrophs.

Savile (43) suggests that obligate parasitism is a fundamental attribute of primitive groups and that more than once saprotrophism may have derived from it. This does not mean that parasitism, especially of facultative saprotrophs, cannot have originated from saprotrophism,but, according to these hypotheses, it is as parasites protected by the host tissues, that fungi left the water. It is this concept that induces Kohlmeyer (23) to think that the most ancient fungi are to be found among the parasites of algae.

These ideas on macro-evolution of fungi are, at present, in active, continuous re-adjustment. Savile (44) has very recently discussed some assumptions by Lewis (24), who, while admitting other possibilities, favours a derivation of biotrophs from obligate saprotrophs, stating "Why start at one extreme rather than with a relatively unspecialized mutualism, from which more specialized mutualism, parasitism and saprophytism repeatedly developed ? ..... The occurrence of Devonian mycorrhizae and the symbiotic origin of eukaryote organelles .... would push biotrophy at some level back 1000 million years; so it is plausible that fungi accompanied the first plants out of the water".

Others who support these ideas stress the following. 1. The parasitic relationship is established especially when the confronting entities (host and parasite) are still in conditions of plasticity, so that the antiquity of the parasite is the same as that

of its hosts. 2. The evolution of parasites, particularly those
which are specialized and obligate, is strictly bound to that of
their hosts. We càn assume that now and then adapted phenotypes,
particularly if specialized, may be subjected, through the host,
to an ecological isolation followed by genetic barriers and speciation.

Recent ideas on the evolution of parasitic fungi may not always be applicable to animal parasites. Nevertheless, Meeuse (31)
has recently proposed three possible ways in which special host-
parasite relationships could have developed. 1. Through exploit-
ation, that is a change from a situation of mutual benefit to dependence of one life-form, the parasite, on the other. 2. Development of a certain degree of interdependence and mutual benefit from
a primarily one-sided parasite-host relationship favouring co-evolution. 3. The "tolerance-adaption principle" of several authors
which involves biochemical interactions as the origin of evolution.
Most of the taxonomically investigated cases of host-parasite relationship seem to be referable to this last pathway (31).

Very few fungal species have been fully investigated in respect
of their evolution and natural affinities. In mycology, the species is still mainly an instrument of identification. Therefore,
the species depends on morphological criteria, supported by meta-
genetic data, and as such is not well suited to a discussion on
speciation. Ramsbottom (42), referring to rusts, stated that
"taxonomical work at its best leads only to certainty of identification" and not much has changed since then. Fischer and Bolton
(13), referring to smuts, noted that nature has made individuals
and populations, and that man has made both species and genera. In
Deuteromycetes, studies on the origin of conidia have given us very
interesting ideas on the development of these structures but still
principally related to morphogenetic criteria.

In the past, particularly for tendentially monophagous obligate
parasites such as species of *Peronospora*, host ranges have been
stressed and are still used. In another complex case, the genus
Fusarium, only a few common morphological characters could be considered by Snyder and Hansen (46) at the species level. Perhaps,
the most frequent tendency today, especially for fungi with few
morphological characters such as the smuts or rather large variability such as *Cercospora* spp., is to consider jointly morphology
and specificity at the level of host families (14) and even sometimes of genera (8). One must accept reality, however, i.e. that
the situation is very different for different groups of fungi.

Gustavsson (15) has suggested that biologic species should be

based not on host species but on relationship between parasite and
host genes. These genic relations, however, are so complex that,
at present, the largest possibilities of their investigation can be
admitted for intraspecific categories (20); and, even here, they
are not always readily applicable. We must consider that it is not
the "naked gene" that is exposed to natural selection but the phenotype,"the manipulation of the entire genotype" (29).

Taxonomy, as Johnson (20) notes, is necessarily confined to
phenotypes,especially morphological, and this author prudently
commends that rigid and permanent limits of pathogenicity should
not be applied to species. Morphology, in fact, is sufficiently
static, whereas behaviour, which is fundamentally important for
parasites, is not easily limited in time and space, because of the
dynamism of host-parasite relationships especially for agricultural
plants.

Visible characteristics associated with pathological behaviour
such as ecto- or endo-parasitism, material position of receptacles,
and so on, have long been used in taxonomy of genera and higher categories, but,all things considered up to the present we have been
unable to give to parasitism the role in species taxonomy that, at
least indirectly, it must have had in speciation. This has, however, been attempted. Thus, Nelson (36) has re-arranged, according
to their morphology, interfertility and other characters such as
pathogenicity, about 10 000 collections of *Helminthosporium* in five
types or species. A few cases of complete intra-type isolation
were considered as possible initial divergent speciation. Further
work along these lines with other pathogens would be most useful.

At a certain point the roads taken by taxonomists and pathologists have diverged. In recent decades, the activity of pathologists has been directed especially to the intra-specific field, to
understanding parasitism and pathogenicity and to identifying the
components of fungal populations and the factors controlling their
behaviour.

Pathogenic capabilities, as such, are rather marginal in taxonomic nomenclature. *Formae speciales* are the highest category in
which classification explicitly considers adaptation to different
hosts, because varieties, forms and cultivars are more or less related to morphology. Special forms, however, do not follow the
rules of botanical or horticultural nomenclature, and notwithstanding some divergent opinions (2), it is perhaps better that for the
time being they remain in a kind of limbo. Special forms should be
used for adaptation at host species level although their meaning
remains rather wide and indefinite. Recently, for example, Joffe

and Palti (18, 19) have given reasons against their acceptance in
F. solani.

Below the special forms, are physiologic forms or races, which
are commonly used for adaptation at the host cultivar level.

Stakman (48), many years ago, stressed that "the determination
of physiologic forms is quite as definite and precise as is the determination of morphologic species" and that, in rusts, they "are as
definite and constant entities as are the species and varieties of
host plants in which they grow". More recently (50) the statement
was made that in *Puccinia graminis* var. *tritici* the physiologic race
is "the only taxon that is recognized". And this can be accepted
in some cases. Esser (12) stressed that when heterogenic incompatibility is present, as in *Podospora anserina,* the race and not the
species becomes the unit of evolution. However, mutations and
nuclear exchanges, heterocaryosis, homogenic incompatibility and,
generally speaking, heterothallism are important instruments of
adaptation, but they are not favourable to genetic isolation and
speciation. Gene flow is essentially a retarding element in evolution (29).

Difficulties multiply with increase of our knowledge and they
sometimes induce a certain perplexity. In this connexion, Kernkamp
(22) for rusts, and Armstrong and Armstrong (1) for *Fusarium oxysporum,* have advocated the greatest uniformity of methods and test
materials and perhaps the establishment of a World Central Identification Laboratory (22).

Nature, however, can give us variants *ad infinitum* and race
investigation, when asked to do more than it can, may even appear as
an "academic gymnastics" (22). In the end, notwithstanding attempts
to analyse precisely their behaviour (3), races are mainly "collections" of different genotypes (40). Thus, *F. solani* f. sp. *cucurbitae*
includes two races which constitute diverse independent mating populations (28) which do not cross, race 1 and some isolates of race
2 being in one group whereas the rest of race 2 belongs to the other.
In this case, the same behaviour in pathogenicity and virulence might
be a convergence of independent populations as a result of selection
by the host.

As is known, these difficulties with races have in some cases
encouraged the use of genes and the development of refined research
tools and materials such as isolines. This has been very useful for
some fields of investigation, but it is evident that it is the total
developmental system (epigenotype), the totality of interactions
among genes which finally the host must face, in relation to climate
and more generally to environment.

As suggested in previous pages, specificity, that is, limitation in the range of hosts, isolates the specialized parasite from
the outer environment in a way somehow recalling the isolation of
free-living organisms due to geographical or ecological barriers.
This is especially true for obligate parasites. The host is the
real environment, the biotope of the parasite; but here the comparison with ecotypes becomes difficult. The host is alive and can,
by mutation, recombination, selection, react continuously to changes
in the parasite. With cultivated plants, man multiplies unendingly
the capabilities of our crops and, through suitable agronomic prac-
tices introduces them into marginal ecological conditions and in
innumerable ways influences their nutrition and physiology in
general. Balanced polymorphisms of host and parasite which are thought to have been so important for micro-evolution over long periods
of time, and other very interesting models of co-evolution, of hosts
and their parasites, are thus disturbed (39). Agricultural systems
are continuously unbalanced by man and instability is perhaps their
main character. Parasites of agricultural plants are, therefore, in
some aspects, also cultivated. In many cases, the parasite, when
we do not breed hosts against it, is encouraged to grow and is protected from competition. We might, therefore, reverse a statement
by Ramsbottom (42) and say that modern horticultural conditions are,
for parasites, similar in many ways to those of pure culture. In
more than one case, organisms now of little pathogenic importance
are given the opportunity of assuming it. This induced condition
is new and it is similar to that of the "opportunistic pathogens"
of human medicine. These are micro-organisms, up to now considered
as innocuous contaminants, which, because of the increasing use of
new instrumentation, antibiotics and drugs, are able to multiply
extensively in the tissue and cause disease (47).

Yarwood (56) points out that in some cases pathogens of agricultural plants under intensive cultivation may be considered as organisms out of place, because their relations with cultivated plants
are different from those with their wild hosts in natural ecosystems.
Indeed, there are examples of extreme tolerance, especially in spontaneous hosts (carriers) not damaged by parasites; and this is true,
not only for virus diseases but also for those caused by cryptogamic
parasites. The tracheomycoses caused by *Fusarium* and *Verticillium*
spp. seem to offer good examples of this behaviour. Most weeds infected by *Verticillium albo-atrum sensu lato* do not show symptoms
(17, 30, 55); and some special forms of *Fusarium oxysporum* such as
albedinis, batatas and *lycopersici* have been found in carriers (1,
5, 21) where they are localized mainly in the lower part of plants,
but sometimes spread to the tops. Carriers certainly have some importance in the conservation of these parasites.

If it is true that "violent damage .... from an infection rep-
represents a primitive state of evolution of host-parasite associations" (56), the condition of a carrier might reflect a very long
established relationship, which is almost an apparently well balanced condition of "indifferent" parasitism. We might also regard
carriers as occasional hosts to which parasites have extended secondarily, possibly also as a result of the intermittent occurrence
of the specific host, when the crop is under rotation (21).

These considerations pose the question as to whether there are
at present indications of heritable adaptations of a parasite to a
new host plant. After Marshall Ward (53), "adaptative parasitism"
was intensively studied at the beginning of this century. Later
studies have generally shown that, if the fungi tested are genetically homogeneous, there is little evidence of heritable adaptations
to hosts (49). Results for adaptation to pesticides are rather different, especially in restricted environments, but such adaptation
to a non-adaptable counterpart, the chemical, is also simpler to
identify.

Recently, Coyne and Schuster (10) and Coyne *et al.* (9) have
confirmed that it is possible to isolate considerable populations
of *Xanthomonas phaseoli* from bean plants *(Phaseolus vulgaris)* tolerant to it and have reminded us that Wellhausen (54) and Lincoln
(25) found that the virulence of a bacterial population increased
through mutation and selection during passage through a tolerant
host. Coyne and Schuster (10) also draw the practical conclusion
from these observations that tolerance might be broken down following the emergence of more virulent strains and races, and that, in
order to reduce this possibility, seed of tolerant varieties should
be saved from plants which are free from bacteria.

Literature relating to fungi is complex to summarize. Matta
(27) has recently reviewed many aspects in this field, and I shall
limit myself to just a few comments on some recent tests on successful inoculation of powdery mildews on non-compatible hosts.

It is well known that previous infection of a host by a parasite may cross-protect or predispose the same plant towards another
parasite. The literature on this subject is rather large. Some
aspects of predisposition have been summarized, among others, by
Brokenshire (4), who predisposed wheat to *Septoria nodorum* by previously inoculating it with *Erysiphe graminis* f. sp. *tritici*. When
a host was previously inoculated with virulent cultures, Moseman
and Greeley (33) and Moseman *et al.* (34) obtained both infection
and propagation of *E. graminis* f. sp. *hordei* on two resistant barley varieties and on wheat, and propagation of *E. graminis* f. sp.
tritici on barley. They stressed the importance of this fact for
the survival of these fungi and, having found no change in genes

conditioning pathogenicity, hypothesized that the first inoculation
altered the physiological environment within the host. This line
of research has been developed by Japanese scientists (35, 37, 38,
51).

The technique followed by Tsuchiya and Hirata (51) and by others
consisted in removing young colonies of the barley powdery mildew
growing on barley leaves and then secondarily inoculating the leaves
with powdery mildews from wheat, from *Agropyron ciliare* var. *minus*,
and with 49 powdery mildews from Dicotyledons. The haustoria of
E. graminis f. sp. *hordei* remained in the epidermal cells and sur-
vived for many days. Although growth was generally inferior to
that of *E. graminis* f. sp. *hordei*, 30 other species produced conidia on barley, and another 15 species grew close to the colonies
of the barley mildew. The secondarily inoculated fungi also differentiated the majority of their first haustoria in cells containing haustoria of the barley mildew, or in adjacent epidermal cells.
In the secondarily inoculated fungus, haustoria and even conidia
with characteristic morphology could be seen (37).

Ouchi *et al.* (37, 38) used the word *accessibility* for the first
access of a parasite into host tissues and consider it important in
understanding evolutionary aspects of parasitism. From different
points of view, this has also been considered by several other authors (41, 45).

At this point let us remember the pioneer work of Miyoshi (32),
Massee, Petri and many other early plant pathologists on induction
and, indirectly, specificity of parasitism. They were able to form-
ulate hypotheses which are still plausible, provocative and lively,
despite their very limited facilities and the lack of the vast amount
of information and data now available to us. Their epoch is past
but their foresight and acute intuitions are fortunately still with
us.

REFERENCES

1.   ARMSTRONG, G.M. and ARMSTRONG, J.K. (1948). Nonsusceptible
          hosts as carriers of wilt *Fusaria*. *Phytopathology*, 38,
          808 - 826.

2.   ARMSTRONG, G.M. and ARMSTRONG, J.K. (1968). Formae speciales
          and races of *Fusarium oxysporum* causing a tracheomycosis
          in the syndrome of disease. *Phytopathology*, 58, 1242 -
          1246.

3.   ARMSTRONG, G.M. and ARMSTRONG, J.K. (1974). Races of *Fusarium
          oxysporum* f. sp. *pisi*, causal agents of wilt of pea.
          Phytopathology, 64, 849 - 857.

4. BROKENSHIRE, T. (1974). Predisposition of wheat to *Septoria* infection following attack by *Erysiphe*. *Trans. Br. mycol. Soc.*, 63, 393 - 397.

5. BULIT, J., LOUVET, J., BOUHOT, D. and TOUTAIN, G. (1967). Recherches sur les fusarioses. I. Travaux sur le Bayoud, fusariose du Palmier Dattier en Afrique du Nord. *Annls. Epiphyt.*, 18, 213 - 239.

6. BULLER, A.H.R. (1931). *Researches on fungi*. Vol. IV. Longmans, Green and Co., London (Hafner Publ. Co., New York, 1958), 329 pp.

7. CHADEFAUD, M. (1972). Les cycles des champignons comparés à ceux des algues. *Mém. Soc. bot. France*, 333 - 368.

8. CHUPP, C. (1953). *A monograph of the fungus genus Cercospora*. Ithaca, New York, 667 pp.

9. COYNE, D.P., SCHUSTER, M.L. and HILL, K. (1973). Genetic control of reaction to common blight bacterium in Bean *(Phaseolus vulgaris)* as influenced by plant age and bacterial multiplication. *J. Am. Soc. hort. Sci.*, 98, 94 - 99.

10. COYNE, D.P. and SCHUSTER, M.L. (1974). Inheritance and linkage relations of reaction to *Xanthomonas phaseoli* (E.F. Smith) Dowson (common blight), stage of plant development and plant habit in *Phaseolus vulgaris* L. *Euphytica*, 23, 195 - 204.

11. DE BARY, A. (1881). Zur Systematik der Thallophyten. *Bot. Ztg.*, 39, 1 - 17 and 33 - 36.

12. ESSER, K. (1974). Breeding systems and evolution. *In : Evolution in microbial world*. (CARLILE, M.J. and SKEHEL, J.J., Eds.), 87 - 104. *24th Symp. Soc. gen. Microbiol.*, Cambridge Univ. Press.

13. FISCHER, G.W. and HOLTON, C.S. (1957). *Biology and control of the smut fungi*. Ronald Press, New York, 622 pp.

14. FISCHER, G.W. and SHAW, G.C. (1953). A proposed species concept in the smut fungi with application to North American species. *Phytopathology*, 43, 181 - 188.

15. GUSTAVSSON, A. (1959). Studies on the nordic Peronosporas. II. General account. *Op. bot.*, 3, 1 - 61.

16. HANSEN, H.N. (1938). The dual phenomenon in imperfecti fungi. *Mycologia,* 30, 422 - 455.

17. ISAAC, I. (1967). Speciation in *Verticillium. Ann. Rev. Phytopath.,* 5, 201 - 222.

18. JOFFE, A.C. and PALTI, J. (1972). *Fusarium* species of the *Martiella* section in Israel. *Phytopath. Z.,* 73, 123 - 148.

19. JOFFE, A.C. and PALTI, J. (1974). *Fusaria* isolated from field crops in Israel and their pathogenicity to seedlings in glasshouse tests. *Z. PflKrankh. PflPath. PflSchutz.,* 81, 196 - 205.

20. JOHNSON, T. (1968). Host specialization as a taxonomic criterion. *In : The Fungi* (AINSWORTH, G.C. and SUSSMAN, A.S., (Eds.)., Vol.III, 543-556. Academic Press, New York, London.

21. KATAN, J. (1971). Symptomless carriers of the tomato *Fusarium* wilt pathogen. *Phytopathology,* 61, 1213 - 1217.

22. KERNKAMP, M.F. (1965). Pathogenic specialization in relation to taxonomy. *Phytopathology,* 55, 821 - 822.

23. KOHLMEYER, J. (1975). New clues to the possible origin of Ascomycetes. *Bioscience,* 25, 86 - 93.

24. LEWIS, D.H. (1974). Micro-organisms and plants : the evolution of parasitism and mutualism. *In : Evolution in microbial world.* (CARLILE, M.J. and SKEHEL, J.J., Eds.). 367 - 392. *24th Symp. Soc. gen. Microbiol.,* Cambridge Univ. Press.

25. LINCOLN, R. (1940). Bacterial wilt resistance and genetic host parasite interactions in maize. *J. agr. Res.,* 60, 217 - 239.

26. MARTIN, G.W. (1968). The origin and status of fungi. *In : The Fungi.* (AINSWORTH, G.C. and SUSSMAN, A.S., Eds.), Vol.III, 635-648. Academic Press, New York, London.

27. MATTA, A. (1971). Microbial penetration and immunization of uncongenial host plants. *A. Rev. Phytopath.,* 9, 387 - 410.

28. MATUO, T. and SNYDER, W.C. (1973). Use of morphology and mating populations in the identification of *formae speciales* in *Fusarium solani. Phytopathology,* 63, 562 - 565.

29. MAYR, E. (1971). *Populations, species and evolution.* Belknap
     Press, Harward Univ. Press, Cambridge, Mass., 2nd printing,
     453 pp.

30. McCAIN, A.H., WILHELM, S. and RAABE, R.D. (1967). Plants res-
     istant or susceptible to *Verticillium* wilt. California
     Agricultural Extension Service, Axt. 40, 5 pp.

31. MEEUSE, A.D.J. (1973). Co-evolution of plant hosts and their
     parasites as a taxonomic tool. *In : Taxonomy and ecol-
     ogy.* (HEYWOOD, V.H., Ed.), 289 - 316. Systematics
     Association. Academic Press, London, New York.

32. MIYOSHI, M. (1894). Ueber Chemiotropismus der Pilze. *Botan.
     Z., 52, 1 and 24.*

33. MOSEMAN, J.G. and GREELEY, L.W. (1964). Predisposition of
     wheat by *Erysiphe graminis* f. sp. *tritici* to infection
     with *Erysiphe graminis* f. sp. *hordei. Phytopathology,*
     54, 618 pp.

34. MOSEMAN, J.G., SCHAREN, A.L. and GREELEY, L.W. (1965). Prop-
     agation of *Erysiphe graminis* f. sp. *tritici* on barley
     and *Erysiphe graminis* f. sp. *hordei* on wheat. *Phytopath-
     ology,* 55, 92 - 96.

35. NAITO, H. and HIRATA, K. (1969). Fusion of hyphae and conidial
     germ tubes in powdery mildew fungus of barley, *Erysiphe
     graminis* f. sp. *hordei* (in Japanese). *Niigata agric.
     Sci.,* 21, 29 - 36. (*In* : TSUCHIYA and HIRATA, 1973).

36. NELSON, R.R. (1965). Assessing biological relationships in
     the fungi. *Phytopathology,* 56, 823 - 826.

37. OUCHI, S., OKU, H., HIBINO, C. and AKIYAMA, I. (1974). Induct-
     ion of accessibility and resistance in leaves of barley
     by some races of *Erysiphe graminis. Phytopath. Z.,* 79,
     24 - 34.

38. OUCHI, S., OKU, H., HIBINO, C. and AKIYAMA, I. (1974a). Ind-
     uction of accessibility to a nonpathogen by preliminary
     inoculation with a pathogen. *Phytopath. Z.,* 79, 142 -
     154.

39. PERSON, C.O. (1966). Genetic polymorphism in parasitic systems.
     Nature, Lond., 212, 266 - 267.

40. PERSON, C.O. (1968). Genetical adjustment of fungi to their
     environment. *In : The Fungi.* (AINSWORTH, G.C. and
     SUSSMAN, A.S., Eds.), Vol. III, 395 - 415, Academic
     Press, New York and London.

41. RAGGI, V. (1964). Affinità sistematica e affinità biologica
     nel caso di parassiti obbligati specializzati. *Phytopath.*
     medit., 3, 135 - 155.

42. RAMSBOTTOM, J. (1940). Taxonomic problems in fungi. *In : The*
     new systematics. (HUXLEY, J., Ed.), 411 - 434. The
     Systematics Association, London. (Reprint, 1971).

43. SAVILE, D.B.O. (1968). Possible interrelationships between
     fungal groups. *In : The Fungi.* (AINSWORTH, G.C. and
     SUSSMAN, A.S., Eds.), Vol. III, 649 - 675, Academic Press,
     New York and London.

44. SAVILE, D.B.O. (1975). Review of : CARLILE, M.J. and SKEHEL,
     J.J. *Evolution in the microbial world.* *Mycologia,* 67,
     201 - 202.

45. SEMPIO, C. and CAPORALI, L. (1957). Sur la pénétration et la
     diffusion de l'*Uromyces appendiculatus* dans le haricot
     et dans d'autres espèces ayant différentes affinités
     systématiques avec l'hôte spécifique. *C.r. IV Congr. int.*
     Lutte ennemis plantes, Hamburgm 1, 117 - 120.

46. SNYDER, W.C. and HANSEN, H.N. (1940). The species concept in
     Fusarium. *Am. J. Bot.,* 27, 64 - 67.

47. SPAULDING, E.H.(1974). Introduction. *In : Opportunistic*
     pathogens. (PRIER, J.E. and FRIEDMAN, H., Eds.), 13 -
     15, Macmillan, London.

48. STAKMAN, E.C. (1929). Physiologic specialization in plant
     pathogenic fungi. *Leopoldina,* 4, 263 - 289.

49. STAKMAN, E.C. and HARRAR, J.G. (1957). *Principles of plant*
     pathology. Ronald Press Co., New York, 581 pp.

50. STAKMAN, E.C., STEWART, D.M. and LOEGERING, W.Q. (1962). Id-
     entification of physiologic races of *Puccinia graminis*
     var. *tritici.* *U.S.D.A., ARS E617,* 1 - 53 (revised 1962).

51. TSUCHIYA, K. and HIRATA, K. (1973). Growth of various powdery
     mildew fungi on the barley leaves infected preliminarily
     with the barley powdery mildew fungus. *Ann. phytopath.*
     Soc. Japan, 39, 396 - 403.

52. VAN DER PLANK, J.E. (1968). *Disease resistance in plants.*
     Academic Press, New York and London, 206 pp.

53. WARD, H., MARSHALL (1902). On the question of "predisposition"
     and "immunity" in plants. *Proc. Camb. phil. Soc.,* 11,
     307 - 328.

54.   WELLHAUSEN, E.J. (1937). Effect of the genetic constitution
        of the host on the virulence of *Phytomonas stewarti*.
        Phytopathology, 27, 1070 - 1089.

55.   WOOLLIAMS, G.E. (1966). Host range and symptomatology of
        Verticillium dahliae in economic, weed and native plants
        in interior British Columbia. *Can. J. Pl. Sci.*, 46, 661 -
        669.

56.   YARWOOD, C.E. (1967). Pathogens as organisms out of place.
        Phytopath. Z., 58, 305 - 314.

THE PHENOMENON OF SPECIFICITY IN PLANT DISEASE

PERCY W. BRIAN

Botany School, University of Cambridge

Cambridge, U.K.

INTRODUCTION

In this introductory lecture I propose to make a broad survey of the field, thus hoping to open the way for later contributors who will provide the necessary detail. I shall deal mainly with fungal diseases, with some mention of bacterial diseases, but I shall not deal with viral diseases at all.

LEVELS OF SPECIFICITY

Specificity in the relations between host and parasite exists at several levels.

1. There is specificity involved in the distinction between pathogens in general and absolute non-pathogens. It is very difficult to account in physiological or biochemical terms for this basic distinction. Saprophytes appear to have all the necessary enzyme systems to enable penetration of a host to take place, yet they are unable to penetrate and colonize plants.

2. Even pathogens with a very wide host-range, such as *Botrytis cinerea, Armillaria mellea* and *Pythium* spp., appear to be unable to infect some plant species. This distinction between host and non-host is again a form of specificity.

3. Some pathogens of wide host range, for example *Corticium solani*, exist in the form of several races specialised in the sense that they each attack only a restricted range of host species.

4.   There is the greater specificity of pathogens restricted to a
single species, or to a few closely related species, such as
Plasmodiophora brassicae, *Phytophthora infestans*, *Colletotrichum
lindemuthianum* and most of the rusts, powdery mildews and downy
mildews.

5.   Finally, specificity of the type mentioned above may be even
more marked than appears at first sight, in that either the pathogen species or *forma specialis* may consist of physiologic races
each restricted to a few genotypes of a host species, as, for
instance, in the well-studied case of *Puccinia graminis* f. sp.
tritici on wheat. It is in this ultimate type of specificity that
we encounter, on the genetic level, the well-known gene-for-gene
situation in host-parasite relations.

It is very doubtful whether these different levels of specificity represent the same phenomenon and it is equally doubtful
whether they can be explained in similar terms.

While it is true that extreme specificity is characteristic
of certain taxonomic groups, notably the *Erysiphales* and *Uredinales*,
three related points should also be noted. 1. Marked specificity
can be found in pathogens from virtually all taxonomic groups of
fungi, presumably having evolved independently on several occasions.
2. High specificity is most well-known in diseases of cultivated
plants and may be to some extent characteristic of disease in such
plants as compared with wild hosts. 3. Taxonomically closely related pathogens may differ widely in their specificity; one may
compare the wide host-range of *Botrytis cinerea* with the restricted
host-ranges of *B. fabae* or *B. tulipae*, or the wide host-range of
Phytophthora cryptogea with the narrow host-range of *P. fragariae*.
One must admit, however, that these apparent contrasts may to some
extent be consequences of imprecise or inconsistent taxonomy.

These variations in width of specificity should not prevent
us from realising that as a general rule most fungal pathogens show
a high degree of host specificity and that in such important pathogens as the rusts and mildews extreme specificity is the rule. I
suspect that it is with this type of specificity that we shall be
mainly concerned at this meeting.

A special case of specificity, and one that will test any
suggested explanation of the material basis of specificity, is that
of the heteroecious rusts with two unrelated hosts involved in the
life cycle - a form of double specificity. Though in such cases
relations with one of the hosts usually shows more specificity than
with the other, any explanation of specificity will eventually have
to accommodate this phenomenon.

HOST SPECIFICITY AND PARASITE NUTRITION

Having attempted to define what we mean by specificity in plant disease, let me approach the topic from another direction.

Lewis (6), developing an earlier scheme of de Bary (2), has recently classified fungi in five groups on a nutritional-ecological basis.

Group 1. Ecologically obligately saprophytic saprotrophs. These are obligate saprophytes, with no capacity for parasitic or mutualistically symbiotic relations with plants; they are the absolute non-pathogens mentioned earlier.

Group 2. Ecologically facultative symbionts, whose nutrition is necrotrophic in the parasitic phase but otherwise saprophytic. These are facultative parasites which cause a good deal of cell damage or cell death, the parasite drawing nutritionally upon such dead or dying cells.

Group 3. Ecologically obligately symbiotic necrotrophs. This group includes a considerable number of plant pathogens which have lost the capacity to compete with saprophytes in the absence of a host, but which retain the necrotrophic habit within the host.

Group 4. Ecologically facultative symbionts whose nutrition is biotrophic in the symbiotic phase but otherwise saprotrophic. This group need not concern us here as it is mainly composed of certain mycorrhizal fungi, such as those involved in the ectotrophic or sheathing mycorrhizae of forest trees, which are also capable of independent saprophytic existence.

Group 5. Ecologically obligately symbiotic biotrophs. This group includes all obligate parasites and the vesicular-arbuscular mycorrhizal fungi. These fungi draw nutritionally upon living cells.

Looking at this classification it is noticeable that the least specific plant pathogens fall clearly into group 2 and that specificity reaches its ultimate expression in the obligate parasites of group 5. This strong correlation between obligate biotrophy and high specificity in host-parasite relations is perhaps not surprising, since both are a measure of a closely integrated host-parasite relationship; it can be argued that, in a sense, specificity is an almost inevitable consequence of biotrophy.

But there are exceptions to this generalisation, two of which are worth mentioning. 1. While lack of specialization of parasitism is associated with necrotrophic relationships, quite a number

of fungal pathogens with necrotrophic nutrition, at least in later
stages of pathogenesis, show a good deal of specialization ;   this
is typical of some representatives of group 3.   Perhaps this is not
surprising since there is a continuous gradation from Lewis' group
2, through group 3 to group 5.   2. There are certain biotrophs
which show very little specialization.   I have in mind the very wide
host-range of the fungal partners (*Endogone* spp.) in vesicular-
arbuscular endogenous mycorrhizae, and such obligately biotrophic
parasites as *Synchytrium macrosporum* (5).

    These exceptions need explaining.

MECHANISMS DETERMINING SPECIFICITY

    When considering the mechanisms underlying specificity, we
should first note that specificity is exhibited very early in the
infection process, the latest point, and frequently the most
critical point, being the moment of contact between the pathogen
cell-wall and the plasmalemma of a penetrated cell.   This is also
the characteristic point at which the hypersensitive response is
exhibited, the hypersensitive response being characteristic of the
most highly developed forms of host specialization.

    Thus, in more extreme specialization the phenomenon is deter-
mined very quickly, early in the process of infection.   We are
still uncertain, except for a few isolated examples, where the
'challenge' comes which excludes an absolute non-pathogen from a
plant, or which exclude an unspecialized pathogen from a non-host
species.

    Mechanisms underlying specificity will be the main topic of
many of the following lectures.   I see my function as being to
point out *possible* mechanisms, most of which will be dealt with in
more detail in subsequent lectures.   Let me make one point first;
specificity may appear to be determined mainly by host characters,
or mainly by pathogen characters, but in fact in most cases it is
by a combination of or an interaction between the two.

    There are many possible causes of specificity;   I shall men-
tion only some.

1.   An antagonistic root- or leaf-surface microflora might exclude
certain pathogens, the nature of the antagonistic microflora being
at least partially determined by host characters, such as root- or
leaf-exudates.   Timonin (9) analysed one such situation on flax
roots.   Evidence for a role for the leaf-surface microflora in
determination of resistance is beginning to accumulate.   It is
doubtful, however, whether such a mechanism would have much spec-

ificity or would have more than occasional significance.

2.   Surface microflora might have infection-promoting properties
in relation to some pathogens;   currently some attention is being
given to pollen grains as promoters of infection, but once again it
seems doubtful if highly specific effects can be expected.

3.   Many pathogens enter a host by penetration of the cuticle.
Nevertheless it could be possible that the host-range of a pathogen,
or the number of pathogens which can infect a particular plant, may
be limited by chemical or physical characteristics of the cuticle
in relation to the cutinolytic enzymes produced by a potential
pathogen. I know of no evidence for the existence of such a situation but it seems to me that it could be a basis of resistance
of non-hosts to inappropriate pathogens.

4.   Albersheim *et al.* (1) have drawn attention to the immense
variety of cell wall polymers which could exist and have pointed
out that this multiplicity of polymers, especially if associated
with a similar multiplicity of polymer-splitting enzymes produced
by potential pathogens, could in itself explain the most complex
and highly specific host-pathogen relations. We shall need to
assess the evidence that this is in fact a real situation;   certainly there is a good deal of evidence against the hypothesis.

5.   Specificity could be determined by selective inhibitors
produced by the host;   this is perhaps the hypothesis to which we
shall devote most attention. This broad category could include
either 'pre-existing' inhibitors or phytoalexins induced by infection. The specificity might be introduced in one or more of several
ways, by specific toxicity of the host product, by specific degradation of the toxin by some pathogens, or the process of induction
in the case of phytoalexins might be specific in nature. We shall
need to assess the evidence for each of these specificity-determin-
ing mechanisms. In this context 'inhibitors' will include not only
antibiotic metabolic products which inhibit growth or metabolism of
potential pathogens, but will also include inhibitors of cell-wall
degrading enzymes or proteolytic enzymes produced by potential
pathogens, and host enzymes, such as chitinases, which lyse pathogen cell-walls.

6.   Similarly, we shall need to consider the possible role of
selective stimulants, or essential substrates for a pathogen,
produced by some potential hosts, which can consequently be infected,
but not by others. This possibility, which at first sight may seem
rather remote, has recently been made more real by the discovery of
the role of choline and betaine as specific stimulants determining
infection of flowers of wheat by *Fusarium graminearum* (8). Another
example of infection being determined by stimulants is afforded by

the selective stimulation of germination of sclerotia of *Sclerotium cepivorum* by root-exudates from some species of *Allium* (3), though in this case the soil microflora is also involved (4).

7.   The only certainly established mechanism of specificity is in those parasitic associations where infection is dependent upon killing of host tissue by host-specific toxins released by the pathogen, as in diseases caused by such pathogens as *Helminthosporium victoriae, H. sacchari* and a few other species of fungus (11). This situation is at present known only in a few host-pathogen combinations but further research may show it to be more common. It is arguable whether, in such cases, the specificity lies in the toxin or in the receptor; here is a case where it is the interaction between host and pathogen characters that is the determinant.

8.   Finally, we have the widespread phenomenon of hypersensitive response. This is a reaction which in some respects is the reverse of the reaction to host-specific toxins mentioned above. In the latter case the toxin causes necrosis, which is an obligatory precursor to infection. In the hypersensitive response infection also leads to a rapid but localized necrosis which is followed, more or less rapidly, by termination of the infection. We are still very ignorant concerning the hypersensitive response and two major questions remain unanswered. 1. What causes cell death? If a toxin is involved it must be host specific; so far the only evidence that we have in this connection comes from the work of van Dijkman and Sijpesteijn (10) on *Cladosporium fulvum* infection of tomato. 2. Why does cell necrosis lead to inhibition of the pathogen and thus to resistance? In some cases this appears to be explicable in terms of rapid triggering of phytoalexin production - note the combination of a specific trigger with a relatively non-specific inhibitor - but in other cases, for example in rust infections of wheat, there is no evidence yet for involvement of a phytoalexin.

Assuming that specific toxins are involved in eliciting the necrosis in the hypersensitive response, we need to consider the possible nature of the toxins that might be involved in a complex situation such as that of *Puccinia graminis* f. sp. *tritici* on wheat. There must be a recognition system based on a family of related toxins and a family of related receptors. Our thoughts naturally turn to a large molecule such as a protein, a poly-saccharide, or a glycoprotein, for though simple molecules may sometimes show marked specificity (e.g. helminthosporoside, the host-specific toxin of *Helminthosporium sacchari*) it is difficult to see how the necessary large number of variants can occur in a small molecule. Inevitably our thoughts travel to the analogous situation of immune responses in mammals, with more foundation to certain less elaborate immune responses in insects (7), and perhaps even more realistically to incompatibility relations in flowering plants, where pollen wall proteins react with receptors on or in the stigma, or to specificity in the *Rhizobium*-legume system where there is some evidence for the involvement of lectins. All this is sheer speculation at present, but there is a great likelihood that we shall be looking for specific molecules in the pathogen cell-wall, especially the haustorial wall, and in the host plasma-lemma.

THE EVOLUTION OF SPECIFICITY

Finally, we need to consider briefly the evolution of specificity. At first sight the capacity to invade many hosts would seem to be a superior character to extreme specificity, but it is undeniable that extreme specificity is characteristic of the most successful pathogens. It seems probable that pathogen specificity and host specificity have evolved in parallel from situations of lower specificity on several occasions, and, as I have suggested above, biotrophy - a characteristic of close host-parasite relations - probably inevitably entails specificity. But we need to consider why biotrophy in the vesicular-arbuscular mycorrhizae, perhaps the ultimate development of biotrophy, is not associated with any marked degree of host specialization.

REFERENCES

1. ALBERSHEIM, P., JONES, T.M. and ENGLISH, P.D. (1969). Biochemistry of the cell wall in relation to infective processes. *A. Rev. Phytopath.*, 7, 171 - 194.

2. BARY, A. DE (1887). *Comparative morphology and biology of the fungi, mycetozoa and bacteria*. Clarendon Press, Oxford, 522 pp.

3. COLEY-SMITH, J.R. and HOLT, R.W. (1966). The effect of species of *Allium* on germination in soil of sclerotia of *Sclerotium cepivorum* Berk. *Ann. appl. Biol.*, 58, 273 - 278.

4. COLEY-SMITH, J.R., KING, J.E., DICKINSON, D.J. and HOLT, R.W. (1967). Germination of sclerotia of *Sclerotium cepivorum* Berk. under aseptic conditions. *Ann. appl. Biol.*, 60, 109 - 115.

5. KARLING, J.S. (1974). Further induced infectivity by *Synchytrium macrosporum*. *Bull. Torrey bot. Club*, 101, 311 - 316.

6.   LEWIS, D.H. (1973). Concepts in fungal nutrition and the
       origin of biotrophy. *Biol. Rev.*, 48, 261 - 278.

7.   SALT, G. (1970). *The cellular defence reactions of insects.*
       University Press, Cambridge, 118 pp.

8.   STRANGE, R.N., MAJER, J.R. and SMITH, H. (1974). The isolation
       and identification of choline and betaine as the two major
       components in anthers and wheat germ that stimulate
       Fusarium graminearum in vitro. *Physiol. Pl. Path.*, 4,
       277 - 290.

9.   TIMONIN, M.I. (1941). The interactions of higher plants and
       soil microorganisms. III. Effect of by-products of plant
       growth on activity of fungi and actinomycetes. *Soil
       Sci.*, 52, 395 - 413.

10. VAN DIJKMAN, A. and KAARS SIJPESTEIJN, A. (1973). Leakage of
       pre-absorbed ^{32}P from tomato leaf disks infiltrated with
       high molecular weight products of incompatible races of
       Cladosporium fulvum. *Physiol. Pl. Path.*, 3, 57 - 67.

11. WOOD, R.K.S., BALLIO, A. and GRANITI, A. (Eds.) (1972).
       Phytotoxins in plant diseases. Academic Press, London,
       New York, 530 pp.

CONTRIBUTIONS

HEATH, M.C.   Specificity determining interactions during rust
     infection.

MANNERS, J.G.   The growth of *Erysiphe graminis* on normally inap-
     propriate or resistant hosts.

OUCHI, S.   Localization and irreversibility of induced susceptib-
     ility.

SMITH, I.M.   Host specificity in vascular wilt diseases.

STAUB, T.   Development of fungi on host and non-host plants.

SUMMARY OF POINTS FROM CONTRIBUTIONS AND DISCUSSIONS
BY
E. W. B. WARD

Chairman and Discussion Leader

The nature of the hypersensitive response and its relevance

to specificity of interactions between fungi and bacteria and higher plants was of major concern to many participants. Problems arise partly from definition of the response because it means different things to different authors and it varies in detail in different hosts, in different tissues of the same host and in the same tissue at different stages of development. Basically, hypersensitivity involves death of one host cell, or a small number of host cells to which the pathogen causing the response is confined. There are incompatible interactions however, in which death of protoplasts does not occur and it is possible that death is not a necessary part of the reaction that determines specificity. Problems of definition again arise due to difficulties in establishing when death of cells of host and of pathogens occurs. Evidence for death of hyphae, particularly in dead host cells in a hypersensitive response is often lacking. This, however, may not be important for resistance where the fungus is contained and does not spread to colonize other cells.

In a paper on vascular wilt diseases in tomato, Smith observed that the species specificity of *Fusarium oxysporum* and *Verticillium* spp. and much of the varietal specificity in the resistance of host plants to these species is presumably complex in expression and genetic control. The best models for examining specificity are isoline pairs differing in single resistant genes. Explanations of specificity based on sensitivity of leaves to symptom development do not readily account for observations that fungal growth is much more restricted in an isoline of tomato with the *Ve*-gene for resistance to *V. albo-atrum*, unless there is feedback to the fungus from leaves to stem or unless the factor acting in the leaves has a primary action at the infection site. Resistant roots were found to react hypersensitively to *V. albo-atrum* whereas susceptible roots gave a compatible interaction. Comparable interactions were difficult to demonstrate in vascular tissue although there were clear differences in phytoalexin production. The difficulties are due evidently to the lack of direct contact of the fungus with the living cell and to the impossibility of defining a clear point of divergence in infection behaviour that is consistently recognized when different methods of inoculation are used.

In discussion it was pointed out by Graniti that the mal secco disease of Citrus caused by *Phoma (Deuterophoma) tracheiphila* provides another example of the influence of inoculation method on apparent specificity in a vascular disease. Only a few species e.g. lemon are susceptible to leaf infection (by conidia through stomata) under natural conditions, but all known species become severely diseased when infection occurs through wounds in stems or roots.

In response to questions, Smith stated that differences in phytoalexin production in resistant and susceptible tomato plants are

large after direct inoculation of the vascular tissue provided high
inoculum levels were used. Comparable differences had not been demonstrated in root inoculated plants. Differences in growth rate of
the pathogen in root inoculated susceptible plants were apparent
early and persisted. In the resistant variety there were few vessels with hyphae from the beginning. In the susceptible variety
numbers continued to increase for three to four weeks.

The stages during rust infection at which specificity may be
determined were discussed in a paper by Heath. An initial if not
completely effective stage occurs on the surface of the host where
location of stomata by germ tubes tends to be less efficient on
non-hosts than on hosts. Where this is by-passed and in the infection of resistant varieties, specificity must be determined by
events following leaf penetration. During susceptible interactions
there are few major ultrastructural changes in host or parasite until sporulation occurs. It was assumed during a large number of
ultrastructural comparisons of incompatible responses to cowpea
rust that any response of plant or parasite at a particular stage
in infection indicates an interaction between the two organisms.
There are a number of stages during the infection process where
such interactions may occur. These may also occur in the susceptible
response but without any resulting ultrastructural changes. These
stages can be regarded as "switching points", an incorrect response
at only one of which may result in resistance. Susceptibility however, would demand the "correct" response at every point, and hence
is a complicated process, as its rarity suggests. Non-host responses
characteristically occur before the formation of the first haustorium and may represent the types of resistance overcome during the
evolution of the parasite towards its host. Varietal resistance is
usually expressed after the formation of the first haustorium and
probably represents gene-for-gene interactions "put back" once the
host's initial resistance has been overcome. Haustorium formation
in the non-host combination *Phaseolus vulgaris*/cowpea rust (*Uromyces
phaseoli* var. *vignae*) is apparently prevented by the deposition on
the adjacent cell wall of osmiophilic material which probably contains phenolic substances and can be induced by exudates of infection hyphae grown on oil-containing collodion membranes. The active
material(s) appears to be of mol. wt. 15 000 to 150 000.

Durbin supported Heath's views by referring to the importance
of early "switch points" in the elimination of *Septoria* spp. from
incompatible plants. These pathogens rarely penetrate but where
they do an early hypersensitive response occurs. If they are art-
ificially introduced early barriers are by-passed and they become
pathogenic. Tani pointed out that in crown rust of oats resistance
is determined ten hours before the first ultrastructural changes can
be detected.

Staub described observations by light and scanning electron

microscopy of the development of barley and cucumber powdery mildews on their hosts and on their non-hosts, cucumber and barley respectively. On barley, germination of *Erysiphe graminis* reached 90% but on cucumber it was only about 45%. The germ tubes left clear trails on leaves, indicating interaction with the cuticle wax. Development stopped only after penetration of epidermal cells and formation of haustoria initials. The penetrated cucumber cells collapsed and died with no visible response in neighbouring cells. *Erysiphe cichoracearum* germinated equally well on barley and cucumber but did not affect wax on barley leaves as did *E. graminis*. It did not penetrate epidermal cells of the non-host plant or cause cotton-blue staining halos beneath the germ tubes as developed in the compatible combinations.

In discussion Wood suggested that differences in germination of powdery mildew conidia on different hosts may not be significant for specificity because such effects could presumably be nullified by increasing the size of the inoculum. Heath stated that the situation may be less simple in the rusts where germ tubes respond to specific surface stimuli that direct them towards stomata whereas on non-hosts they may be actively directed away. Day made the interesting suggestion that if conidia of the two powdery mildews could be distinguished in mixed inoculations, it could then be determined whether occurrence of a non-host reaction affects adjacent compatible interactions. Ouchi stated that conidia and appressoria of *E. graminis hordei* and *Sphaerotheca fulginea* can be distinguished as can appressoria of *E. graminis hordei* and *E. graminis tritici* and Manners stated that species specific antigens and immunofluorescent dyes may also be used to distinguish species.

In a paper on the induction of susceptibility to powdery mildews Ouchi reported that cells conditioned to be non-accessible by either an incompatible race or non-pathogenic species subsequently reject any powdery mildew. Cells conditioned to be accessible are subsequently unable to recognize normally incompatible or non-pathogenic mildews as foreign entities and permit the co-existence of those mildews in the same cell. The primary cellular recognition and conditioning mechanisms are active responses, are localized and irreversible.

Manners reporting work by Laurence stated that in wheat or barley inoculated with single spore isolates of races of mildew from barley or wheat respectively, sporulation did not occur but there was more development when the pathogen possessed virulence genes or the host lacked major resistance genes. When, in successive inoculations a compatible race was followed by an inappropriate mildew after removal of surface mycelium, seven wheat races grew on barley and oats, six barley races grew on wheat and oats, and an oat isolate grew on wheat and barley. All colonies were identified by culturing

from single spores. When a host was inoculated simultaneously with
pairs of races, development was as if they were independent and although hyphal fusions occurred, spores were only of one or the other
type.

In the discussion of these two papers on induced susceptibility
to powdery mildews it was pointed out that the effects persisted
for at least four days in the experiments of Ouchi and for seven
days in those of Manners. Both speakers emphasized that subsequent
isolation and testing demonstrated that the infection that developed
originated from the secondary avirulent inoculum. Ouchi considered
the possibility of heterocaryosis to be small because all mycelium
of the primary fungus was removed from the leaf surface with wet
cotton-wool prior to the second inoculation; this almost certainly
excluded hyphal fusion. With regard to the time sequence for the
induction of susceptibility Ouchi and Manners agreed that some degree of haustorial formation by the primary fungus was essential.
Ouchi stated that establishment of the compatible race was inhibited in leaves treated with $3.5 \times 10^{-6}M$ cycloheximide within 15 h
following inoculation but not later. Kuć suggested that there was
general agreement between the results for mildews and his work with
Phytophthora infestans and potatoes where prior inoculation with
a compatible race suppressed accumulation of rishitin and related
terpenoids in response to subsequent treatment with a cell-free
sonicate of cell walls from a compatible or incompatible race. Kuć
emphasized that use of the term 'induction' with reference to susceptibility or resistance is not intended to imply the involvement
of protein synthesis.

TISSUE AND ORGAN SPECIFICITY IN PLANT DISEASES

ANTONIO GRANITI

Istituto di Patologia vegetale, University of Bari

Italy

INTRODUCTION

In plant pathology, it is common to refer to generalized (tobacco mosaic), local (apple scab, corn smut), and localized (wheat bunt) diseases, according to the site of infection, the rate and path of spread of the pathogen in the host, and the site of symptom expression. Systemic diseases such as tracheomycoses are more or less limited to one tissue system, although infection often spread throughout the plant. The propensity of parasites for specific organs or tissues of susceptible hosts is usually called electivity (sometimes, selectivity).

In more general terms, we recognize a preferred habitat in the plant on the part of many parasites and we refer to these parasites as radicicolous, caulicolous, foliicolous and so on. The same can be said for internal structures such as meristems, teguments, cortex and parenchymas.

During their parasitic life most plant pathogenic micro-organisms live inside host tissues, sometimes within host cells; only a few develop on the host surface with haustoria within the epidermis. In all cases, the host serves both as food and habitat for the parasite.

In the manifestation of preferences, taxonomic relatedness is not always a primary factor. Sometimes, even related *taxa* of obligate parasites may behave differently. Thus, in general, *Erysiphaceae* are ectophytic but *Leveillula* spp. invade principally mesophyll tissues from which conidiophores arise through stomata. Such differences are sometimes still recognized as important taxonomic characters,

27

especially in *Coelomycetes*, although in many cases other criteria, such as the mode of differentiation of condia and conidiophores are now used (27, 38).

Related parasites are known to attack the same plant organs in areas geographically distant but climatologically similar (*Sclerotinia fructigena* and *S. fructicola* in Europe and North America, respectively). Where the climate is different, it is perhaps not surprising that parasites occupying the same ecological niche are unrelated. Thus, *Corynebacterium sepedonicum* and *Pseudomonas solanacearum*, cause similar vascular wilts of potatoes but the former is restricted to the northern regions of Europe, America and Asia, whereas the latter occurs mainly in tropical and warm temperate regions.

Sometimes, in the same place, but in different seasons, two non-related species alternate. On tomatoes, for example, in the Mediterranean, two imperfect fungi, *Verticillium dahliae* and *Fusarium oxysporum* f. sp. *lycopersici*, have a similar functional position in the specific habitat of the vascular system but the latter usually prevails during the summer.

To understand why plant parasites are differently distributed within a host, both the requirements of the parasite and the variability of the internal environment of the plant must be considered. As each parasite has particular nutritional and environmental requirements, its appearance and distribution within the host can be expected to occur in seasons when and at sites where the various environmental components are within the limits required for its growth and multiplication. Since these sites are not uniformly distributed in the host, infection and colonization also are not uniform. From this point of view, the site and mode of infection are often determinant factors. For example, if *Corynebacterium michiganense* enters the host through roots or through wounds on stems, it soon becomes systemic, extending to placentary tissues and seeds. However, it produces only local epicarp infections if it penetrates through small lesions on the fruit surface. Similarly, seedling infection of maize and sorghum from oospores of the downy mildew *Sclerospora sorghi* results in a systemic disease, whereas infection on leaves from conidia remains local and results in stippled necrotic areas on the leaves.

The case just mentioned also illustrates how a particular biological phase of a parasite may have a restricted distribution in the host. Other examples are that oospores of the grape downy mildew, *Plasmopara viticola*, are formed especially in the small, polygonal spots on leaves characteristic of late infections, and the various phases in the life cycles of rust and other parasitic fungi may develop on different parts of the main or alternate hosts.

The entrance of a parasite into its host may depend upon casual occurrences. Certain parasites take advantage of lesions of any kind on the plant surface such as wounds, broken trichomes, leaf or flower scars, and points of emergence of secondary roots. Others are intro- duced by vectors such as insects, mites, nematodes and fungi, or are carried by pollen into a particular organ or tissue otherwise inac- cessible to them. Still others enter hosts through intact walls of delicate organs (tender leaves, flowers or young fruits). Penetra- tion of parasites into the host may also occur through natural open- ings such as stomata, hydathodes, lenticels and nectaries or through special organs such as root hairs and flower stigmas. Many, however, enter the host directly by mechanical or chemical means. Specific structures such as appressoria and hyphal penetration pegs may be differentiated by fungi to pierce the cuticle and outer wall of epid- ermal cells. During penetration, enzymic degradation of certain components of the protective covering of the plant as well as mech- anical forces may be involved.

When more than one type of parasite is present in the same organ or tissue, competition for nutrients and for multiplication sites may lead to a territorial partition of the substratum on the basis of "first arrived - first served". In this way, competition may ex- clude the overlapping of parasites with similar nutrient requirements, such as cellulolytic wood-rotting fungi. Antibiosis, cross protec- tion and related phenomena favour spatial segregation and even exclu- sion of the weaker type or of the later invader. Sometimes, invasion of certain tissues or organs by a parasite may prevent another from differentiating reproductive structures. On citrus twigs, the trach- eiphilic "mal secco" pathogen *Phoma (Deuterophoma) tracheiphila* extends from the xylem to the bark and forms pycnidia only in areas free of *Glomerella cingulata*, a weak pathogen which readily invades the bark of mal secco-infected shoots, and produces a profusion of acervuli. Thus, distribution patterns of different parasites on the same host are usually separated. If they overlap, co-existence may be determined by the specific requirements of each parasite. Assoc- iation of two or more parasites with different requirements and path- ogenicity may initiate a sort of metabiotic relationship, although later the successor organism may prevail. Superinfection by secondary pathogens or succession by saprophytes often complicates the picture of diseases caused by specific primary parasites, for example *Phytophthora* spp.

It is thus clear that many parasites, with their ability to attack preferentially certain parts of the plant and with their restricted colonization of the host, exhibit a marked degree of spec- ialization toward tissues and organs where they find suitable envir- onmental conditions and no adverse reactions.

Obviously, the host-parasite association is not a mere nutrit- ional relationship. Besides supplying the parasite with food and water, the host shelters it from harmful radiations and desiccation, maintains stable internal physical and chemical conditions and reacts

against other potential invaders. Thus, a facultative parasite can
escape microbial competition and unfavourable conditions which it
otherwise would have to face, e.g. in the soil, and becomes segregated in the living plant. Its isolation within the host is increased either by specific adaptation to an appropriate internal environment, sometimes spatially limited, in which it may spread, or
through specific responses of the host. These may result in restriction of the parasite to certain sites either by morphological and
chemical barriers or by anomalous growth of surrounding tissues as,
for example, galls. The speed and extent of these reactions are
often a function of tissue responsiveness to infection stimuli.

Conditions in the plant which lead to a restricted colonization
of the parasite may refer only to certain stages of growth. Resistance to parasites may vary with differentiation of tissues, maturity
of organs and age of the plant. Moreover, during the development
of the plant, its inner environment changes and consequently, especially when the plant flowers, the expression of parasitism may be
fundamentally affected. In cereals, for example, smut fungi are
present throughout the growing plant and yet cause little or no disease until heads are formed ("organotropic diseases", Gäumann, 1950).
Thus, the differentiation of reproductive organs and the stimulus
of related physiological processes are apparently needed for inducing
smut fungi to manifest their pathogenicity.

ELECTIVITY AS A SPECIFIC FEATURE OF PLANT PATHOGENS

In the stages of infections by various pathogens, certain specific features may be recognized.

Detection of virus aggregates in a particular plant tissue or
organ does not exclude the possibility that low concentrations of
the virus may occur in other parts of the plant; moreover it does
not necessarily mean that the virus particles replicate there. Multiplication of tobacco ring spot virus, for example, does not occur
or is negligible in root tips of French beans to which the virus
is transported and where it accumulates (1). Usually, transport
through phloem speeds up the process by which a viral infection at
first local later becomes generalized through the plant. However,
phloem transport is not the only avenue of spread of virus particles
systemically within a host, because viruses with icosahedral, tubu-
lar, filamentous, and bacilliform particles have been observed within
xylem vessels (14) where presumably they move in the transpiration
stream.   These viruses seem to multiply in the immature tracheary
elements where they persist during differentiation. In certain diseases caused by Reoviruses, such as clover wound tumor and Fiji disease of sugar cane, tumors formed on leaves, stems or roots are
usually associated with the phloem. The spherical particles of

maize rough dwarf virus are abundant in the tumors, but seldom are
found elsewhere in the plant (31).

Plant mycoplasmas are generally restricted to the phloem, both
in sieve tubes and in phloem parenchyma, where their characteristic
pleomorphic bodies can be easily demonstrated.

Rickettsia-like organisms, such as the pathogen of phony disease of peach, and some Gram-positive bacteria, such as the agent
of Pierce's disease of grape (2), seem to thrive only in the xylem
where they are usually seen.

Particular plant tissues may be specific for certain bacteria.
Secondary meristems or undifferentiated tissues are usually required
for hyperplastic responses to the olive knot pathogen, *Pseudomonas
savastanoi* which otherwise lives epiphytically on the plant surface
(13). Frost injuries, by splitting the bark of tender shoots, often
expose the cambial face of the woody cylinder to bacterial infection
and results in the formation of knots (4). These also occur in
regular series along the branches at sites where leaf scars were infected. Development of knots is thought to be associated with production by the bacterium of auxin and cytokinins which would induce,
in succession, hypertrophy, hyperplasia and vascular differentiation
(37). Once the bacterium has died, growth of the knot stops, presumably because diffusion of these growth factors ceases.

Abnormal differentiation of specific organs is associated with
infection by certain phytopathogenic bacteria. Thus, *Agrobacterium
rhizogenes* induces susceptible plants to produce a mass of roots at
the site of infection which gives the name "hairy root" to this disease. The ability of the pathogen to stimulate the host to differentiate roots seems to be associated with the production of a root-
inducing factor which can be extracted from the cultures of the
pathogen and reproduces the syndrome of the disease (24). Soon
after inoculation, localized areas of hyperplastic growth become
apparent. If these areas are located within 200 - 300 μm of the
vascular elements, they continue to enlarge, ultimately giving rise
to roots. If, however, they occur at a greater distance, they generally cease to develop after several days (25).

Fungi display a number of topological relations with their
hosts ranging from the highly specific intracellular growth of those
belonging to the *Plasmodiophoromycetes* and *Chytridiomycetes* to the
superficial growth of some *Ascomycetes (Meliolales)*.

Electivity of plant pathogenic fungi for reproductive organs,
flowers, fruits and seeds, has been analysed by Ciccarone (1959) in
relation to infection, colonization and sporulation which may occur
only or mainly in these organs ("fructicolous pathogens"). Ergots

of cereals and grasses (*Claviceps* spp.) and similar diseases caused
in warm regions by *Balansia* spp. are well known examples. Fungi
belonging to the *Ustilaginaceae*, *Tilletiaceae*, and to other families
or classes are able to co-exist with the host after infection and
become generalized within the plant during a long period of incubation. Symptoms appear only when the parasite sporulates in the reproductive organs of the host. The path of infection varies with
the species involved (embryo or flowers, seed coats, pericarp and
husks, seedling epicotyls, stem and other organs). A high degree
of electivity is shown by several species infecting or sporulating
on certain flower parts (anthers, stigmas, ovaries and petals)
or attacking only one type of flower of dioecious plants. However,
the electivity shown by these parasites for reproductive organs is
not absolute. In certain environments, *Ustilago tritici*, *Tilletia*
spp. and *Claviceps purpurea* may also sporulate on vegetative parts.

Infection of reproductive organs by parasites which usually
invade other parts of the plant may be conditioned by chemical factors in certain floral structures. Strange and coworkers (1974)
found that infection of wheat heads by *Gibberella zeae* usually occurs
after anthesis through the extruded anthers. Headblight does not
occur or is rare in emasculated plants exposed to infection by
conidia or ascospores (36). Two fungal growth stimulants, choline
and betaine which are produced by and accumulate in the anthers,
seem to promote the initial growth of the parasite thus permitting
further invasion of the wheat heads.

Specificity of pathogens for certain organs is obvious for
diseases characterized by abnormal growth proliferation of buds,
shoots or flowers, such as witches' brooms on plum and birch trees
following infection by *Taphrina insititiae* and *T. betulina* respectively.

One of the most striking cases is a disease of cacao caused by
Calonectria rigidiuscula f. sp. *theobromae* (conidial state, *Fusarium
decemcellulare*). This fungus causes various hyperplastic conditions
in cacao the most common being the so-called "green-point galls" and
"flowery galls". Green-point galls are made up of a large number
of small vegetative buds crowded on galls on twigs and stems. Flowery galls and other flower cushion disorders are characterized by
massive flower proliferation. The ability to induce bud or flower
proliferation is associated with the heterothallic clones of the
fungus. Seed inoculation with isolates from green-point galls results in bud proliferation and gall formation at the cotyledon sites
but it fails if flower isolates are used as inocula. The green-
point gall inducing isolates of the fungus are morphologically indistinguishable from the isolates from flower proliferation and
both groups are interfertile. A few crosses between two mating types
with the ability to induce either flower or bud proliferation gave

a progeny of green-point gall inducers (15, 40). The inherent ability to cause either bud or flower proliferation has been found to
be associated with the production of different growth regulating
substances by various clones of the fungus (30). IAA is produced in
culture both by the homothallic clones and the heterothallic isolates which induce flower or bud proliferation. The latter group
(green-point gall inducers), however, produce cytokinins in addition
to IAA (29). This may contribute to the differential responses of
infected plants.

Wood rot and wood inhabiting fungi are perhaps the best known
examples of facultative parasites electivity of which for woody
tissues makes them one of the most common and devastating groups of
micro-organisms for forest, ornamental, and fruit trees. Various
types of wood decay such as brown, dry, white and soft rots are
distinguished. The species of wood-rotting micro-organisms responsible for each type of decay, even if taxonomically unrelated, possess a common ability to attack and preferentially degrade a particular class of the basic components of woody tissues,cellulose, hemi-
celluloses and lignin (26).

The ability to degrade native cellulose differs significantly
among the various classes of wood-attacking fungi. The most efficient cellulolytic species belong to the *Basidiomycetes* and, to a
lesser extent, to the *Ascomycetes* and *Deuteromycetes,* whereas only
a few of the lower fungi produce cellulases. The arrangement of
cellulose molecules and microfibrils in cell walls and their encru-
station with substances such as lignin may prevent cellulases from
reaching and hydrolysing their specific substrate. Increasing cellulose breakdown by degradation of encrusting substances enables
brown rot fungi to rot wood more efficiently than white rot fungi
although both are good producers of cellulases.

Apart from a high degree of lignification of cell walls, other
factors such as low moisture and low nitrogen content, and the
presence of inhibitory or toxic substances, account for the natural
resistance of heartwood to rotting micro-organisms. In this context,
terpenoids, flavonoids, tropolones and stilbenes are especially
important in the conifers whereas tannins and other phenolics have
major roles in angiosperms.

Sapwood of living trees is usually resistant to attack by wood-
rotting fungi but with age the non-functional wood can be extensively
degraded. Most of the very old olive trees in warm Mediterranean
countries have a hollow trunk which is often reduced to a ring of
peripheral tissues. The heartwood of living carob trees is also
easily rotted. Very few of the fungi which degrade the ligno-
cellulosic complex of non-functional woody tissues can attack living
sapwood. However, they become pathogenic if products of their metabolism diffuse to functioning vessels or reach living parenchyma.

Summer wilt ("apoplexy") of grapevines, for instance, is thought to
be caused by the sudden spreading through the plant of toxic compounds produced in the heartwood by the tinder fungus *Phellinus
(Fomes) igniarius* and other associated fungi. A chronic condition
of the same disease, the so-called "black measles" of white grapes
(6, 7, 17), is probably caused by slow diffusion of metabolites from
the rotted wood to fruits and leaves.

The electivity of tracheiphilic fungi and the origin of vascular
wilts has been discussed by various authors in the last decade (9,
10, 3, 42, 39, 23). Restriction of the pathogens to the functioning xylem is, however, only a phase, even if a main one, in the
parasitic life of these fungi. In fact, *Ceratocystis fagacearum,
C. ulmi, Cephalosporium diospyri* and other species may invade wood
parenchyma and medullary rays and *C. ulmi* may do so before it invades tracheary elements (32, 41).

Nowadays, the view that vascular fungi are weak parasites unable to colonize living host parenchyma which can react actively to
their infection, is being questioned. Many species of tracheomycotic
fungi can penetrate host leaves through stomata (*Verticillium dahliae,*
Cirulli, 1974) or through intact rhizoderm (*V. albo-atrum* and *V.
dahliae,* Isaac, 1966), as do non-tracheiphilic parasites. Similarly,
once they have entered, they have to degrade middle lamellae and
breach mechanical or structural barriers such as the endodermis
before reaching the vascular elements. On their way, they may be
impeded or arrested by host reactions, such as cell wall lignification and lignituber formation. The vascular system offers a very
suitable place for these fungi to avoid such reactions in ground
parenchyma and also permits rapid advance of mycelium through the
plant.

Adaptation of vascular fungi to conditions in the xylem is
made easy by the availability of low concentrations of nutrients in
the xylem sap and by their ability to degrade and utilize some components of the walls of vessels. Fragments of high-molecular
compounds are consequently released into the transpiration stream
together with viscous metabolites which are sometimes produced by
these fungi. These materials together with production of tyloses
and gum, all contribute to vascular dysfunction (9, 10). Also,
fungal metabolites such as growth factors and phytotoxins may diffuse
through the plant via the vascular system. They can affect plant
growth, disturb the water balance or permanently damage the integrity of cells in leaf tissues so that the latter lose their osmotic
properties and collapse. Hence, because of their electivity for
tracheary elements, the potential pathogenicity of vascular fungi
is fully expressed as a wilt disease.

SUBCUTICULAR FUNGI AS PARASITES RESTRICTED TO
SPECIFIC HOST TISSUES

A single example will be considered to illustrate in detail the complex range of factors which affect the interplay of host and parasite and restrict the latter to a well defined part of the plant.

Certain fungal pathogen which belong to the *Pseudosphaeriales* (*Venturia* and the related imperfect states *Spilocaea, Fusicladium, Pollaccia*) infect leaves, shoots and fruits of many perennial plants, especially *Rosaceae*, and grow, at least initially, beneath the cuticle. The peculiar location of these parasites (also, some species of *Taphrina*) in the host led Ducomet (1907) to consider these "sub-cuticular fungi" as intermediate between ectophytic and endophytic forms. Certain host-parasite relations have been studied in detail for the apple scab pathogen, *Venturia inaequalis*, and a few other species; their sub-cuticular habitat has been generally interpreted in terms of nutritional and ecological adaptation.

Olive leaf spot caused by *Spilocaea (Cycloconium) oleagina* has also been extensively studied in attempts to explain why this fungus remains for so long in a marginal position between the cuticle and the cellulosic part of the epidermal cell wall. In the upper epidermis of an olive leaf, the outer periclinal wall is considerably thickened by a continuous process of cutinization. Four superimposed layers can be distinguished (18, 19) : 1. an inner, non-cutinized, hydrophilic stratum or cellulosic layer, mainly of cellulose and pectin, which at its edges extends towards the anticlinal walls; 2. a thick, lenticular-shaped, slightly hydrophilic inner cuticular layer, in which cutinization is confined to a number of lipoidal islets, or drops, scattered in the pecto-cellulosic matrix which forms a permeable, acidic web; 3. a heavily cutinized outer cuticular layer, neutral or subalkaline, in which wax and partially polymerized, unsaturated lipids prevail; at the edges of the cell wall this layer extends downward with cuticular peg-like protrusions which are thrust down between adjacent epidermal cells; 4. outwardly, the cell wall is covered by a thick cuticle, appearing as a continuous sheet, including wax.

The pathogen forms round, flat, sub-cuticular colonies growing parallel to the leaf surface. Starting from the point of penetration of the infection hypha, a hyaline, septate mycelium develops into branched radiating ribbons of mono-layered strands just beneath the cuticle. It invades only the outer cuticular layer extending to the protrusions of this layer between adjacent epidermal cells. The constituents of this layer are utilized by the advancing hyphae as sources of food. Conidiophores arising at right angles from the sub-cuticular mycelium pierce the cuticle by mechanical and enzymic action.

The results of the investigations seem to suggest that the peculiar location of *S. oleagina* within the outer cuticular layer results from the concomitance of a number of favourable conditions of the substratum and of certain unfavourable or adverse events (20) occurring in leaf tissues after infection. These are described below.

Favourable conditions of the substratum. The outer cuticular layer of the epidermal cell appears to offer an appropriate medium for growth of the pathogen. All constituents seem to be enzymatically degraded and the products utilized by the pathogen. Pectolytic, cellulolytic, and lipolytic enzymes produced by *S.oleagina* are active at the pH values in the outer cuticular layer (22). Olive cutin is readily hydrolysed by the fungus which is able to grow on it as the sole carbon source (34). Thus, once the infection hypha has pierced the cuticle and reached the hydrophilic part of the cell wall through which it can obtain water and solutes, the parasite can freely spread tangentially as immersed, radiating colonies. Neither mechanical nor chemical barriers against the fungus are present in the outer cuticular layer. A continuous water, and presumably, solute supply diffuses to the cuticular layer through the permeable, non-cutinized parts of the epidermal cell wall. Outwardly, effective protection of the monolayered colonies of *S. oleagina* from sunlight and desiccation is provided by the thick cuticle.

Unfavourable and adverse events in leaf tissues. The healthy olive leaf possesses a natural resistance to macerating enzymes which is conferred chiefly by phenolic compounds. On infection, this is enhanced in the diseased, but not invaded tissues underlying and around the scab spots (21). In susceptible olive leaves, infection by *S. oleagina* seems to trigger a series of reactions leading to an increased synthesis of phenolics including the major phenolic glucoside of the olive leaf, oleuropin. This bitter compound, which can reach a concentration of 2%, can be mobilized and hydrolysed by a β-glycosidase, thus liberating the water-soluble aglycone which inhibits the pectolytic enzymes of *S. oleagina*. In order to invade the epidermis and the mesophyll from its sub-cuticular position, the scab fungus must degrade the middle lamellae or penetrate the cell walls. However, it is prevented from breaking down the hydrophilic layers of cell wall and lamellae so long as the phenolic inhibitors of its pectolytic enzymes are present in an active form. Thus, the pathogen grows in a sub-cuticular position away from tissues where a host defense reaction may occur. The efficiency of this response seems to depend on rapid hydrolysis and translocation of the aglycone to the cell walls rather than simply in increased synthesis of oleuropin. Thus, it is conceivable that these events proceed actively soon after infection while the parasite is engaged in the slow degradation of the host cuticle. As the sub-cuticular growth of the fungus progresses and the phenolic inhibitors are inactivated by oxidation and polymerization, more of the glucoside is presumably hydrolysed so that a sufficient level of phenolic aglycone is present in the host tissues.

In old scab spots formed on leaves and fruit, other barriers, such as suberization, tend to exclude any further activity of the parasite other than new sub-cuticular growth at the margin of the spots. In senescent leaves, the defense reaction outlined above is attenuated or stopped, the products of oleuropin degradation are inactivated, and the pathogen eventually invades the mesophyll.

CONCLUSION

We have seen that many concurrent factors are involved in processes determining whether an infection will result in limited invasion of the host. These factors can be referred to two broad classes.

1.   Those endowing a parasite with pathogenicity towards a plant tissue or a system of tissues, and conferring on it the ability to overcome plant defense mechanisms operating locally.

2.   The inherent properties of various plant tissues which in various ways affect the course and outcome of the infectious process and confine the pathogen and/or its effects to a limited portion of the host.

Biochemical and physiological research on these lines is needed for most plant pathogens showing electivity for host tissues and organs, a field which appears to have been largely neglected by plant pathologists. Thus, we can repeat for plant diseases what Dubos (1954) has stated for medicine, that "the study of tissue factors which are responsible for arresting the process of infection, or for allowing it to evolve into overt disease, is still in the most primitive state".

REFERENCES

1.   ATCHINSON, A. and FRANCKI, R.I.B. (1972). The source of tobacco ringspot virus in root-tip tissue of bean plants. *Physiol. Pl. Path.*, 2, 105 - 111.

2.   AUGER, J.G., SHALLA, T.A. and KADO, C.I. (1974). Pierce's Disease of Grapevine : Evidence for a Bacterial Etiology. *Science, N.Y.*, 184, 1375 - 1377.

3.   BECKMAN, C.H. (1964). Host responses to vascular infection. *A. Rev. Pl. Path.*, 2, 231 - 252.

4. CICCARONE, A. (1950). Alterazioni da freddo e da rogna sugli ulivi, esemplificate dai danni osservati in alcune zone pugliesi negli anni 1949 - 1950. *Boll. Staz. Patol. veg. Roma,* Ser. 3, 6 (1948), 141 - 174.

5. CICCARONE, A. (1959). Reproduction is affected. *In : Plant Pathology. An Advanced Treatise.* (HORSFALL, J.G. and DIMOND, A.E., Eds.), Vol. I, 249 - 312. Academic Press, New York and London.

6. CHIARAPPA, L. (1959) Wood decay of the grapevine and its relationship with black measles disease. *Phytopathology,* 49, 510 - 519.

7. CHIARAPPA, L. (1959a). Extracellular oxidative enzymes of wood-inhabiting fungi associated with the heart rot of living grapevines. *Phytopathology,* 49, 578 - 583.

8. CIRULLI, M. (1974). Infezioni locali da *Verticillium dahliae* Kleb. su foglie di Pomodoro. *Phytopath. medit.,* 13, 23 - 26.

9. DIMOND, A.E. (1970). Biophysics and biochemistry of the vascular wilt syndrome. *A. Rev. Pl. Path.,* 8, 301 - 322.

10. DIMOND, A.E. (1972). The origin of symptoms of vascular wilt diseases. *In : Phytotoxins in plant diseases.* (WOOD, R.K.S., BALLIO, A. and GRANITI, A., Eds.), 289 - 306. Academic Press, London and New York.

11. DUBOS, R.J. (1954). *Biochemical determinants of microbial diseases.* Harvard University Press, Cambridge, Mass., 152 pp.

12. DUCOMET, V. (1907). *Recherches sur le développement de quelques champignons parasites à thalle subcuticulaire.* Guillemin et Voisin, Rennes, 287 pp.

13. ERCOLANI, G.L. (1971). Presenza epifitica di *Pseudomonas savastanoi* (E.F. Smith) Stevens sull'Olivo, in Puglia. *Phytopath. medit.,* 10, 130 - 132.

14. ESAU, K. and HOEFERT, L.L. (1971). Cytology of beet yellows virus infection in *Tetragonia.* II. Vascular elements in infected leaf. *Protoplasma,* 72, 459 - 476.

15. FORD, E.J., BOURRET, J.A. and SNYDER, W.C. (1967). Biologic specialization in *Calonectria (Fusarium) rigidiuscula* in relation to green point gall of cocoa. *Phytopathology,* 57, 710 - 712.

16. GÄUMANN, E. (1950). *Principles of plant infection*. Crosby
     Lockwood, London, 543 pp.

17. GRANITI, A. (1960). Il "mal dell'esca" della vite in Puglia.
     Italia agric., 97, 543 - 550.

18. GRANITI, A. (1962). Osservazioni su *Spilocaea oleagina* (Cast.)
     Hugh. I. Sulla localizzazione del micelio nelle foglie
     di Olivo. *Phytopath. medit.*, 1, 157 - 165.

19. GRANITI, A. (1965). Osservazioni su *Spilocaea oleagina* (Cast.)
     Hugh. III. Struttura submicroscopica della parete epider-
     mica fogliare dell'Olivo sana o invasa dal fungo. *Phyto-
     path. medit.*, 4, 38 - 47.

20. GRANITI, A. (1965a). Defence mechanisms in plants infected by
     scab or leaf-spot fungi. *In : Atti del Seminario di
     studi biologici*. (QUAGLIARIELLO E., Ed.), Vol. 2, 217 -
     228. Adriatica Editrice, Bari.

21. GRANITI, A. and DE LEO, P. (1966). Osservazioni su *Spilocaea
     oleagina* (Cast.) Hugh. IV. Resistenza alla macerazione
     enzimatica come probabile reazione di difesa delle foglie
     di Olivo al parassita. *Phytopath. medit.*, 5, 65 - 79.

22. GRANITI, A., DE LEO, P. and BAGORDO, F. (1962). Osservazioni
     su *Spilocaea oleagina* (Cast.) Hugh. II. Attività enzima-
     tiche del fungo in relazione al suo insediamento nelle
     foglie di Olivo. *Phytopath. medit.*, 2, 20 - 36.

23. GRANITI, A. and MATTA, A. (1969). Indirizzi attuali degli studi
     sulla patogenesi delle malattie vascolari. *Annls. Phyto-
     path.*, 1, No. hors-série, 77 - 120.

24. HOPKINS, D.L. and DURBIN, R.D. (1971). Induction of adventit-
     ious roots by culture filtrates of the hairy root bacterium,
     Agrobacterium rhizogenes. *Can. J. Microbiol.*, 17, 1409 -
     1412.

25. HUISINGH, D. and DURBIN, R.D. (1968). Histological observations
     on root formation induced by *Agrobacterium rhizogenes*.
     Phytomorphology, 18, 334 - 338.

26. HUDSON, H.J. (1972). Fungal saprophytism, Arnold, London. 68
     pp.

27. HUGHES, S.J. (1953). Conidiophores, conidia and classification.
     Can. J. Bot., 31, 577 - 659.

28. ISAAC, I. (1966). Vascular disease pathogens with particular
     reference to the genus *Verticillium*. *J. Indian bot. Soc.,*
     45, 209 - 231.

29. LERARIO, P. (1975). Produzione di sostanze regolatrici di
     crescita da parte di *Calonectria rigidiuscula* f. sp.
     theobromae. *Phytopath. medit.* (In press).

30. LERARIO, P., GRANITI, A. and REYES, H. (1972). Production in
     culture of plant growth regulating substances by *Calonect-
     ria rigidiuscula* f. sp. *theobromae,* the incitant of cacao
     galls. *Inftore bot. ital.,* 4, 209 - 211.

31. MILNE, R. and LOVISOLO, O. (1974). Viruses that cause grass
     tumors. *New Scient.,* 113, 252 - 253.

32. OUELLETTE, G.B. (1962). Studies on the infection process of
     Ceratocystis ulmi (Buism.) C. Moreau in American elm trees.
     Can. J. Bot., 40, 1567 - 1575.

33. PETRI, L. (1930). Lo stato attuale delle ricerche sul "mal
     secco" dei limoni. *Boll. Staz. Patol. veg. Roma,* n.s.,
     10, 63 - 107.

34. SPARAPANO, L. and GRANITI, A. (1975). Cutin degradation of two
     scab fungi, *Spilocaea oleagina* (Cast.) Hugh. and *Venturia
     inaequalis* (Cke.) Wint. (In press).

35. STRANGE, R.N., MAJER, J.R. and SMITH, H. (1974). The isolation
     and identification of choline and betaine as the two com-
     pounds in anthers and wheat germ that stimulate *Fusarium
     graminearum in vitro*. *Physiol. Pl. Path.,* 4, 277 - 290.

36. STRANGE, R.N. and SMITH, H. (1971). A fundal growth stimulant
     in anthers which predisposes wheat to attack by *Fusarium
     graminearum*. *Physiol. Pl. Path.,* 1, 141 - 150.

37. SURICO, G., SPARAPANO, L., LERARIO, P., DURBIN, R.D. and
     IACOBELLIS, N. (1975). Cytokinin-like activity in extracts
     from culture filtrates of *Pseudomonas savastanoi*. *Experi-
     entia* (In press).

38. SUTTON, B.S. (1973). Coelomycetes. *In : The Fungi.An advanced
     treatise*. (AINSWORTH, G.C., SPARROW, F.H. and SUSSMAN,
     A.S., Eds.), Vol. IVA, 513 - 582, Academic Press, New York
     and London.

39. TALBOYS, P.W. (1968). Water deficits in vascular diseases.
     In : Water deficits and plant growth. (KOZLOWSKI, T.T.,
     Ed.), Vol. 2, 255 - 311. Academic Press, New York and
     London.

40. THOMAS, D.L. and SNYDER, W.C. (1970). Relationships among
       clones of *Calonectria (Fusarium) rigidiuscula* isolated
       from gall diseases of *Theobroma cacao*. *Phytopathology*,
       60, 1542 (Abstr.).

41. WILSON, C.L. (1965). *Ceratocystis ulmi* in elm wood. *Phyto-
       pathology*, 55, 477.

42. WOOD, R.K.S. (1967). *Physiological Plant Pathology*. Blackwell,
       Oxford and Edinburgh, 570 pp.

CONTRIBUTIONS

MOREAU, M., DUBOUCHET, J. and CATESSON, A.-M.   Variation of iso-
     peroxidases in carnation tissues after infection by *Phialophora
     (Verticillium) cinerescens*.

SUMMARY OF POINTS FROM CONTRIBUTIONS AND DISCUSSIONS
BY
W. C. SNYDER

Chairman and Discussion Leader

     The study of specificity in plant disease is plant pathology.
In a broad sense all diseases are specific. Some plant pathogens
possess narrow host ranges, others possess broad ones. No pathogen
is known to attack all species of plants; in fact resistance is the
common situation. Simply stated, a disease is the product of an
interaction between a pathogen and a host in a given environment.
Our concern is with the detailed phenomena of specificity. What
chemical-physical phenomena take place during the initial challenges
of a micro-organism to a plant and what physiological-biochemical
interactions ensue ?

     Each corner of the disease triangle is subject to great var-
iability. There is a multiplicity in the kinds and syndromes of
disease but susceptibility is not the rule. There is no denial
that a large variety of plants live on earth but there can be no
agreement on the number of species that exist. Species are named
by men, and who would expect scientists to agree upon each others'
species delimitations ? Yet the degree of specificity recognized
depends upon whether a "splitter's" or a "lumper's" concept of
species is followed.

     It has been shown by Ciccarone that the taxonomy and nomen-
clature of pathogens are in a state of flux. Mycologists, for example, do not agree on the classification or the nomenclature of

many pathogens, and even if they did the name itself is always subject to change. In the early literature, members of a species were
thought of as identical. Later, after the concept of variability in
organisms became established it was recognized that a species is a
group of similar but genetically different individuals which resemble each other more than they do members of other groups. Eventually we will swing to the more realistic concept of larger, highly
variable species in all biology.

Use of a single culture or clone of a pathogen to study specificity is to consider an individual to be representative of a
normal, genetically variable species. Thus one investigator's
species may be another investigator's clone. It is essential for
each to identify clearly the test organism(s) with which one is
working.

The situation is, in fact, more complicated than this. Most
fungi when kept in culture mutate, some more than others. In a
rust fungus the mutant may represent another race, while a mutant
of a pathogenic *Verticillium* or *Fusarium* or *Colletotrichum* species
may be much less virulent than the wild type from which it arose.
Some of these mutants would not survive in nature. Therefore, the
findings from a study of such a mutant would not necessarily be
representative of what happens in nature.

The question raised by a realization of the confusion and general lack of agreement on the delimitation and nomenclature of plant
pathogens is whether we shall set up one world center to identify
all species for us, and to decide on one system of taxonomy and no-
menclautre ? The answer certainly is an emphatic no, that is, if we
wish to promote progress and change for the better.

Genetic variability is perhaps better recognized among higher
plants, than among micro-organisms. Those who breed plants and collect their progenies, rapidly become aware of the enormous genetic
diversity that may exist in species. To test the interaction of a
pathogen with the genetically variable individuals of a host species
by field exposures in different regions of the world is good. But
the test could be more meaningful if the fungus were bred too, as
the host plant is, to reveal its fuller potential.

It is generally accepted that a pathogen may be selective for
the host tissue or organ of the species which it attacks. For example, *Venturia* spp. invade the cuticular tissue of their hosts
where they seem to be limited during the life of cuticularized structures such as leaves and fruit.   Upon death the pathogens may enter
other tissues and there initiate the perfect state. As pointed out
by Graniti, this is in contrast to a pathogen of cacao, *Calonectria
(Fusarium) rigidiuscula* f. sp. *theobromae*, clones of which may infect

flower cushions, twigs and branches and may cause cankers on the trunk. Infected seed is attacked below ground at germination, and the fungus invades the cotyledons and eventually the meristematic tissue underlying them. Leafy galls develop at the site of the coty-ledons if the proper clone is involved, and later, when the flower cushion is invaded. A different but interfertile clone of the path-ogen causes excessive flower proliferation when it infects the meris-tematic tissue of the flower cushions. At the death of infected twigs and branches perithecia develop abundantly both in the tree and in infected debris on the ground. It is evident that both mating types of the fungus are abundantly present in nature, and that the various clones of this heterothallic pathogen infect differentially the meristematic tissues of host organs. The pathogen however is highly selective for *Theobroma cacao*, yet the production of leafy galls, or of flower proliferations depends upon the genetic capab-ilities of the particular infecting thallus when environmental con-ditions are suitable.

Rhizoctonia (Thanetephorus) solani has a broad host range, and a rather selective specificity for cortical tissues of the hypocotyl of young seedlings. This species is highly variable genetically. Some clones attack roots deep in the soil, and others attack the internal tissues becoming systemic in the vascular tissues of such plants as wheat and pepper. In Central America *R. solani* may prod-uce enough basidiospores on the above-ground stems of the bean plant to cause leaf spots on the leaves. These effects are greatly influ-enced by the environment but they reflect the great versatility and genetic variability of the pathogen.

This brings us to the powerful influence of environment upon specificity. Each pathogen has its range of temperature, humidity, light, nutrition, etc. required for the disease cycle and its life cycle.

Wilt diseases caused by *formae speciales* of *Fusarium oxysporum*, possess chlamydospores for survival in soil, mycelium for the in-vasion of susceptible roots, micro-conidia for movement in the xylem, and macroconidia produced on the surface of infected, above-ground parts for water-borne spread. Under a moist, temperate condition aster plants infected by *Fusarium oxysporum* f. sp. *callestephi* dev-elop masses of pink Fusarium spores in sporodochia emerging from the cortex of the stem. These conidia when washed into the soil rapidly convert to chlamydospores. Another such pathogen, *Fusarium oxysporum* f. sp. *albedinis* attacks the data palm in the Sahara desert. There the macroconidia of the Fusarium state are not seen, nor do they function in the disease cycle. This Fusarium is specific for its host yet the Fusarium state is excluded from its full cycle by the environment.

Fusarium spp. and perhaps most soil-borne pathogens require an

outside source of nutrients in order to germinate and to attack the
host plant. This is true of the soil-borne chlamydospores of the
Fusaria causing wilts and root rots. It appears also to be true of
Fusarium (Gibberella) roseum 'Graminearum' which attacks the roots,
stems and heads of wheat. Choline and betaine have been reported to
be the nutrients in anthers which enable spores of the pathogen to
cause the scab disease of wheat.

When *Fusarium (Gibberella) moniliforme* occurs on maturing rice
plants in Japan, the grain may become pink with macroconidia. Yet
the same fungus on the same variety grown in Taiwan causes no apparent infection of seed, but stems of infected plants may become blue-
black with the abundant production of the Gibberella state which is
less often seen in Japan. The climatic differences of nearby areas
can dramatically affect fungus cycles and the expression of diseases.
In what way do slight differences in the environment so strongly influence disease development, and the specificity of disease for host
organs and tissues ?

Thus, in studies aimed at the intimate phenomena which take
place between host and pathogen as expressed in disease specificity,
we must continually be aware of the enormous variability in the pathogen, the host, and the environment. Studies involving one clone
of the pathogen and one clone of the host, in one fixed set of environmental conditions may allow analysis of a phenomenon. But until
the result can be related to what happens in nature it is difficult
to know if the finding represents a basic biological phenomenon or
an artefact.

Plant pathology is concerned with the nature of disease and its
control. The pressure is now for new approaches to more efficient
control of diseases of agricultural crops around the world. Meaningful analyses of specificity in disease can provide the fundamental knowledge we need for a better understanding of the nature of
disease and for providing the new approaches so urgently needed for
disease control.

Professor Moreau summarized her work with Dubouchet and Catesson
which showed that the same isoperoxidases are present in healthy
carnation plants and in carnations infected by *Phialophora cinerescens*
but that infection does lead to increases in the amounts of these
enzymes. The effects of infection on the ultrastructure of cell
walls of xylem were also described as were the effects of filtrates
from cultures of the pathogen on the peroxidase activity of extracts
from roots of *Lens culinaris*.

GENETICS OF HOST-PARASITE INTERACTIONS

ROY JOHNSON

Plant Breeding Institute

Cambridge, U.K.

INTRODUCTION

Accentuation of Specificity in Host-Parasite Interactions

Most plant pathogens, whether obligate parasites (biotrophs) or facultative saprophytes (necrotrophic pathogens) have a host range limited to few species. For host and parasite to survive and evolve together it is necessary for equilibrium, either dynamic or stable, to be established between them, such that each is capable of adequate reproduction. Examination of wild plant communities reveals, as for *Solanum* species and *Phytophthora infestans,* that their resistance may comprise genetic elements which show pronounced specificity to different strains of the pathogen and other elements which apparently do not show such specificity. During development of cultivated crops several processes have occurred which have tended to accentuate the effects of specificity. Among these I believe that the following are of importance.

Geographic separation of host and pathogen. Most plant species are resistant to most plant pathogens so that in the long term separate evolution of a host and pathogen might be expected to lead to a loss of compatibility between them. In the short term during which certain plant species have been transported to areas where some of their existing pathogens are absent the process has often been the reverse of this, and greater compatibility between host and pathogen has developed. Thus the arrival of *Phytophthora infestans* in Europe caused severe epidemics of late blight in the 1840's on potato varieties which had been selected in the absence of the disease. Similar examples include maize rust in Africa (32) and coffee rust in Brazil in 1969 (17) both of which caused serious

epidemics on crops which had for many years been selected for yield
and quality in the absence of the pathogens.

Methods of using race-specific resistance in plant breeding.
In attempting to control undesirable susceptibility to disease in
crops, plant breeders have sought varieties with high levels of resistance, preferably controlled by one or a few dominant genes,
making transfer in breeding simple without incorporating other undesirable characters. In numerous cases this type of resistance
has proved to be race-specific, but this has often only become evident after the breeding programme has been completed. During the
programme the resistance has often been complete, and the attention
of the breeder could be turned to selection for other characters,
some of which may have been derived from plants with high levels of
susceptibility to the pathogen against which the resistance gene is
effective. There is no selection against the susceptible component
derived from such a parent since the presence of the resistance gene
prevents any assessment of its effect. When, later, the pathogen
population acquires virulence against the resistance gene, the background level of resistance is sometimes much lower than in varieties
in which no such protection by a race-specific gene has been provided. Van der Plank (40) called this the 'Vertifolia' effect after
the potato variety of the same name which was initially resistant
to late blight due to the presence of resistance *(R)* genes but was
shown to be more susceptible than varieties without *R* genes to
races capable of overcoming those in 'Vertifolia'.

Genetic uniformity of crops. The effects of specificity are
more noticeable and can be more damaging when they occur in crops
which are genetically uniform. This was dramatically shown in the
epidemic caused by *Helminthosporium maydis* race T on maize in the
United States in 1970 (17) which occurred on what was probably the
most extensive genetically uniform substrate ever presented to a
plant pathogen. There were, however, many earlier examples particularly in cereals where genetic uniformity of varieties played an
important part in the generation of epidemics such as those of stem
rust in the United States in 1916, 1935, 1953 and 1954.

It should be remembered, however, that genetic uniformity alone
does not create specificity in disease resistance but makes it more
apparent and enhances the spread of pathogen strains with specific
virulence. It may also create an environment suitable for selection
of pathogen strains with narrower host ranges but increased pathogenicity on selected hosts.

The case of Race T of *H. maydis* on maize in the United States
in 1970 is particularly important for plant breeders as well as
for those interested in mechanisms of resistance. The cytoplasmic
character which led to increased susceptibility to the pathogen was
incorporated for its ability to cause male sterility; it was not

expected that this would influence the ability of any pathogen to cause disease on maize, and there was no apparent reason, therefore, to fear the uniformity which was consciously created.

Above I have pointed out three ways in which naturally occurring specificity in resistance and susceptibility to disease in plants is thrown into prominence by the methods of production of plant varieties and crops which bring about genetic changes and increased genetic uniformity. Further aspects of these problems are discussed by Wolfe (43). I will now consider in more detail some aspects of specificity and its genetic control which raise interesting challenges for the plant breeder and I hope also for those interested in the mechanisms of specificity.

SPECIFICITY AND ITS GENETIC CONTROL

I will discuss these topics under the following three categories.

1. Specificity in host range.

2. Degrees of specificity in host-parasite interactions.

3. The role of specific genes in induced resistance.

I will refer mainly to rust diseases of cereals and will illustrate some points with results from recent work on *Puccinia striiformis* which causes yellow rust on cereals and grasses.

Host Range

Early investigations of cereal rusts suggested that species could be sub-divided according to their ability to infect different host species. Such divisions were found not to be rigid, however, and there have been many reports that wild and cultivated grasses can be hosts to rust fungi that can also attack cereals. Sanford and Broadfoot (34) showed that *P. striiformis* which could infect wheat was present on various *Agropyron* species and on *Hordeum jubatum* in Canada; several other grass species were found to be infected by *P. striiformis* in the United States (16). D'Oliveira and Samborski (11) showed that there was a complex relationship between the various strains of *Puccinia recondita* both in their ability to infect uredial hosts of *Aegilops, Agropyron, Triticum* and *Secale,* and their aecial hosts which include species of *Thalictrum* in the Ranunculaceae and *Anchusa* and *Echium* in the Boraginaceae.

Recently there have been several descriptions of the genetic control of specialization into *formae speciales* in rusts and powdery mildews. Cotter and Roberts (8) showed that hybrids between

Puccinia graminis f. sp. *avenae* and f. sp. *agrostidae* were non-virulent on many oat varieties which had no known genes for resistance
to the oat-attacking form, but that known genes for resistance did
not condition resistance to the hybrids. Some varieties of oats
were more susceptible to the hybrids than to the parental race 7 of
f. sp. *avenae*. Eshed and Dinoor (12) tested the host range of six
formae speciales of *Puccinia coronata* in Israel on more than 100
native grass species. They found variations of resistance in grass
samples within species but also found several host species which
were susceptible to more than one form of the rust pathogen. In
crosses between *formae speciales* some progeny with a wider host
range than that of the parents were found. Hiura (15) hybridized
Erysiphe graminis f. sp. *agropyri* with f. sp. *tritici* and found that
hybrid cultures were non-virulent on many wheat varieties which were
susceptible to the wheat attacking parental strain. He suggested
that the f. sp. *agropyri* possessed many genes for non-virulence on
wheat and also noted that those isolates with low virulence on *Agropyron* tended to show virulence to many wheat cultivars, and *vice
versa*. When mixtures were grown on *Agropyron,* cultures with virulence on many wheat varieties tended to decrease.

There have been several investigations of the genetic control
of the specialization of *P. graminis* into *formae speciales* attacking
wheat and rye. Using sexual hybrids produced on barberry, Green
(14) showed that some F_1 and F_2 cultures of hybrids between *P. graminis* f. sp. *tritici* and *P. graminis* f. sp. *secalis* could infect
winter wheat or rye and some could infect both wheat and rye. Some
of the latter group were tested on wheat stem rust differential varieties and could only infect 'Einkorn'. Although these hybrids had
a wider host range than their parents the infection types were lower than those of the parental cultures on their respective host
species. It was noted that barley was more susceptible to the F_1
cultures than was either wheat or rye.

In more extensive tests, using more varieties of wheat and rye,
Sanghi and Luig (35) showed that sexual or somatic hybrid cultures
between wheat- and rye-attacking forms of *P. graminis* could reveal
genes for resistance in wheat varieties which were usually regarded
as susceptible to *P. graminis* f. sp. *tritici*. They also showed that
some genes for resistance to *P. graminis* in wheat, such as *Sr-8,*
apparently gave resistance to all the hybrid cultures tested whereas *Sr-11,* which gave resistance to the parental wheat-attacking culture did not give resistance to all the hybrid cultures. They suggested that two systems of genes were present in wheat, one providing specific resistance to rye stem rust, but not interacting with
wheat stem rust, and another providing resistance to strains of
wheat stem rust and sometimes interacting with rye rust.

Somatic hybrids between wheat- and rye-attacking forms of *P.
graminis* were reported by Bridgmon and Wilcoxson (5) who showed that

the hybrids had increased host ranges which included both wheat and
rye, although infection types were lower on the wheat than were
those produced by the parental wheat-attacking culture. Luig and
Watson (23) discussed the role of wild and cultivated grasses in
permitting formation of somatic hybrids between ff.spp.*tritici* and
secalis of *P. graminis*. *Hordeum leporinum* and *Agropyron scabrum*
were both susceptible to these two *formae speciales* and allowed the
production of somatic hybrids which could attack both wheat and
rye cultivars. It was also reported that *P. graminis* f. sp. *secalis*
could attack certain barley cultivars, that the rust from *A. scabrum*
was more pathogenic on barley and that *P. graminis* f. sp. *tritici*
was the most pathogenic of the three on barley. These authors suggested that their findings had significance for breeding for resistance especially when genes for resistance are being transferred
to wheat from rye or a grass species. Green (14) also referred to
this problem but thought it possible that the rusts had evolved
from more primitive forms with wide host ranges and relatively low
pathogenicity to more specialised forms with narrower host ranges
and greater pathogenicity. If resistance on a sufficiently broad
base could be incorporated into crops, using such sources as rye
and grasses, he suggested that it might be possible to cause a reversal of the evolution.

     In practice transferring genes to wheat from other species has
been difficult and those so far transferred have not achieved the
objective of producing reduced pathogenicity in the pathogen. Resistance to *P. recondita* was transferred from *Aegilops umbellulata*
to wheat by Sears (36) but a culture of *P. recondita* highly virulent on the cultivar, 'Transfer',was found by Samborski (33). Resistance to *P. striiformis* transferred from *Aegilops comosa* to wheat
(31) was effective against European races of the pathogen but not
to the Kenyan population (R. Little, personal communication). A
chromosome from rye, carrying resistance to *P. graminis, P. recondita, P. striiformis* and *Erysiphe graminis* was transferred to wheat
by Riebesel and has become very widely distributed in European
wheat cultivars (26, 45). Races of *P. recondita* capable of overcoming the resistance have appeared in Romania (27), and in races
of *P. striiformis* in the Netherlands (R.W. Stubbs, personal communication). In most cases the levels of susceptibility of the attacked varieties was high.

     All these results suggest that the differences between *formae
speciales* in rusts and mildews are due to increasing numbers of
specific genes for non-virulence in the pathogen and increasing
numbers of specific genes for resistance in the hosts. They also
raise an interesting question about the specificity of susceptibility.
Despite the very specific compatible and incompatible relationships
which occur between a *forma specialis* and its host species, compatible relationships with other species can exist, as in the ability of wheat-attacking and rye-attacking rusts to infect species of

barley. This evidence, and the fact that resistance has declined
in crop varieties selected in the absence of pathogens suggests to
me that whereas susceptibility of a host species to a pathogen is
obviously highly specific, resistance is even more so within the
close relationship established between the host and pathogen. If
it is assumed that the pathogen must induce a state favourable to
its growth by specific activity it seems surprising that whereas,
for example, a rust can achieve this on hosts as diverse as wheat
and barley it cannot do so for all wheat varieties. It therefore
seems more likely that resistance results from a specific interaction between the host and pathogen and that susceptibility results
from the absence of such an interaction. In this context it is not-
able that, as pointed out by Brian (4), the vesicular arbuscular
mycorrhizae have a wide host range despite their biotrophic growth
habit. This again indicates that susceptibility to a biotrophic
organism need not be a highly specific condition. I suggest that
this difference between the pathogens and the symbiotic mycorrhizae
in specificity of relationship with the host could result from effects of selection during evolution. A pathogen must, by definition,
be detrimental to the host. There will therefore be selection for
any mechanism in the host which provides resistance, and for any
mechanism in the pathogen overcoming such resistance. In the case
of a symbiotic relationship there is no advantage to the host in
developing resistance and when such resistance occurs it will not
provide a selective advantage for the host. Indeed, it may be detrimental. There will, therefore, be no development of corresponding
systems for resistance in the host and the ability by the pathogen
to overcome the resistance. This again suggests greater specificity
in resistance mechanisms than in the condition of susceptibility.

Degrees of Specificity in Host-Parasite Interactions

As already mentioned the degree to which specificity is manifest
in host-parasite relations appears to have been accentuated in crop
plants by several factors. One of the greatest challenges to plant
breeders is to reverse this trend. How this should be done is a
matter for much speculation and often heated argument.

The idea that resistance to pathogens could be divided into
distinct components which could be labelled as race-specific (vertical) or non-race-specific (horizontal) probably originated from
work on potatoes for resistance to the late blight pathogen *Phytophthora infestans*. Much original work was reviewed by Thurston (39)
who gives many other terms used to describe horizontal resistance.
Black and Gallegly (3) defined what they called 'field resistance'
as all forms of inherent resistance that plants possess with the
exception of hypersensitivity as controlled by R genes. The R genes
were resistance genes derived from *Solanum demissum* and gave resistance which was expressed as small chlorotic or hypersensitive
flecks. The components of this 'general' resistance were described

under such categories as the following.

1. The chance that a spore will successfully penetrate a leaf.
2. The extension rate of mycelium in a leaf. 3. The rate of
sporangial production (42). It was, apparently, assumed by some
workers that these characters would not show race-specificity.

   Evidence from work with cereals does not support such a hypothesis. It is clear that rate of spore production, or such characters as slow rusting are not, themselves, a guarantee of non-spec-
ificity. Browder (6) showed that the wheat cultivar 'Bulgaria 88',
which was described as slow rusting by Caldwell *et al.* (7) possessed
two genes giving a low infection type (resistant) to *P. recondita,*
one of which was *Lr-11* and the other a previously unknown but race-
specific gene. Browder (6) pointed out that a gene conditioning a
visually measurable low infection type may have a 'slope-flattening'
effect on disease development similar to that attributed by Van der
Plank (41) to horizontal resistance.

   Working with *P. striiformis* we showed that slight differences
in high infection types in two wheat cultivars 'Joss Cambier' and
'Hybrid 46' were race-specific. Isolate WYR 69/10 produced a 3+
reaction (heavy sporulation with slight chlorosis) on 'Joss Cambier'
and a 4 reaction (heavy sporulation, no chlorosis) on 'Hybrid 46',
whereas the reverse was true of isolate WYR 71/2. These slight differences were reflected in large differences in sporulation of seedlings of the two varieties (20, 21). The differences in spore production on seedlings of 'Joss Cambier' were correlated with differences in percentage leaf area infected in the field (Table 1).

TABLE 1. *Production of spores on seedling leaves of wheat*
          cultivar 'Joss Cambier' and per cent leaf area per
          plot infected in field trials after inoculation
          with isolate WYR 69/10 or isolate WYR 71/2 of
                   P. striiformis

Test	Isolates of *P. striiformis*		L. S. D.
	WYR 69/10	WYR 71/2	(P = 0.05)
Spore production mg/100 cm^2 leaf			
Seedling leaves	49	187	36.5
Per cent leaf area per plot infected			
Field (25 June 1972)	23	44	5.0
Field (18 June 1974)	14	24	5.8
Field (19 June 1975)	28	46	5.9

These results and those of Lupton and Macer (25) show that
race-specific resistance to *P. striiformis* in wheat can be conditioned by genes for low infection type from the seedling stage on-
wards as well as by genes (not yet identified) of much smaller visual effect in seedlings, affecting rate of rusting in the field.

Despite these demonstrations of race-specificity there are
examples of wheat cultivars which have been widely grown in Britain
for many years without becoming severely infected with *P. striiformis*. We do not have sufficient evidence to state that their resistance is not race-specific or that it would be permanent and I
therefore prefer to refer to them as having durable resistance,
which simply means their resistance has remained effective for a
long time while the variety was widely grown.

In genetic analysis of one such cultivar, 'Little Joss', (24)
we obtained evidence that resistance appeared to be polygenically
controlled. This was suggested by our failure to obtain discontinuous segregation into discrete classes with different levels of
susceptibility in the F2 from a cross between 'Little Joss' and a
susceptible variety. This result may be surprising in view of the
fact that Biffen and Engledow (2) selected 'Little Joss' as a stable
and true breeding line in F3, from the cross of a susceptible English
cultivar 'Squarehead's Master' with 'Ghirka', a Russian wheat susceptible to yellow rust at the seedling stage, but resistant at
later stages. These results may suggest that 'Squarehead's Master'
already possessed several genes for resistance and only a single
further one was obtained from 'Ghirka' in the selection of 'Little
Joss'. Another possibility is that there are few genes for resistance in 'Little Joss' and that their penetrance is variable. Alternatively it could be suggested that a linked group of genes was
obtained from 'Ghirka' during the breeding programme for 'Little
Joss'. If the linkage was not very tight the genes could have separated by crossing over in segregating generations, giving apparently continuous distribution. This analysis was not pursued further
as the main objective was to show that resistance from cultivars
such as 'Little Joss' could be successfully transferred to other
cultivars with normal breeding techniques.

Arnold *et al.* (1) have presented evidence suggesting that polygenic systems may sometimes evolve in such a way that segregation
of a discontinuous type can be observed in progeny from some crosses
between resistant and susceptible lines as observed in programmes
for breeding cotton for resistance to *Xanthomonas malvacearum*.
Segregation in populations derived from crosses appeared in the
Sudan to be attributable to major genes; in Uganda segregation in
the same populations was evidently under polygenic control. Selection carried out under conditions in Uganda, with apparently continuous variation for virulence in the pathogen, enabled resistant
lines to be selected. When tested in the Sudan they segregated for

resistance equivalent to that caused by known major genes. It was concluded that selection for genes of small effect may have resulted in polygenic blocks which, under appropriate conditions, might show discontinuous segregation, as suggested by Thoday (38) in *Drosophila*.

At Cambridge we have recently been studying some French wheat cultivars which have shown durable resistance to *P. striiformis* (22). 'Cappelle-Desprez' is susceptible at the seedling stage to a small group of races, but in the field it has an intermediate level of resistance which has remained approximately the same for almost 20 years while the cultivar was grown on a high proportion of the winter wheat acreage in Britain. It also has specific resistance to some races of *P. striiformis* and to these it gives a very low infection type (25). Another French cultivar, 'Hybride de Bersee', has shown the same spectrum of resistance and susceptibility in our tests as 'Cappelle-Desprez', with which it has some common ancestors, and it has maintained a moderate to high level of resistance for many years.

The availability of various euploid and aneuploid stocks of 'Bersee', developed at the Plant Breeding Institute, Cambridge, has enabled us to investigate its resistance to *P. striiformis*. A line which is deficient for one chromosone (called 5BS-7BS because its arms are genetically related to the short arms of chromosome 5B and 7B in the cultivar 'Chinese Spring') was shown to be much more susceptible in the field than the euploid (with full chromosome complement). The absence of this chromosome did not change the known race-specific resistance of the cultivar to race 37 E132 of *P. striiformis* (22).

Spore production from infected seedlings of the euploid, the line deficient for chromosome 5BS-7BS (nullisomic 5BS-7BS) and from lines possessing either the 5BS or the 7BS arms of the chromosome (ditelosomic 5BS and ditelosomic 7BS) was measured under various environments using the techniques of Johnson and Bowyer (20).

Under environments with low mean spore yields less sporulation occurred on the euploid and ditelosomic 7BS than on the nullisomic 5BS-7BS and ditelosomic 5BS (Table 2). Under conditions favouring high spore production ditelosomic 5BS was the most resistant line. This suggests that both arms of the chromosome can contribute resistance. This has been confirmed in field trials where both ditelocentric lines are more resistant than the nullisomic but are less resistant than the euploid (R.G. Gaines, personal communication).

Thus, although much of the resistance of 'Bersee' is controlled by one chromosome the resistance is not due to a single gene and may be due to a linked group of genes such as postulated by Arnold *et al.* (1). The question as to whether this resistance will be effective when transferred to other genotypes and especially whether

TABLE 2.   *Uredospores of* P. striiformis *produced per 100 cm^2 of seedling leaf tissue of euploid and aneuploid lines of wheat cultivar 'Hybride de Bersee'*

Experiment	Euploid	Ditelo 7BS	Ditelo 5BS	Nulli 5BS	L. S. D. P = 0.05	Mean
	Spore production mg/100 cm^2 of leaf					
1	26	32	57	49	17	41
2	23	20	83	73	25	50
3	115	99	114	120	ns	112
4	112	116	108	129	22	116
5	141	155	151	145	ns	148
6	329	301	201	290	59	280

it will provide durability must await further investigations.

I have given two examples of resistance of wheat to *P. strii-formis,* the first in 'Joss Cambier' which has shown race-specificity after a few years of exposure in commercial crops, and the second in 'Bersee' and 'Cappelle-Desprez' which has not done so after many more years and, therefore, is described as durable. To the plant breeder the different behaviour of these two types of resistance is of great importance. It is often supposed that the mechanisms of resistance must be different in these two types of situation and clearly there is a difference in durability. I have already discussed, however, the evidence that slow growth of the pathogen in the leaf, and low rates of sporulation are not distinctive characters of resistance described as 'horizontal' or non-race-specific since they can equally well be demonstrated in cases known to be race-specific.

I would like to emphasize further the similarities between the appearance of host responses which are race-specific and those which have not been shown to be race-specific. In seedling tests, euploid 'Bersee' gives a 3+ infection type to several races of *P. striiformis* whereas the nullisomic 5BS-7BS gives type 4. Similarly, 'Cappelle-Desprez' seedlings give a 3+ infection type and seedlings of 'Nord Desprez', a more susceptible sister line, give a type 4. These differences have not been shown to be race-specific but the 3+ reaction of 'Joss Cambier' to isolate WYR 69/10 of *P. striiformis* which was identical in external appearance to the 3+ of 'Bersee' and 'Cappelle-Desprez', is race-specific and is not manifested to isolates

such as WYR 71/2 which cause a type 4 reaction. Thus in one case
the sporulating lesion with chlorosis (3+) is race-specific, but in
the other a reaction with identical appearance is not race-specific,
or, at least, does not rapidly elicit a race-specific response from
the pathogen.

Both the resistance of 'Joss Cambier' to isolate WYR 69/10
of race 104 E137 and the resistance of 'Bersee' and 'Cappelle-Desprez' are expressed in seedlings by reactions showing evidence of
physiological incompatibility. It seems likely that the difference
between these slight manifestations of incompatibility and the
extreme incompatibility expressed in hypersensitive reactions is
merely one of degree. Indeed, even in work on resistance to potato
late blight, where the hypersensitive response was often regarded
as distinct from the other responses, Graham (13) described the
selection of lines from crosses between *Solanum verrucosum* and *S.
bulbocastanum* with all gradations in reaction from hypersensitive
to fully susceptible.

Several workers have suggested the possibility that the accumulation of sufficient race-specific genes for resistance will result
in non-race-specificity or generalised resistance (e.g. 14, 29).
At present there is little evidence to support the hypothesis, although the demonstration by Slesinski and Ellingboe (37) that the
gene *Pm-1* for resistance to powdery mildew in wheat appeared to limit
the uptake of ^{35}S even in clones of the fungus carrying the corresponding gene for virulence may indicate one way in which such genes
could, collectively, produce residual effects sufficient to reduce
the ability of the pathogen to invade the host tissues. On the
other hand there is as yet little convincing evidence that the mechanisms associated with the durable resistance in the cases I have
described are in fact qualitatively different from those associated
with race-specificity.

The fact remains, however, that many degrees of specificity in
resistance exist in wild as well as in cultivated species. In some
cases it is believed that race-specific resistance genes alone provide little protection against disease. Niederhauser (30) suggested
that no tuber bearing potato clones were immune to *P. infestans* in
central Mexico where the pathogen exhibits a great range of race-
specific virulence. Since race-specific genes evidently persist in
wild populations it may be assumed that they do have some value to
the host species. There is, of course, the possibility that their
effects on disease resistance are incidental to some other vital
function which they perform. Nevertheless it has been suggested that
the presence of a mixture of such genes in the host population will
result in a delay in the build-up of an epidemic because a proport-
tion of the mixture of pathogen strains will be non-virulent on some
host plants thus reducing the effective level of inoculum. It has
been suggested that multiline varieties could be used to control

rust diseases in cereals in this way (18).

The Role of Specific Genes in Induced Resistance

We have recently been considering another possible action of race-specific genes in such situations as multiline crops. It has been demonstrated in laboratory studies that inoculation with non-virulent strains of pathogens can induce resistance to normally virulent strains (9, 44). We have been able to demonstrate a similar phenomenon with *P. striiformis* in wheat and with *Uromyces appendiculatus* in beans and we have suggested that the induced resistance might increase the effectiveness of disease control in multiline varieties (19).

Sporulation of virulent races of *P. striiformis* was reduced on seedlings inoculated with non-virulent strains of *P. striiformis* on four wheat cultivars (Table 3). Similar reductions in sporulation in field conditions could reduce the rate of development of disease and conditions suitable for such events could occur in multiline varieties where virulent and non-virulent strains of the same pathogen were present in the same crop mixture.

TABLE 3. *Spore production on seedling leaves of wheat cultivars inoculated with virulent (V) races of* P. striiformis *only, or with non-virulent (NV) and virulent races*

Variety	Leaf surface with NV and V	Days between NV and V	Spore production (mg/cm^2 leaf)		P^a
			V	NV+V	
'Cappelle-	Opposite	0	79.9	42.2	<0.001
Desprez'	Upper	4	85.7	29.8	<0.001
'Maris Beacon'	Opposite	0	115.1	63.0	<0.001
	Upper	0	122.2	53.2	<0.001
	Upper	4	102.1	49.6	<0.001
'Maris Ranger'	Opposite	0	77.9	58.0	<0.01
	Upper	0	99.3	73.0^b	<0.01
	Upper	4	13.3	9.5^b	ns
'Maris Templar'	Opposite	0	178.7	149.3	<0.05
	Upper	0	104.0	84.2	<0.05
	Upper	4	53.5	23.6	<0.001
	Upper	6	133.6	115.9	<0.05

a Probability for significant difference between means
b Both inoculations with non-virulent race in error

One feature which might limit the effectiveness of such induced resistance is its localisation on the leaf. Several examples of chimaeras are known in cereals where adjacent resistant and susceptible sectors of leaves apparently do not interact with each other (10). Our results, however, indicate that resistance induced by a non-virulent race on one surface of a leaf can be effective against a virulent race applied to the opposite surface. It thus seems unlikely that the resistance induced by the non-virulent race depends on its occupation of infection sites so as to make them unavailable for the virulent race. More probably, a fungistatic factor is induced in cells or transmitted to cells some distance away from the infected site.

We are working on trials in the field to test whether such resistance can be effective in delaying disease development. Nelson and Tung (28) suggested that cross protection by race O against race T of *Helminthosporium maydis* in maize may have delayed the build up of race T in areas where race O was common.

CONCLUSIONS

There is a wide variation in both the range and in degree of specificity shown by plant pathogens. Factors such as geographical separation of hosts and pathogens, breeding techniques and crop uniformity may have increased the degree of specificity manifested in cultivated plants. The range of host specificity of some pathogens such as rusts enables them to attack different host species but at the same time to have highly specific interactions within their main host species. This may suggest that susceptibility in the host is a less highly specific response than is specific resistance found in host-variety x pathogen-race interaction. The highly specific interactions between host varieties and pathogen races may have developed through evolutionary selection for resistance in the host and virulence in the pathogen rather than as an intrinsic feature of biotrophism, since symbiotic biotrophs such as the vesicular arbuscular mycorrhizae do not show such specificity.

Disease resistance in agricultural crops varies in its durability and this aspect is of great interest to the plant breeder. Genetic investigations of cultivars with durable resistance show that although resistance may be controlled by several genes these may sometimes be grouped together to act as a unit. Much remains to be learnt of the mechanisms associated with durability of disease resistance. There is no distinctive difference in superficial appearance between some reactions which are clearly race-specific and others which have not been shown to be so. There is a need for these aspects of resistance and for the possible value of known race-specific resistance in disease control to be more fully investigated.

It is possible, for example, that race-specific resistance may play
an active role in limiting disease development by producing an induced resistance response when a host is inoculated with a non-
virulent strain of the pathogen.

ACKNOWLEDGEMENTS

I am grateful to the Director and Governors of the Plant
Breeding Institute and the Agricultural Research Council for allowing me leave of absence to attend the Advanced Study Institute, and
I thank Miss Jean Bower of Trent Polytechnic for obtaining the data
given in Table 3.

REFERENCES

1.   ARNOLD, M.H., INNES, N.L. and BROWN, S.J. (197). Resistance
          Breeding in Agricultural Research for Development : the
          Namulonge Contribution. (ARNOLD, M.H., Ed.), Cambridge
          University Press and Cotton Research Corporation. (In
          press).

2.   BIFFEN, R.H. and ENGLEDOW, F.L. (1926). Wheat breeding inve-
          stigations at the Plant Breeding Institute, Cambridge.
          Res. Monogr. Minist. Agric. Fish., 4, 114.

3.   BLACK, W. and GALLEGLY, M.E. (1957). Screening of *Solanum*
          species for resistance to physiologic races of *Phytoph-
          thora infestans*. *Am. Potato J.*, 34, 273 - 281.

4.   BRIAN, P.W. (1975). The phenomenon of specificity in plant
          disease. This NATO Advanced Study Institute on Specific-
          ity in Plant Diseases.

5.   BRIDGMON, G.H. and WILCOXSON, R.D. (1959). New races from
          mixtures of urediospores of varieties of *Puccinia grami-
          nis*. *Phytopathology*, 49, 428 - 429.

6.   BROWDER, L.E. (1973). Specificity of the *Puccinia recondita*
          f. sp. *tritici* : *Triticum aestivum* 'Bulgaria 88' relat-
          ionship. *Phytopathology*, 63, 524 - 528.

7.   CALDWELL, R.M., ROBERTS, J.J. and EYAL, Z. (1970). General
          resistance ("slow rusting") to *Puccinia recondita* f. sp.
          tritici in winter and spring wheat. *Phytopathology*, 60,
          1287 (Abstr.).

8.   COTTER, R.U. and ROBERTS, B.J. (1963). A synthetic hybrid
          of two varieties of *Puccinia graminis*. *Phytopathology*,
          53, 344 - 346.

9.   CHEUNG, D.S.M. and BARBER, H.N. (1972). Activation of resist-
      ance of wheat to stem rust. *Trans. Br. mycol. Soc.*, 58,
      333 - 336.

10.   DAY, P.R. (1974). *Genetics of host-parasite interaction.*
      W.H. Freeman and Company, San Francisco, 238 pp.

11.   D'OLIVEIRA, B. and SAMBORSKI, D.J. (1966). Aecial stage of
      Puccinia recondita on Ranunculaceae and Boraginaceae in
      Portugal. *Proc. Cereal Rust Conf.*, Cambridge 1964, 133 -
      150.

12.   ESHED, N. and DINOOR, A. (1973). Genetic studies on the spec-
      ialisation of crown rust into pathogenic varieties
      (Formae speciales). *2nd Int. Congr. Pl. Path.*, Minneap-
      olis, Minnesota (Abstr.).

13.   GRAHAM, K.M. (1963). Inheritance of resistance to *Phytophthora
      infestans* in two diploid Mexican *Solanum* species.
      Euphytica, 12, 35 - 40.

14.   GREEN, G.J. (1971). Hybridization between *Puccinia graminis
      tritici* and *Puccinia graminis secalis* and its evolution-
      ary implications. *Can. J. Bot.*, 49, 2089 - 2095.

15.   HIURA, U. (1973). Genetic basis of the host specialization in
      Erysiphe graminis DC. *2nd Int. Congr. Pl. Path.*, Minn-
      eapolis, Minnesota (Abstr.).

16.   HENDRIX, J.W., BURLEIGH, J.R. and TU, JIN-CHANG. (1965). Over-
      summering of stripe rust at high elevations in the pacific
      Northwest-1963. *Pl. Dis. Reptr.*, 49, 275 - 278.

17.   HORSFALL, J.G., *et al.* Committee on Genetic Variability of
      Major Crops, Agricultural Board. (1972). *Genetic vulner-
      ability of major crops.* National Academy of Sciences,
      Washington, D.C., 307 pp.

18.   JENSEN, N.F. (1952). Intra-varietal diversification in oat
      breeding. *Agron. J.*, 44, 30 - 34.

19.   JOHNSON, R. and ALLEN, D.J. (1975). Induced resistance to
      rust diseases and its possible role in the resistance of
      multiline varieties. *Ann. appl. Biol.*, 80, 359 - 364.

20.   JOHNSON, R. and BOWYER, D.E. (1974). A rapid method for meas-
      uring production of yellow rust spores on single seedlings
      to assess differential interactions of wheat cultivars
      with *Puccinia striiformis*. *Ann. appl. Biol.*, 77, 251 -
      258.

21. JOHNSON, R. and TAYLOR, A.J. (1972). Isolates of *Puccinia*
     striiformis collected in England from the wheat varieties
     Maris Beacon and Joss Cambier. *Nature, Lond.*, 238, 105 -
     106.

22. JOHNSON, R. and LAW, C.N. (1975). Genetic control of durable
     resistance to yellow rust *(Puccinia striiformis)* in the
     wheat cultivar Hybride de Bersee. *Ann. appl. Biol.*, 81
     (In press).

23. LUIG, N.H. and WATSON, I.A. (1972). The role of wild and cult-
     ivated grasses in the hybridisation of *formae speciales*
     of *Puccinia graminis*. *Aust. J. biol. Sci.*, 25, 335 -
     342.

24. LUPTON, F.G.H. and JOHNSON, R. (1970). Breeding for mature
     plant resistance to yellow rust in wheat. *Ann. appl. Biol.*,
     66, 137 - 143.

25. LUPTON, F.G.H. and MACER, R.C.F. (1962). Inheritance of resi-
     stance to yellow rust *(Puccinia glumarum* Erikss. and Henn.)
     in seven varieties of wheat. *Trans. Br. mycol. Soc.*, 45,
     21 - 45.

26. METTIN, D., BLÜTHNER, W.D. and SCHLEGEL, G. (1973). Additional
     evidence on spontaneous 1B/1R wheat-rye substitutions and
     translocations. *Proc. fourth int. Wheat Genetics Symp.*,
     Columbia, Missouri; 179 - 184.

27. NEGULESCU, F. and IONESCU-COJOCARU, M. (1974). The outbreak of
     a new form of race 77 of *Puccinia recondita* f. sp. *tritici*
     on wheat cultivar Aurora in Romania in 1973. *Cereal Rusts
     Bull.*, 2, 19 - 22.

28. NELSON, R.R. and TUNG, G. (1973). Cross protection by race O
     against race T of *Helminthosporium maydis*. *Pl. Dis. Reptr.*,
     57, 971 - 973.

29. NELSON, R.R., MACKENZIE, D.R. and SCHEIFELE, G.L. (1970). Inter-
     action of genes for pathogenicity and virulence in *Tricho-
     metasphaeria turcica* with different numbers of genes for
     vertical resistance in *Zea mays*. *Phytopathology*, 60, 1250 -
     1254.

30. NIEDERHAUSER, J.S. (1968). Resistance to *Phytophthora infestans*
     in Mexico. *First int. Congr. Pl. Path.*, London, Abstr.,138.

31. RILEY, R., CHAPMAN, V. and JOHNSON, R. (1968). The incorporat-
     ion of alien disease resistance in wheat by genetic inter-
     ference with the regulation of meiotic chromosome synapsis.
     Genet. Res., 12, 199 - 219.

32. ROBINSON, R.A. (1973). Horizontal resistance. *Rev. appl. Mycol.*,
     52, 483 - 501.

33. SAMBORSKI, D.J. (1963). A mutation in *Puccinia recondita* Rob.
     ex. Desm. f. sp. *tritici* to virulence on Transfer, Chinese
     Spring x *Aegilops umbellulata* Zhuk. *Can. J. Bot.*, 41,
     475 - 479.

34. SANFORD, G.B. and BROADFOOT, W.C. (1932). The relative sus-
     ceptibility of cultivated and native hosts in Alberta
     to stripe rust. *Scient. Agric.*, 13, 714 - 721.

35. SANGHI, A.K. and LUIG, N.H. (1974). Resistance in three com-
     mon wheat cultivars to *Puccinia graminis*. *Euphytica*, 23,
     273 - 280.

36. SEARS, E.R. (1956). Transfer of leaf-rust resistance from
     Aegilops umbellulata to wheat. *Brookhaven Symp. Biol.*,
     9, 1 - 22.

37. SLESINSKI, R.S. and ELLINGBOE, A.H. (1971). Transfer of ^{35}S
     from wheat to the powdery mildew fungus with compatible
     and incompatible parasite/host genotypes. *Can. J. Bot.*,
     49, 303 - 310.

38. THODAY, J.M. (1961). Location of polygenes. *Nature, Lond.*,
     191, 368 - 370.

39. THURSTON, H.D. (1971). Relationship of general resistance :
     late blight of potato. *Phytopathology*, 61, 620 - 626.

40. VAN DER PLANK, J.E. (1963). *Plant diseases : epidemics and
     control*. Academic Press, New York and London, 349 pp.

41. VAN DER PLANK, J.E. (1968). *Disease resistance in plants*.
     Academic Press, New York and London, 206 pp.

42. VAN DER ZAAG, D.E. (1959). Some observations on breeding for
     resistance to *Phytophthora infestans*. *Eur. Potato J.*,
     2, 278 - 286.

43. WOLFE, M.S. (1972). The forced evolution of cereal disease.
     Outl. Agric., 7, 27 - 31.

44. YARWOOD, C.E. (1954). Mechanism of acquired immunity to a
     plant rust. *Proc. natn. Acad. Sci. U.S.A.*, 40, 374 - 377.

45. ZELLER, F.J. (1973). 1B/1R wheat-rye chromosome substitutions
     and translocations. *Proc. Fourth int. Wheat Genetics Symp.*
     Columbia, Missouri, 209 - 221.

CONTRIBUTIONS

BOUCHER, C.   Search for genetic transfer system in phytopathogenic bacteria.

DAY, P.R.   Specificity of killer characters in *Ustilago maydis*.

MANNERS, J.G.   The significance of somatic recombination in *Puccinia striiformis*.

MANNERS, J.G.   Epidemiology and specificity in *Puccinia striiformis*.

SUMMARY OF POINTS FROM CONTRIBUTIONS AND DISCUSSIONS
BY
A. H. ELLINGBOE

Chairman and Discussion Leader

Although Johnson's paper dealt primarily with accentuation and degrees of specificity in host-parasite interactions, most of the discussion related to specificity dealt with the effect of inoculation with one race of *Puccinia striiformis* on the success of a second, later inoculation with a different race. Johnson made the following points in relation to questions from Mansfield, Ward, Ciccarone, DeVay, Brethauer and Ellingboe. There is evidence that virulence, in some rusts, is dominant but in most cases virulence is recessive. In our (Johnson *et al.*) studies on induced resistance, virulent inoculum applied to 'Man's Templar' four days before nonvirulent inoculum had no effect on sporulation of the latter, a white spored race; this does not suggest induced susceptibility. Resistance in wheat to *P. striiformis* has been induced following inoculation with strains of *P. striiformis* which are virulent on barley; others have also induced resistance to rusts by inoculation with non-host species of rust. We have not found evidence of specificity to races of *P. striiformis* in the cross protection we have studied. The mechanism for the resistance we have observed does not operate through effects on spore germination. Based on studies of transfer of resistance and the different reactions of parts of chimeras, it is possible that transmission of resistance from adaxial to abaxial regions of leaves occurs more easily than does transfer across the leaf. Alternatively, genetically resistant tissue may be necessary, and this is lacking in the susceptible tissue in the chimeras. Action of a diffusible substance is suggested by the fact that the effect extends beyond the infected site, if only for a few cells. I understand that there may be two systems of genes in wheat, one effective against *P. graminis* f. sp. *secalis* and the other against *P. graminis* f. sp. *tritici*. If so it should be possible to eliminate those for resistance to *P. graminis* f. sp. *tritici* without affecting resistance to *P. graminis* f. sp. *secalis*. I do not think that genes for virulence on one host species would mutually exclude those for virulence on another unless they were allelic. Such evidence as there is, from work such as that by Hiura on *Erysiphe graminis*, may indicate that hybrid races with virulence genes necessary to infect both host species are poorly competitive on one host compared with races able to infect only that one host.

Short papers presented by Manners showed how genetic differences which affect host-parasite specificity can be expressed. Races of *P. striiformis* have been isolated which differ in such characteristics as incubation period, spore production and sensitivity to temperature, as well as final infection type. He also gave the example of wheat cultivar 'Man's Huntsman' in which an isolate of *P. striiformis* spread only slowly through a field although it sporulated profusely. Johnson's paper pointed out that the idea that genetic systems controlling characters described by Manners are different from the genetic systems affecting infection type is not supported by genetic studies with cereals. The idea of non-specific mechanisms of resistance (i.e. not gene-for-gene relationships) is not supported by detailed genetic studies of naturally-occurring variability of host-parasite interactions.

The idea that "major" and "minor" genes differ in function is probably misleading. Ellingboe defined "major" genes as those where differences between the presence of alternate alleles are large and easy to see and "minor" genes are those where differences are not so large as to be easy to see. The genes differ, therefore, in the ease with which the presence of the two alleles are distinguished. Johnson considered that discontinuous or continuous variation in segregating populations provides a means for separating "major" from "minor" genes. The control of environmental variables, therefore, helps to determine whether genes are "major" or "minor". There is no reason to believe that their gene-for-gene functions are different.

Boucher's paper described attempts to make a mutational analysis of host-parasite interactions. Auxotrophic mutants of *Pseudomonas solanacearum* are usually avirulent on tomato. Prototrophic revertants are usually virulent. Some prototrophic revertants are not virulent. This suggested that the product of the mutated gene may have at least two distinct functions; one, a role in intermediary metabolism, the other in expression of virulence. Sequeira and Person both cautioned that the use of auxotrophic mutants for examining avirulence could reflect only the fact that the mutant strains was no longer prototrophic and, therefore, could not grow because the host did not provide the necessary compounds. Boucher pointed out, however, that although prototrophic revertants had regained their ability to grow in the absence of the supplement, some revertants had not regained pathogenicity. If there were two or more mutations, one affecting prototrophy and one affecting pathogenicity, most revertants should have a common low level of pathogenicity.

The paper by Day presented evidence for the specificity of "killer" factors in three strains of *Ustilago maydis*. Each strain carries a virus-like particle which can be transmitted to strains lacking virus-like particles. The relationship between the presence of virus-like particles to virulence or to resistance in the host is not yet established.

GENE FUNCTIONS IN HOST-PARASITE SYSTEMS

PETER R. DAY

Genetics Department, Connecticut Agricultural Experiment Station
New Haven, Connecticut, U.S.A.

There are two important reasons for trying to explain the nature of host-parasite specificity. The first is to be able to manipulate and improve specific resistance and hence deploy it more effectively against crop plant parasites. The second reason is that an understanding of the mechanism of specificity will increase knowledge of the nature and control of gene action in eukaryotes an end with broad philosophical and practical rewards. My interest in specificity began with work in the early 50's on breeding tomatoes for resistance to the fungal parasite *Cladosporium fulvum* and with the consequences of releasing resistant varieties and seeing that their resistance was race specific.

As plant breeders our definition of resistance was a practical one. Resistance had to restrict pathogen development and multiplication enough to be useful. Our goal was no loss of yield or quality either in the presence or the absence of the pathogen. The products of our breeding programs had to satisfy a variety of requirements other than resistance or no grower would use them.

Pathogen development was curtailed only by the resistant varieties that the growers planted and the cultural and other measures they took to limit the disease. Our first problem was to decide whether the forms of the pathogen present were virulent or avirulent on the varieties grown at the time. Again the definition of virulent was a practical one. It described the ability of an isolate to cause as much damage to a resistant variety as it could cause to a susceptible one. The gene for resistance was unable to restrict pathogen development and multiplication sufficiently to prevent significant yield reduction in heavy infections. While these broad

definitions are characteristic of nearly all host-parasite interactions manipulated by breeding for resistance, intermediate reactions between the extremes of virulence and avirulence or susceptibility and resistance are frequently recognised. Nowadays we
can quantify infection for example by weighing spore yields (9) and
resistance by determining the effects of different levels of infection on host yield (15).   We are also in a much better position to
recognize and handle differences in resistance in the host or of
virulence in the pathogen that are due to genes of small effect.

   Flor's gene-for-gene hypothesis is central to much of our discussion of the genetic basis of specificity. Working with flax
(*Linum usitatissimum*) and its autoecious rust *Melampsora lini* Flor
noted that resistance in flax was dominant and virulence in the rust
recessive and that both showed segregation patterns indicating that
they were often determined by single genes in each species. Table
1 shows the results he obtained in the F_1 and F_2 from crosses of
two flax varieties,'Ottawa 770B'and'Bombay', and from crosses of two

Table 1. *Interactions of race 22 and 24 of flax rust* (Melampsora
      lini) *with 'Ottawa' and 'Bombay', cultivars of flax* (Linum
      usitatissimum). *Also shown are* F_1 *and* F_2 *reactions of both
      host and parasite. Note that since the rust uredial gen-
      eration is scored, and this is dikaryotic, the* F_2 *ratios
      are like those of a diploid organism (From 3,4).*

Rust Race	'Ottawa'	'Bombay'	Flax F_1	F_2			
22	S	R	R	R	R	S	S
24	R	S	R	R	S	R	S
				110	43	32	9

Flax	22	24	Rust F_1	F_2			
'Ottawa'	S	R	R	R	R	S	S
'Bombay'	R	S	R	R	S	R	S
				78	23	27	5

rust races, 22 and 24. Flor concluded that resistance in'Ottawa
770B'and'Bombay'was determined by two independent genes L and N
respectively and that the virulence of race 22 on'Ottawa'was determined by a gene vL independent of a gene vN that determined the virulence of race 24 on'Bombay'· The choice of these particular flax
varieties and rust races as parents is important because, as the
table shows, the two races differentiate the two cultivars and the
two cultivars differentiate the two races. On the basis of this and
other similar evidence Flor concluded that "for each gene conditioning rust reaction in the host there is a specific gene conditioning
pathogenicity in the parasite" (5). This generalisation has since
been shown to hold true for a number of other host-parasite systems
(2, 14).

    One of the simplest methods for representing the interaction,
the so-called quadratic check, is shown in Figure 1. Two alleles at
a locus controlling resistance are shown interacting with two alleles at a locus controlling virulence. The diagram is simplified to
show phenotypes and assumes that dominance is complete or, in other
words, that in the host $R1R1$ and $R1r1$ have equivalent resistant phenotypes and in diploid and dikaryotic parasites $V1V1$ and $V1v1$
also have equivalent avirulent phenotypes. Only one of the four
possible interactions between host and parasite results in avirulence
and resistance. The other three reactions are virulent and susceptible. Similar patterns are generated for a saprophyte by alleles

Figure 1.   *Interaction between resistant* (R1) *and susceptible*
      (r1) *hosts and avirulent* (V1) *and virulent* (v1) *pathogen.*

   − = *no disease*   + = *disease*

for auxotrophy versus prototrophy on media with and without the required supplement and by alleles for drug resistance versus sensitivity on media with and without the drug (12). In each case, of
either avirulence or the failure to grow, specificity is restrictive
and limited to one of the four possible situations. The other three
situations are permissive. In terms of models for the kind of mechanisms involved in determining specificity the two saprophyte examples in general seem to be too simple. Although resistance or susceptibility of plum varieties to the canker fungus *Rhodosticta quer-
cina* depends on their ability to supply the requirement of this parasite for myoinositol (10) no more is known and I know of no other
similar examples.

     In fact race specific resistance appears not to be preformed
but to occur in response to infection by a pathogen. The quadratic
check suggests that resistance, or avirulence, is due to the interaction of the products of the genes for resistance in the host and
for avirulence in the pathogen. Evidently there are molecules that
signal the presence of the pathogen as an invader which the plant
can recognise as non-self. These molecules should behave as inducers
of specific resistance and show the same specificity as the agents
that produce them. Other speakers review evidence that specific inducers may be RNA (13), or polysaccharide (1), or will suggest that
resistance genes may determine membrane recognition sites with properties like those that enable mammalian cells to recognise infectious
agents (7). Elsewhere (2) I have summarised some of the evidence for
the idea that resistance is an induced response and that mutation
to virulence may occur as a result of the loss of a specific inducer.
Here I propose to examine an alternative thesis that avirulence could
also arise as a failure to induce susceptibility.

                          INDUCED SUSCEPTIBILITY

     In some host-parasite interactions the outcome is determined at
a relatively early stage. For example Müller and Börger (12) and
Müller (11) found that potato tubers carrying the gene *R1* for late
blight resistance inoculated with the avirulent race 0 of *Phytophthora infestans* produced materials which they called phytoalexins
that inhibited growth of the virulent race 1 applied a short time
later. Varns and Kuć (18) reversed the order of inoculation and
showed that tuber slices inoculated with virulent zoospores 12 hours
previously failed to accumulate the phytoalexins rishitin and phytuberin or to undergo a necrotic hypersensitive response following
inoculation with avirulent zoospores. Varns and Kuć proposed that
the virulent race and host interact to block, or interfere with the
hypersensitive response. It would be useful to know, perhaps with
the help of markers or conditional temperature sensitive mutants,
whether such an avirulent race can continue compatible growth to the

same extent as the virulent race.

Tomiyama (17) had earlier shown that pre-inoculation of potato petiole tissue with a compatible race of *P. infestans* 15 - 20 hours earlier prevented the death of host cells for up to four hours after inoculation with an incompatible or avirulent race. Normally inoculation with an avirulent race is followed by death of individual host cells over a period of from 10 - 60 minutes. The period of four hours was the time during which the two inocula could still be distinguished.

In *Melampsora lini* when pycnia on a resistant flax plant are fertilised with spermatia from an avirulent race normal aeciospores are formed that cannot re-infect other host plants of the same genotype (6). Although avirulence is dominant it is either not expressed in the fresh dikaryotic growth following fertilisation although other genes controlling dikaryon morphology are expressed or, more likely, the host tissue is unable to respond because of a localised condition of induced susceptibility.

A similar phenomenon is found when segregating egg clusters of Hessian fly *(Mayetiola destructor)* are deposited on resistant wheat plants. The close presence of one or more virulent larvae protects adjacent avirulent larvae permitting their survival.

Zucker (19, 20) has described protein synthesis dependent mechanisms in potato tuber tissue and *Xanthium* leaves that destroy phenylalanine ammonia lyase and other enzyme activities. Susceptibility could well depend on induction of proteolytic enzymes that would prevent a normal resistant response from occurring.

Although many studies have examined differences between compatible and incompatible interactions few have examined the effect of compatible pre-inoculations. More interest has centered on incompatible pre-inoculations as a means of inducing resistance. Johnson has referred in this Institute to his own work on induced resistance to rust diseases in wheat and bean (8) and has made the interesting suggestion that induced resistance may play a role in the resistance of multiline varieties. There is also the possibility that induced susceptibility may play a corresponding role in the susceptibility of multiline varieties.

In some pathogens host specificity depends on toxins and may also involve induced susceptibility. It is possible to represent the specificity of *Helminthosporium victoriae* by a quadratic check (Fig. 2) which is the reverse of Fig. 1. Disease only occurs when Hv toxin is produced and only on an oat variety carrying the dominant gene for sensitivity *Pc-2*. This gene is pleiotropic and also controls resistance to oat crown rust*(Puccinia coronata)*. *H. sacchari,* a pathogen of sugar cane, also produces a toxin given the trivial

	Pc-2	pc-2
hv	+	-
hv^+	-	-

Figure 2. *Interaction between oat cultivars susceptible* (Pc-2) *and resistant* (pc-2) *to* Helminthosporium victoriae *and strains of* H. victoriae *producing* (hv) *and not producing* (hv$^+$) *the pathotoxin victorin.*

+ = *disease*   - = *no disease*

name helminthosporoside. The toxin is responsible for the flares of necrotic tissue that develop distally from points of infection on leaves of susceptible cane. Strobel (16) has claimed that a toxin binding protein present in the plasma membranes of susceptible cane cells can be isolated and taken up by protoplasts prepared from re-sistant clones, or even from tobacco. These protoplasts which are normally insensitive then become sensitive to the toxin. Fortunately for plant breeders and pathologists host resistance to pathotoxins is so far not race specific. No variants of *H. victoriae* or *H. sacchari* are known that are virulent on resistant host varieties.

It is a commonplace observation that plants are compatible hosts for an extremely small proportion of the insects, nematodes, micro-organisms and viruses with which they come into contact. Host-parasite interactions no doubt evolved through a series of adaptive changes in both components. Such changes probably included moves by the parasite to avoid triggering defence reactions or inducing resistance, and countermoves by the host to restore the triggers. It would be astonishing if evolutionary interaction did not also extend to regulation or control by the parasite of host pathways that would otherwise prevent further parasite growth and to promote pathways that would favor parasite growth and development. The ab-normal growth caused by gall-forming parasites such as *Ustilago may-dis*, by some insect larvae, and by the gibberellins of *Gibberella fujikuroi* (gibberellins) are examples of parasite control of growth of the host.

Two types of interaction should concern us. First are the interactions that determine the gene-for-gene kind of specificity. These govern host-parasite growth and development at a late stage of evolution. They are the means of finely tuning the balance between the two components on an evolutionary scale. The single gene effects are manageable by plant breeders which is fortunate because until now they are all that has been available to use against the ancient parasites of our food crops.

The second type of interaction determines whether or not a plant is a host for a parasite. Wheat is not a host for *Phytophthora infestans*. It is true that the resistance mechanisms that restrict *Erysiphe graminis*, *Puccinia graminis*, Hessian fly *(Mayetiola destructor)* and other wheat parasites may also restrict *P. infestans*. But it is very likely that since *P. infestans* evolved as a potato pathogen and not as a wheat pathogen it is unable to induce a compatible or susceptible interaction with a wheat plant. I would further make the generalisation that race specific resistance tends to be induced by products of avirulence genes in parasites and that it is superimposed on an interaction that already has most if not all the elements needed for compatibility. Non-host resistance, on the other hand, is generally due to failure to induce susceptibility and will probably prove to be complex both physiologically and genetically.

Non-host resistance may well be of tremendous importance in the future when it becomes possible to exchange large segments of DNA among organisms that have been genetically isolated for a long time. If such hybrids are viable and reproducible it may be possible to outstrip parasite evolution and establish cultivars with really long-lived resistance. What if such inter-specific and inter-generic plant hybrids combine parasite susceptibilities rather than their resistances ? Limited experience to date suggests this may not happen. Wheats with genes for rust resistance carried on chromosome segments of rye are not known to be susceptible to rye pathogens.

REFERENCES

1.   ALBERSHEIM, P. (1976). (This Symposium).

2.   DAY, P.R. (1974). *Genetics of host-parasite interaction.* Freeman, San Francisco, 238 pp.

3.   FLOR, H.H. (1946). Genetics of pathogenicity in *Melampsora lini*. *J. agric. Res.*, 73, 335 - 357.

4.   FLOR, H.H. (1947). Inheritance of reaction to rust in flax. *J. agric. Res.*, 74, 241 - 262.

5.   FLOR, H.H. (1956). The complementary genic systems in flax
      and flax rust. *Adv. Genet., 8,* 29 - 54.

6.   FLOR, H.H. (1959). Differential host range of the monocaryon
      and dicaryons of eu-autoecious rust. *Phytopathology,*
      49, 794 - 795.

7.   HUGHES, C. (1976). (This Symposium).

8.   JOHNSON, R. and ALLEN, D.J. (1975). Induced resistance to
      rust diseases and its possible role in the resistance of
      multiline varieties. *Ann. appl. Biol.,* 80, 359 - 363.

9.   JOHNSON, R. and BOWYER, D.E. (1974). A rapid method for meas-
      uring production of yellow rust spores on single seedlings
      to assess differential interactions of wheat cultivars
      with *Puccinia striiformis. Ann. appl. Biol.,* 77, 251 -
      258.

10.   LUKEZIC, F.L. and DEVAY, J.E. (1964). Effect of myo-inositol
      in host tissues on the parasitism of *Prunus domestica* var.
      President by *Rhodosticta quercina. Phytopathology,* 54,
      697 - 700.

11.   MÜLLER, K.O. (1953). The nature of resistance of the potato
      plant to blight - *Phytophthora infestans. J. natn. agric.
      Bot.,* 6, 346 - 360.

12.   MÜLLER, K.O. and BORGER, H. (1940). Experimentelle Untersuch-
      ungen über die *Phytophthora* - Resistenz der Kartoffel.
      Arb. biol. BundAnst. Land-u. Forstw., 23, 189 - 231.

13.   ROHRINGER, R. (1976). (This Symposium).

14.   SIDHU, G.S. (1975). Gene-for-gene relationships in plant para-
      sitic systems. *Sci. Prog., Lond.,* 62, 467 - 485.

15.   SIMONS, M.D. (1966). Relative tolerance of oat varieties to
      the crown rust fungus. *Phytopathology,* 56, 36 - 40.

16.   STROBEL, G. (1975). Transfer of toxin susceptibility to plant
      protoplasts via the helminthosporoside binding protein of
      sugarcane. *Biochem. biophys. Res. Commun.,* 63, 1151 -
      1156.

17.   TOMIYAMA, K. (1966). Double infection by an incompatible race
      of *Phytophthora infestans* of a potato plant cell which
      has previously been infected with a compatible race. *Ann.
      phytopath. Soc. Japan,* 32, 181 - 186.

18. VARNS, J.L. and KUČ, J. (1971). Suppression of rishitin and
        phytuberin accumulation and hypersensitive response in
        potato by compatible races of *Phytophthora infestans*.
        Phytopathology, 61, 178 - 181.

19. ZUCKER, M. (1968). Sequential induction of phenylalanine
        ammonia-lyase and a lyase-inactivating system in potato
        tuber discs. *Pl. Physiol., Lancaster*, 43, 365 - 374.

20. ZUCKER, M. (1971). Induction of phenylalanine ammonia-lyase
        in *Xanthium* leaf discs. *Pl. Physiol., Lancaster*, 47,
        442 - 444.

CONTRIBUTIONS

CIRULLI, M. and CICCARESE, F.   Interaction between TMV isolates,
    temperature, allelic conditions and combination of the *Tm* re-
    sistance genes in tomato.

ELLINGBOE, A.   Use of temperature sensitive mutants in host :
    parasite studies.

SUMMARY OF POINTS FROM CONTRIBUTIONS AND DISCUSSIONS
BY
C. PERSON

Chairman and Discussion Leader

Discussion of Day's paper centred on the induced-susceptibiltiy
hypothesis. In reply to a question by Hadwiger, Day did not extend
the hypothesis to include organisms which, though normally not regarded as pathogens, can grow in toxin-killed tissues; he did,
however, extend it to include those pathogens which do, in fact, kill
tissue in advance of their growth, and regarded this form of pathogenicity as an extreme example of induced susceptibility. Replying
to a question by Sivak, who had cited Müller's finding that *Phytophthora infestans* can grow in tissues of non-solanaceous hosts, Day
agreed that for the conditions of Müller's experiment this could
also be regarded as an example of induced susceptibility.

The need for postulating inducers of susceptibility was questioned by Albersheim who suggested that the presence of a potential
invader may be signalled to the plant by molecules which are non-
specific, and that non-pathogens may simply fail to outgrow the resistance response which had been non-specifically involved. In reply
Day briefly described the kind of evidence which would support the
hypothesis of induced susceptibility, and referred to experiments

in which a disease reaction established by an initial inoculation
is reversed by a second, following inoculation. It was his opinion
that the hypothesis of induced susceptibility would be strongly supported if in this type of experiment it were found that initially
established compatible reactions are generally not reversible. Day
had inferred from Albersheim's question that for resistance which is
specific (as it is in gene-for-gene relationships) the host has responded to specific molecules which have signalled the presence of
an avirulent pathogen which cannot outgrow the specifically-induced
host response. Albersheim's question had implied, therefore, that
two kinds of resistance response were needed: 1. elicited by non-
specific molecules (produced by non-pathogen and pathogen alike)
which only pathogens can outgrow and 2. elicited by specific molecules (produced only by pathogens) which may or may not evoke a
resistance response, depending on whether the host is in possession
of an appropriate "matching gene" for resistance. It was Day's
view that it was not necessary to postulate two kinds of resistance
response: that induced susceptibility need only imply that the capacity of the host to respond to signals was impaired, and that un-
impaired responses, which result in resistance to non-pathogen and
pathogen alike, may be identical.

The discussion included a comment by Keen that the preponder-
ance of published data supported the idea that it was resistance,
rather than susceptibility, which was induced. Keen cited the fact
that a number of abiotic agents (including heat, narcotics, inhibitors of protein synthesis) could convert normally incompatible
reactions to compatible, and questioned the existence of similarly
pursuasive evidence for conversions in the opposite direction. Day
agreed that further supporting evidence should be looked for, and
suggested that the search should include further studies of systemic
fungicides, since those which fail to limit fungal growth *in vitro*
apparently act *(in vivo)* by blocking the induction of susceptibility.

Wheeler commented on the fact that the quadratic check showing
disease reactions involving victorin were the opposite of those obtained by Flor (see Fig. 1, 2 of Day's paper). He did not view this
as an anomaly, particularly since it is a classic assumption in plant
pathology that disease development involves a two-fold requirement:
1. virulence on the part of the pathogen, and 2. susceptibility
on the part of the host.

The discussion ended with a comment by Rahe, the substance of
which was as follows : that at the present time phytoalexins appear
to be more common in non-obligate than in obligate parasitic systems;
that examples of induced resistance caused by non-specific stimuli
also appear to be more common in non-obligate than in obligate systems; that, if valid, these generalizations would suggest that
non-obligate and obligate parasites operate in ways that are fundamentally different; for example they may differ in whether it is

resistance or susceptibility which is induced, or in whether compatibility or incompatibility serves as the general background in which disease development may take place.

The genetic discussion included a report by Ellingboe of a new line of genetic experimentation which has yielded promising preliminary results. Ellingboe's rationale for these experiments was as follows :

1.   in his view the simplest interpretation of the genetics of specific interaction between host and parasite is one that invokes specificity for the incompatible reaction of the quadratic check;

2.   that the patterns of variability, which are generated in gene-for-gene systems occur as a consequence of specific incompatibility superimposed on a background of compatibility;

3.   that if these two points are valid it would logically follow that specific incompatibility operates either to prevent establishment of a compatible relationship, or to destroy a relationship which would otherwise have been compatible;   and

4.   that distinction between these two possibilities may be made possible through studies involving temperature-sensitive mutants of the pathogen. (For temperature-sensitive mutants it is generally accepted that the mutant gene directs the formation of a product which functions normally at the "permissive", or normal temperature for growth, and abnormally, through loss of three-dimensional integrity, at the "restrictive" temperature, which is usually near the upper temperature limit for growth. Such mutants have been usefully employed, in temperature-shift experiments, for "pinpointing" the time of gene action).

It was Ellingboe's expectation that three types of temperature-sensitive mutation should be inducible in the parasite :

1.   those that are temperature sensitive both in the host and on fully-supplemented agar medium;   this type of mutant would be useful in studying the ontogeny of interactions between the host and parasite, since parasitic growth could be stopped at various stages following infection;

2.   those that do not grow *in vitro* (on fully-supplemented medium) but do grow in the host at the restrictive temperature;   this type of mutant would be useful in determining, by means of analysis of macromolecular biosynthesis taking place at the site of the lesion, what is transferred from host to parasite;   and

3.   those that do not grow in the host but do grow *in vitro* (on fully-supplemented medium) at the restrictive temperature;   this

type of mutant should involve genes that are critical to the establishment of a compatible relationship. (Ellingboe pointed out that a mutant of a gene the normal function of which was to control incompatibility might be expected to give compatibility at the restrictive temperature).

Ellingboe's studies involve two different pathogens, *Phyllosticta maydis* and *Colletotrichum lindemuthianum*. Among the six temperature-sensitive mutations obtained thus far, mutants of types 1 and 2 were both recovered. Mutants of type 3 have not yet been looked for.

Cirulli described host-parasite specificity as influenced by temperature regimes, strains of tomato TMV, and combinations and allelic conditions of the *Tm* resistance genes of tomato. Strains of TMV were described which induced mottling, systemic necroses and local necrotic lesions depending on the host differentials. The activity of the *Tm-1*, *Tm-2* and *Tm-2^a* resistance genes was temperature-dependent. The incubation period was longer and the number of isolates that induced susceptible reactions (mottling or systemic necroses) on near-isogenic tomato differentials was higher at 26° and $30^\circ C$ than at 17° and $22^\circ C$. The genetic factors *Tm-2* and *Tm-2^a* secured protection against more TMV isolates than did the *Tm-1* gene. Moreover their behaviour at 26° and $30^\circ C$ suggests that this is an example of gene cumulative action with two alleles for resistance producing twice the effect of one allele. A gene-for-gene TMV strain classification was best accomplished when differentials possessed resistance genes in the homozygous condition and at 26°C.

CELL SURFACE MEMBRANES OF ANIMAL CELLS AS THE SITES OF RECOGNITION

OF INFECTIOUS AGENTS AND OTHER SUBSTANCES

COLIN HUGHES

National Institute for Medical Research

Mill Hill, London, U.K.

INTRODUCTION

Contacts between the cell surface and extracellular substances such as peptide hormones or antigens are of prime importance in the activation of target cells and the triggering of a defined biological response, for example antibody production in sensitized lymphocytes. Similar contacts are important in intercellular communication and play an integral role in the complex differentiative processes associated with embryogenesis and tissue reorganization. The cell surface must display a large variety of complementary sites that are recognized by other cells and by biologically active effector substances.

The surface membrane is asymmetric functionally in the sense that the recognition molecules involved in such contacts occur on the outer side of the membrane. Perturbation of these sites by contacts with other cells or with extracellular regulatory substances is followed by a complex series of events by which the extracellular signal is passed through the membrane to direct cytoplasmic processes. These later processes although, of course, extremely diverse in biological consequences, nevertheless are likely to incorporate many points of similarity. Thus, the binding of effector substances to the cell surface often has profound effects on the intracellular levels of cations, particularly calcium, and of cyclic nucleotides. In the latter case presumably adenyl cyclase and guanylcyclase are activated or inhibited. The exact mechanisms by which manipulation of the intracellular levels of cations or cyclic nucleotides brings about a terminal biological event are, however, presently unknown. These aspects have been reviewed recently by Peterson (22) and Cuatrecasas (8).

In addition to its critical role in these largely beneficial processes the cell surface also carries sites of recognition for agents such as viruses, bacterial toxins and lectins. The initial event of binding to the specific cell surface receptor is followed in many cases by entry of all or part of the agent into the cell cytoplasm, perhaps by endocytosis, to re-direct or inhibit cellular metabolism. A secondary event in viral replication is to promote synthesis of virus specific messenger RNA which directs the synthesis of virus specific proteins and replication of virus nucleic acid genome.

In this lecture I shall discuss briefly the organization of cell surface membrane components and, in particular, the role that carbohydrates may play in receptor activity at the surface. I shall discuss how in certain cases the susceptibility of a cell to an effector substance or an infectious agent can be correlated with the presence of a specific cell surface receptor, whereas cells that lack the receptor are unaffected. Finally, the biochemical mechanisms by which normally susceptible cells may become resistant to one particular agent, the toxic principle (ricin) of *Ricinus communis* seeds, are discussed. Throughout this paper, "cell" means a mammalian or, rarely, an avian cell. It should be clear however from the other contributions to this Institute that similar mechanisms may well be operating in plants in their reactions to pathogens and toxins.

MEMBRANE CARBOHYDRATES

There has been remarkable progress recently in the characterization of the surface membranes of animal cells. It now seems clear that the membrane is asymmetric structurally as well as functionally (Fig. 1). The complex mixture of lipid and protein molecules and the carbohydrate moieties are distributed differently at the two faces of the membrane. Ultimately the arrangement of these components at the surface must be correlated with membrane function.

It has been known for many years that the carbohydrate components of plasma membrane are located exclusively on the extracellular face of the membrane (13, 29) in positions that are potentially accessible to other cell surfaces as well as to external stimuli. The carbohydrate is distributed between the two main classes of substances, the sphingoglycolipids and the membrane glycoproteins (Fig. 2). The carbohydrate units of these molecules are built up by polymerization of between two and five different monosaccharides into oligosaccharide chains consisting of as many as twenty or thirty sugars. The structural diversity possible in such chains is immense. A simple trisaccharide, for example can exist in several hundred forms differing in ring size (pyranose or furanose), anomeric configuration

External Surface

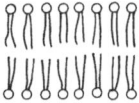 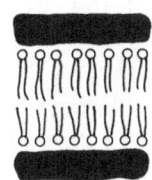 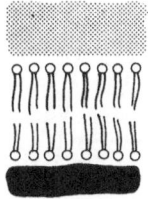 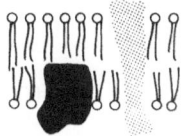

Cytoplasmic Surface

a   b   c   d

Figure 1.   *Some surface membrane models.*

(a)   *Most models incorporate a bilayer of lipid in which the polar head groups of these amphiphathic molecules (represented by the open circles) are exposed to the aqueous environment and the long hydrophobic fatty acid chains are sequestered within the interior of the bilayer from which water is excluded.*

(b)   *Danielli and Davson suggested that the lipid bilayer may be covered on each side by a layer of protein, the interaction between protein and lipid being mainly electrostatic.*

(c)   *However, histochemical studies showed clearly that carbohydrate containing substances were associated only with the external face of the membrane whereas the cytoplasmic face contained carbohydrate free proteins.*

(d)   *Proteins and glycoproteins are integrated into the lipid bilayer and interact hydrophobically within the nonaqueous interior of the bilayer. Carbohydrate free proteins are present at the cytoplasmic face of the membrane; glycoproteins may span the membrane. Carbohydrate is present only at the external surface in the latter case; the portion of the molecule emerging from the cytoplasmic face is devoid of carbohydrate.*

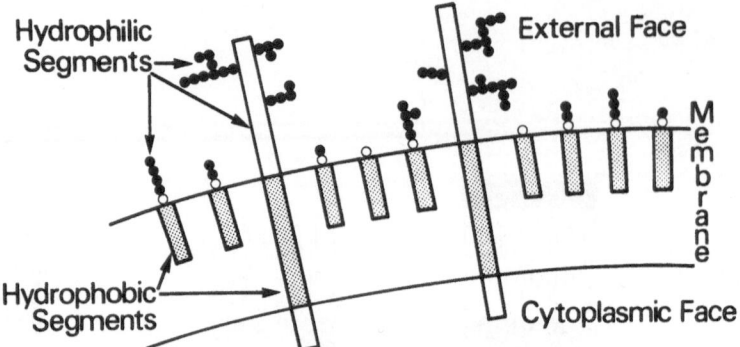

Figure 2. *Idealized picture of the organization of sphingoglyco-*
          lipids and glycoproteins on the surface of the plasma
          membrane. Different kinds of glycolipids and glycopro-
          teins with varying degrees of complexity in carbohydrate
          structure are integrated into the lipid bilayer by hydro-
          phobic interactions with the ceramide unit and fatty acid
          chain of the glycolipids or short peptide segments of the
          glycoproteins. In all cases the carbohydrates are external
          to the lipid bilayer in contact with the extracellular
          space. Each closed circle represents a monosaccharide
          unit and these are polymerized to form complex oligosac-
          charide chains.

(α or β) and types of glycosidic linkage between the residues. It
is by no means certain at present to what extent this extreme diversity of carbohydrate structure is meaningful or utilized in biology. It does however provide the cell with a potentially useful
array of structurally distinct sites that may function in the recognition processes discussed in the Introduction.

     Very little is in fact known about the structure of cell sur-
face carbohydrates, particularly the complex carbohydrate chains of
membrane glycoproteins. The best characterized of the latter is the
major glycoprotein of the human erythrocyte membrane, glycophorin
(Fig. 3). This molecule consists of a single polypeptide chain containing about 130 amino acids and substituted with short carbohydrate chains in the peptide segment nearest to the amino terminus.
There are two basic types of carbohydrate chain. Type A chains are
linked to asparagine residues of the polypeptide and contain fucose,
sialic acid, galactose and mannose, in addition to N-acetylglucosamine. The second type B unit is joined O-glycosidically to serine or threonine residues and contains N-acetylgalactosamine, galactose and sialic acid (34). Each of these sugars except sialic
acid are ubiquitous constituents of glycoproteins from plants, animals and micro-organisms. Sialic acid (Fig. 4) is found in almost

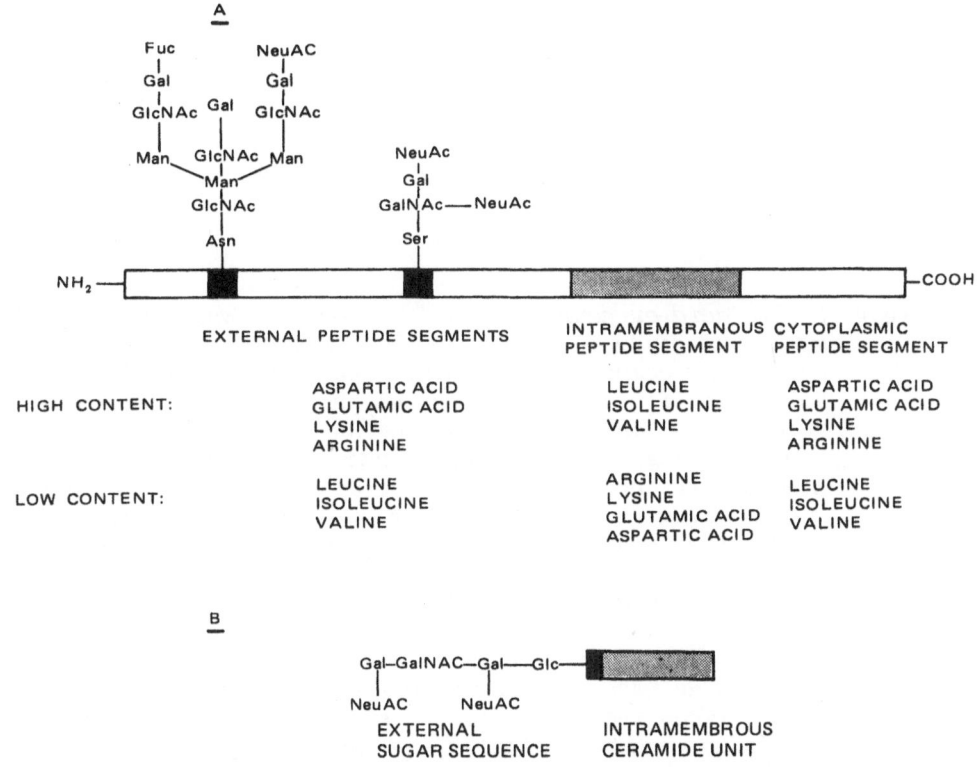

	EXTERNAL PEPTIDE SEGMENTS	INTRAMEMBRANOUS PEPTIDE SEGMENT	CYTOPLASMIC PEPTIDE SEGMENT
HIGH CONTENT:	ASPARTIC ACID GLUTAMIC ACID LYSINE ARGININE	LEUCINE ISOLEUCINE VALINE	ASPARTIC ACID GLUTAMIC ACID LYSINE ARGININE
LOW CONTENT:	LEUCINE ISOLEUCINE VALINE	ARGININE LYSINE GLUTAMIC ACID ASPARTIC ACID	LEUCINE ISOLEUCINE VALINE

Figure 3. *(a) Glycophorin, the primary membrane glycoprotein of human erythrocytes. The single polypeptide chain contains a short hydrophobic peptide sequence of about 23 amino acids that anchors the whole molecule into the lipid bilayer of the membrane. A peptide sequence rich in charged amino acids extends away from the cytoplasmic face of the membrane. All of the carbohydrate projects from the opposite (external) side of the membrane.*

(b) Idealized picture of a sphingoglycolipid. The ceramide unit contains two extended hydrocarbon chains that integrate into the interior of the membrane and interact with fatty acid chains of other membrane lipids. The carbohydrate units are substituted on to the polar head group and extend away from the membranes.

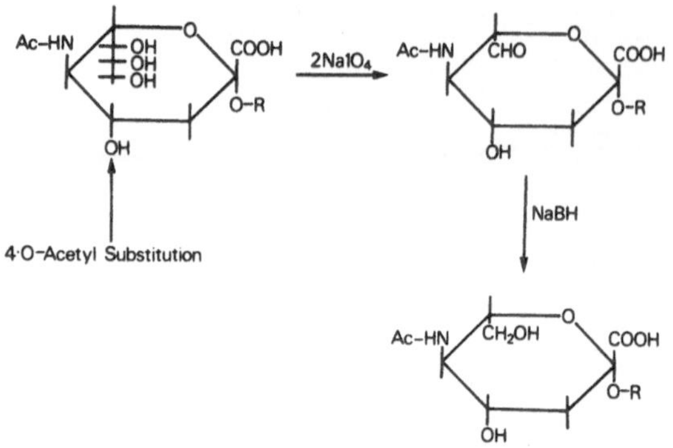

Figure 4. *Sialic acid or neuraminic acid, a common constituent of*
mammalian glycoproteins and sphingoglycolipids. The
basic ring compound containing nine carbon atoms is us-
ually substituted with an acetyl group on the amino acid
function to form N-acetyl neuraminic (NeuAc). More rarely
carbon atom 4 is acetylated in addition. The sugar is
usually present at the terminals of oligosaccharide chains
and is released into the free form by neuraminidase, an
enzyme that hydrolyses the linkage between sialic acid
and the rest of the glycoprotein or sphingoglycolipid
molecule (represented by R).

Sialic acid terminal residues of membrane glycoprot-
eins act as receptors for myxoviruses such as influenza
virus. When the glycoprotein is oxidized by sodium met-
aperiodate and then reduced, an analogue of sialic acid
is formed containing seven carbon atoms instead of nine.
This modified form of sialic acid, which remains attached
to the glycoprotein, is no longer a myxovirus receptor.

all mammalian cells and is present in some bacteria. It plays an
extremely important role as a receptor in myxovirus infections and
will be discussed later.

    The carbohydrate units of glycoproteins and glycolipids grow
biosynthetically (Fig. 5) first by attachment to the polypeptide
of the appropriate sugar, for example, N-acetylglucosamine or N-
acetylgalactosamine for chain types A and B of glycophorin, followed
by chain elongation. This is accomplished by the sequential addition of monosaccharides to the non-reducing terminus of the growing
oligosaccharide chain, catalysed by a battery of specific glycosyl
transferases. The carbohydrate chains of the sphingoglycolipids
(Fig. 3 for idealized structure) are assembled in similar fashion
although in this class of substance the initial reaction is the
attachment of a glucose residue to ceramide (Fig. 5).

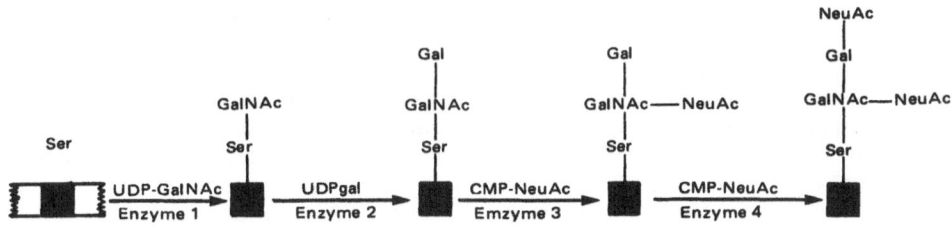

Figure 5. *Biosynthesis of carbohydrate chains. The assembly of one of the chains of glycophorin (Fig. 3) is shown. Each sugar is added stepwise to serine residues of the polypeptide moiety of the glycoprotein. The biosynthesis of a sphingoglycolipid is similar except that the initial step is addition of a glucose unit to ceramide. Each sugar addition is catalysed by a specific glycosyl transferase enzyme.*

The enzymes responsible for attachment of sugar residues to the growing chains of sphingoglycolipids and glycoproteins are extremely specific. This is true not only for the sugar nucleotide precursor but often almost equally so for the acceptor substrate in the growing oligosaccharide chain. The genetic control of carbohydrate structure in these compounds lies in the regulation of synthesis of each specific transferase enzyme, the "one gene - one glycosidic linkage" hypothesis. Therefore it seems likely that *in vivo* the assembly of complex carbohydrate chains may be strictly governed by the relative activities of the glycosyl transferases at the biosynthetic sites and the relative concentrations of the various sugar nucleotide precursors utilized by these enzymes. The heterogeneity of carbohydrate structure produced, which is perhaps the most striking feature of glycoproteins and sphingoglycolipids, although very large, may nevertheless be accurately maintained. If this were not so the degeneracy of surface carbohydrate structure resulting from inaccurate biosynthetic assembly would raise serious doubts about the roles played by carbohydrates in surface receptor reactions. We can now briefly examine the evidence for such roles.

CARBOHYDRATE RECEPTORS AND HORMONES

There are two types of evidence that surface membrane carbohydrate molecules are involved in hormone action. First, the specific receptors for insulin and glucagon have been isolated and shown
to be glycoproteins (8). Secondly, treatment of normally responsive
target cells with glycosidases to remove surface sugars alters dra-
stically the response of the cells to the effector substance. For
instance, recent results in Cuatrecasas' laboratory have shown that
while insulin binding to specific receptors is unaffected by neuraminidase treatment of liver or fat cells (sialic acid groups are
not involved in hormone action) subsequent treatment with β-galact-
osidase rapidly inactivates the receptor sites (galactose groups
are important in binding insulin to the cell surface).

Presumably the binding of insulin or glucagon to the galactose
containing membrane receptor activates another membrane component,
e.g. adenyl cyclase, in a later event in the overall triggering
process, although exactly how this is done is obscure.

CARBOHYDRATE RECEPTORS FOR BACTERIAL TOXINS

Similar initial events seem to occur in the action of some
bacterial toxins on cells, most clearly in the case of cholera toxin
(20). The acute diarrhoea of clinical cholera is due to the action
of a specific protein component of the *Vibrio cholerae* on the epithelial cells of the small intestine. This toxic protein combines
with specific receptors present at the surface of responsive cells
(6, 31) and stimulates the membrane-bound adenyl cyclase so that
intracellular levels of cyclic AMP are greatly increased (6, 7).
It is thought that the flow of salts and fluid into the lumen of the
intestine which are typical clinical symptoms are mediated by the
changes in intracellular cyclic nucleotide levels (2).

Cholera toxin binds specifically to the sphingoglycolipid, GMI.
This lipid has a carbohydrate structure similar to the pentasaccha-
ride unit of the glycolipid shown in Fig. 3 except that the terminal sialic acid residue linked to galactose in the structure is
missing. Toxin receptor activity is dependent on carbohydrate structure since the glycolipid containing two sialic acid residues
shown in Fig. 3 is inactive and so are glycolipids with shorter oligosaccharide chains lacking galactose or N-acetylgalactosamine residues.

Therefore the GMI structure is the primary or sole membrane
component responsible for cholera binding to cells. Thus, incubation of cholera toxin with brain glycolipids, which include GMI, blocks
the biological effects of the toxin on fat cells and on the small
intestine (33). The therapeutic use of such preparations to reverse

cholera toxin action is an interesting possibility. However, it appears that in some cases addition of exogenous glycolipids actively increases the cholera toxin effect, for example on lipolysis in fat cells (9). Presumably, some of the glycolipid becomes associated with the surface membrane of cells to increase effectively the numbers of receptors for cholera toxin. Conceivably, the use of fragments of the GMI molecule, e.g. the carbohydrate moiety might afford several advantages. The binding of toxin would not be affected appreciably but the ceramide unit required for integration into the membrane would be missing.

Other bacterial toxins which have very potent biological effects on certain mammalian tissues also appear to require sphingoglycolipid membrane receptors. Tetanus toxin for example, binds di- and tri-sialoglycolipids very strongly (32) and the activity of botulinus toxin can be blocked with tri-sialoglycolipids (27).

GLYCOPROTEIN RECEPTORS AND SPECIFICITY IN VIRUS REPLICATION

The first step in viral replication involves interaction of virus particles with the surface of the host cell (14). Specific receptor sites are present on the host cell surface membrane and in several cases these have been characterized and found to be glycoproteins. I shall consider two classes of viruses, namely the myxoviruses and the adenoviruses.

For the purposes of the present discussion the animal viruses may be classified by the presence or absence of a viral envelope. In both classes viral adsorption involves complementary structures, namely the receptor sites on the host cell surface and the structural component of the virus responsible for recognition of the specific receptor sites. In the enveloped viruses such as the myxoviruses these structural components must be part of the viral envelope, i.e. they are membrane components. In a non-enveloped virus such as adenovirus the attachment to the host cell involves a structural protein analogous to the tail fibres of T4 phage and the mode of attachment to the host cell is similar (24).

Myxoviruses

The ability of myxoviruses to combine with and agglutinate erythrocytes from many species has been known for many years (11). The attachment to erythrocytes is of course the only event likely to occur in these cells. In the full lytic cycle in nucleated permissive host cells, some mechanism perhaps akin to phagocytosis takes place following attachment of the virus to the cell surface, and the virion or a partially degraded virion penetrates into the cell in which it multiplies (9).

Presumably, similar receptor sites are present in non-permissive cells such as erythrocytes and in cells capable of supporting a lytic infection. Most work has been done, however, using erythocytes. It is now well established that the receptor site includes sialic acid residues. Thus, the haemagglutination of erythrocytes by myxovirus is inhibited by soluble glycoproteins containing sialic acid. By using this assay, two important features of the receptor activity have been found. In the first place, a high concentration of terminal residues of sialic acid is necessary for maximal binding to the virus and hence maximal inhibition of erythrocyte agglutination induced by the virus. Thus, colominic acid, a linear polymer of sialic acid from a bacterium is inactive in preventing haemagglutination (28) presumably because of the very low content of terminal sialic acid residues. Secondly, an intact sialic acid terminal residue is necessary. For example, selective oxidation of the side chain of sialic acid (Fig. 4) in a variety of glycoproteins abolishes their capacity to inhibit viral haemagglutination (30). These results show therefore that the polyhydroxy 3-carbon fragment of sialic acid plays an important role in binding of myxovirus to glycoproteins bearing these residues.

In addition to an intact sialic acid terminal residue, some strains of influenza virus show greater specificity requirements. Thus the binding of the Asian (A2) strain is strongest to sialic acid residues bearing an O-acetyl group in addition to the usual N-acetyl group (21); Fig. 4 shows the structure of this receptor.

On the other hand there appears to be relatively little specificity for the sugar penultimate to a terminal sialic acid residue in the glycoprotein receptor for myxovirus. As mentioned previously a large variety of soluble glycoproteins will compete effectively for virus and prevent haemagglutination. These inhibiting glycoproteins may have very different sugar sequences; sialic acid could be attached to different sugars for example and yet exhibit similar inhibiting activities. This conclusion suggests that there is no single membrane glycoprotein receptor for the myxovirus in most permissive host cells. Sialic acid is distributed in these cells among a variety of different carbohydrate chains, e.g. type A and B chains of glycoproteins (Fig. 3) as well as in sphingoglycolipids. Each of these components may be available at the cell surface for virus binding. Indeed, since sialic acid is ubiquitous in cell surface membranes there is a high concentration of putative receptors for myxovirus adsorption in most cells. Only a few insect cells, e.g. of mosquito appear to lack sialic acid and are unable to bind the virus (5). Since the adsorption of the virus to most tissue is feasible, the resistance of some cells to myxovirus infection cannot be related to a failure of the virus to attach to the cells but rather to some later event, probably internally, in viral replication. These mechanisms of resistance to viral infection are, however, outside the scope of the present lecture.

Adenovirus

In my next example I should like again to illustrate how the susceptibility of cells to a virus, in this case the adenovirus, may reside at least in part in the specificity of interaction of the virus with a cell surface receptor. In certain cases there seems to be a correlation between susceptibility or resistance to infection and the presence or absence of specific receptors at the surface of the challenged cells. The adenovirus receptors are few in number, about 10^4 - 10^5 sites per cell (4, 14, 23), compared to the several million sialic acid receptors present on most cells and, far from being a mixture of glycoprotein species carrying an important sugar determinant necessary for viral attachment, these receptors represent an unique glycoprotein species.

Adenoviruses are non-enveloped viruses 70 to 80 nm in diameter with icosahedral symmetry (Fig. 6). They are common inhabitants of the respiratory and intestinal tracts of man and many animals, mainly causing mild respiratory tract diseases and certain ocular infectious enteritis, rashes, lymph node involvement, intussusception and myocarditis. More recently the oncongenic capacity of the adenoviruses has attracted considerable attention since this appears to be a case in which a human pathogen has been linked to an ability to form experimental tumours (26). The 240 molecules making up the outer surface of the virion surrounding the nucleocapsid core are composed of three proteins separable by polyacrylamide gel electrophoresis of whole virus disaggregated in sodium dodecyl sulphate. The most numerous protein forming the outer faces of the virion is the hexon of molecular weight 120 000. The twelve apical units are complex structures consisting of a base plate protein, the penton, of mole-

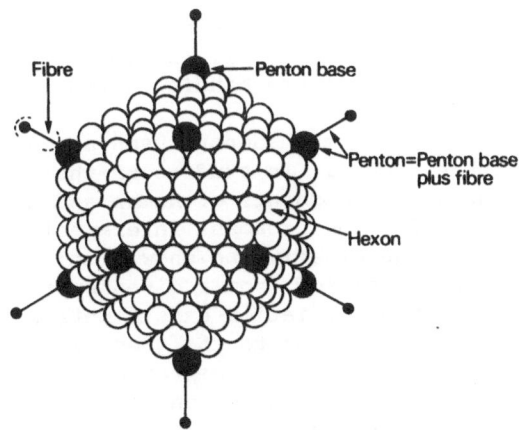

Figure 6.   *Schematic representation of adenovirus showing the disposition of the major capsid proteins.*

cular weight 70 000 to which is attached non-covalently the fibre,
a protein of molecular weight 62 000. The fibre forms a long pro-
jection emanating from each apex of the intact virion.

The adenoviruses are classified into different types antigenically, i.e. the surface viral proteins of individual types show
structural differences that can be detected immunologically, and by
host cell range. The human virus for example, is subdivided into
about 20 types, all of which will grow well in humans and some pri-
mates but not in most laboratory animals. The virus also replica-
tes in some human cell lines in culture, particularly lines of epi-
theloid origin (KB or Hela cells) and to a lesser extent in fibroblasts. Similarly, virus isolated from monkeys is antigenically
distinguishable from the human types, grows less well in human cells
than in monkey cells and very poorly or not at all in other animal
cells.

How is this type of specificity explained at the molecular
level ? Does the initial event in viral adsorption play a role in
defining this specificity ? The interactions involve two components.
The host cell receptor and the structural protein of the virion responsible for recognition of and attachment to the specific recept-
or. We shall consider each of these components in turn.

Viral structural proteins. The structural protein of the virion
responsible for attachment to host cell receptors is the fibre.
This is shown by the binding of radioiodinated ^{125}I fibre but not
labelled hexon or pentonbase proteins to isolated plasma membranes
(4, 14). The binding of radioiodinated fibre to isolated membranes,
for example of human epithelial cell lines, is saturated when high
concentrations of fibre are added to a given amount of membrane
(Fig. 7) and is reversible in the sense that radioiodinated fibre
is rapidly displaced from binding sites on the plasma membrane by
addition of excess cold fibre (Fig. 8). By contrast hexon or penton
base does not compete for the fibre binding sites (14).

As previously mentioned, the adenoviruses are divided into serologically identifiable groups. In studies of the immunological
relationship of structural components of human and non-human adenovirus types, hexons display the broadest interspecies reactivity.
In contrast, fibres diverge greatly and show little cross-reactivity
within the various types isolated from a given species, e.g. humans
(19). Since this structural variation exists between fibre proteins
of different adenoviruses it leads to the question of whether these
proteins recognize the same surface receptor of the host cells that
support their infection. Some results bearing on this point are
shown in Fig. 8. The fact that adenovirus type 5 fibre can displace
only a proportion of ^{125}I type 2 fibre from binding sites on the KB
cell surface (Fig. 7) suggests either that these fibre proteins bind
to the same receptor sites with different affinities or that they

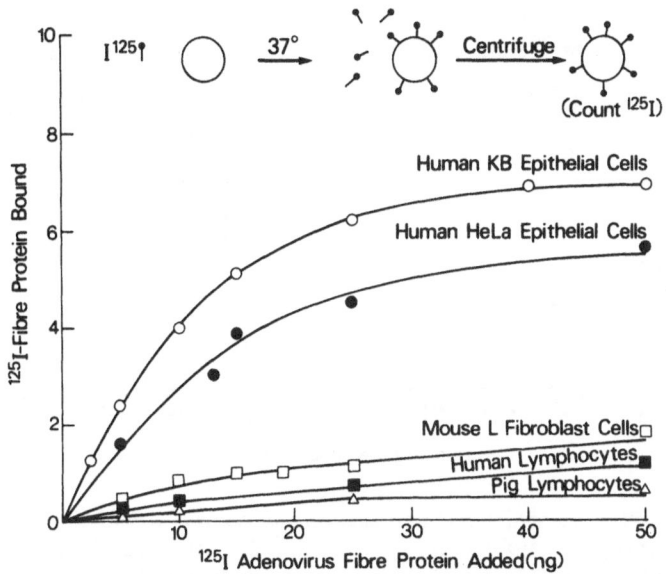

Figure 7. *Adenovirus receptors : binding of adenovirus type 5 fibre*
structural protein to isolated plasma membranes of sever-
al cell lines. The fibre protein which is responsible
for attachment of the virus to cells during infection was
isolated in crystalline form from disrupted virus, iod-
inated with ^{125}I and incubated in increasing amounts
with set amounts of each membrane preparation. After in-
cubation at $37°C$ for one hour, the proportion of viral
protein bound to the membrane receptors was determined
by centrifuging down the membranes, removing the unbound
viral protein left in the supernatant and counting the
radioactivity in the pelleted membranes. Representative
values are shown for separate experiments.

bind to different but adjacent sites on the cell surface. There is
no evidence that the former possibility is true : the dissociation
constant for the interaction of either type of adenovirus with KB
cells is about 10^{-8}M. Therefore, we conclude that type 5 fibre recognizes a receptor site that forms only a part of the type 2 receptor site. These receptor sites are chemically distinguishable.

Host cell receptors. Attempts have been made to isolate these
from KB epithelial cells by affinity chromatography on columns of
the fibre protein insolubilized by covalent linkage to Sepharose
(16). The receptors are solubilized from the membranes by extraction with a fluorocarbon solvent and passed through the column. The
majority of proteins and other components pass unretarded through
the column. The putative receptors bind tightly to the insolubilized fibre molecules and are eluted by increasing the salt concent-

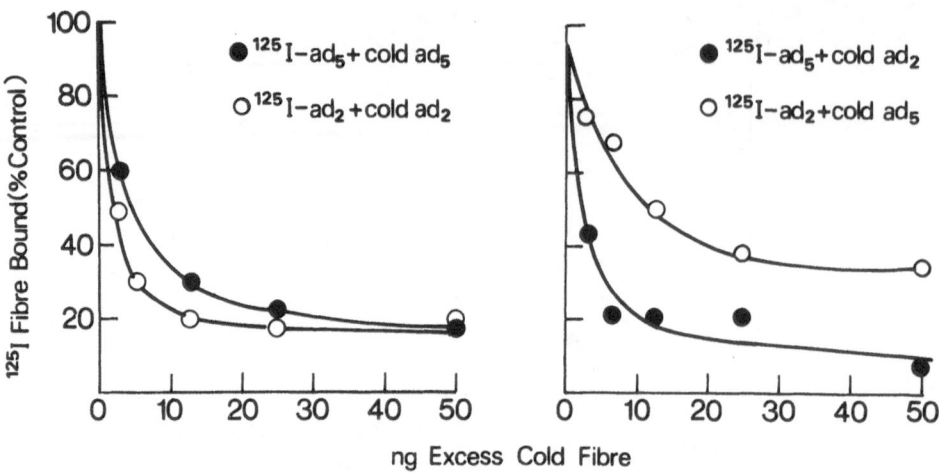

Figure 8. *Serologically distinguishable fibre structural proteins*
from adenoviruses type 2 and 5 were labelled with ^{125}I
and bound to isolated plasma membranes of human KB epith-
elial cells. The unbound fibre was washed away and fibre
bound to specific membrane receptors was then displaced
by incubation at $37^{\circ}C$ with a large excess of either hom-
ologous or heterologous non-radioactive fibre. The ex-
periments show : (a) at least 80% of the bound ^{125}I fibre
is displaced by cold type 5 fibre; (b) binding of type
5 ^{125}I fibre is also reversed by type 2 fibre about as
efficiently as homologous cold fibre; (c) binding of
type 2 ^{125}I fibre is not reversed as efficiently by
addition of type 5 cold fibre as by homologous fibre.
The types 2 and 5 fibres therefore probably recognize
different but closely related surface receptors.

ration of the eluting buffer. When the membrane components isolated
in this manner are examined by polyacrylamide gel electrophoresis
in sodium dodecyl sulphate they are seen to migrate as a glycoprotein species of approximate molecular weight 70 000.

Unlike the myxovirus receptors, however, the sialic acid part
of these molecules may have little to do with virus binding. Thus,
removal of these residues has no effect on receptor activity (3, 14).
This does not exclude of course the importance of other sugar residues of the receptor glycoproteins in virus binding.

The receptors for type 5 adenovirus can be demonstrated on the
surface of a variety of human cells capable of supporting a lytic
infection (14). By contrast, human lymphocytes, in which the adenovirus does not grow, binds fibre protein very poorly (Fig. 7).

Several non-human cells, e.g. mouse fibroblasts, which similarly are non-permissive for adenovirus replication, also carry few specific receptors in the cell surface membrane. In the very limited range of cells examined therefore, there appears to be a clear correlation between permissiveness for adenovirus infection and the presence of specific fibre binding proteins. Cells lacking the specific receptors do not support growth of adenovirus.

It seems likely, of course, that non-permissiveness may stem from a variety of causes only one of which is associated with the presence or absence of a specific surface receptor. Nevertheless, we can conclude that the binding of a virus to its specific surface receptor is an obligatory, although certainly not the only requirement for successful infection. This observation is not limited to the adenoviruses. The classical example is that of poliovirus which adsorbs to and infects mammalian cells but does not adsorb to and does not infect chick cells (1, 12). The chick cells nevertheless can support poliovirus multiplication when infected with RNA extracted from the virus particles because the normal adsorption process is by-passed.

RICIN TOXICITY AND RECEPTOR SPECIFICITY OF CELL SURFACES

In this final section I shall describe some recent work relating to other mechanisms by which cells may render themselves resistant to a noxious agent acting in an initial step at the cell surface. The agent in this case is the plant lectin, ricin, from seeds of castor bean (*Ricinus communis*). It has been known for a century, perhaps longer, that this substance is extremely toxic to animals. The LD_{50} in mice for example, is about 100 ng. Cultured animal cells are also killed readily by low concentrations of the toxin.

The strong toxicity of ricin to cultured cells is easily prevented by certain glycoproteins or oligosaccharides containing galactose. Lactose, for example, at millimolar concentrations increases the number of cells surviving exposure to ricin. In view of these findings, it seems likely that ricin acts on cells by a two step process (Fig. 9). First ricin binds to galactose groups at the cell surface and it is this step that is inhibited by lactose. Subsequent steps require entry of the bound lectin into the cells probably by endocytosis followed by cytoplasmic events (inhibition of protein synthesis) leading to cell death (18, 25).

These conclusions concerning the mechanism of action of ricin are consistent with the structure of this protein. The native lectin of molecular weight 60 000 is composed of two polypeptides each of molecular weight 30 000. After separation, the two chains carry quite different biological activities. Chain A ("the effectomer")

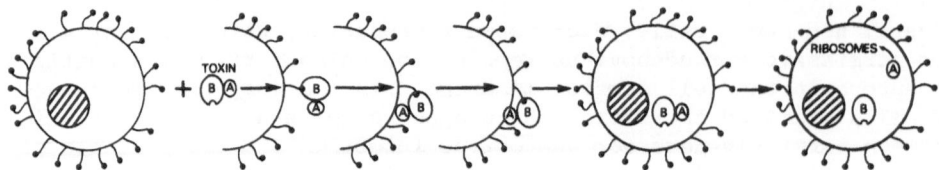

Figure 9. *Schematic structure of the toxic component of* Ricinus
communis *seeds and its interactions with cells. The toxin consists of two polypeptide sub-units linked by one or
more disulphide bonds. The B chain (mol. wt. 34 000) binds
to galactose or N-acetylgalactosamine containing receptors
on the cell surface. The complete toxin molecule enters
the cell perhaps by pinocytosis. The sub-units dissociate
and the A chain inactivates ribosomal protein synthesis.*

inhibits protein synthesis strongly in cell free systems, but is
totally ineffective in whole cells. Chain B ("the haptomer") has
no effect on protein synthesis either *in vivo* or *in vitro*. It does
carry the galactose binding site, however, since this chain binds
to agarose supports, e.g. Sepharose, which contains D-galactose.

                          Toxin Resistant Cells

     Several groups have isolated clones from mutagenized or non-
mutagenized cells in culture resisting the cytotoxic effect of
ricin (10, 15, 17). Resistant cell lines of baby hamster kidney
fibroblasts BHK retain a considerable degree of their resistance
even when cultured for several months in the absence of the toxin
(Fig. 10) and appear therefore to exhibit stable phenotypes with
low reversion rates. Among the resistant cells isolated by this
means, the class blocked in the adsorption of ricin and lacking galactose receptors at the cell surface (Table 1) are most relevant to
the present discussion. I shall therefore not discuss clones that
are resistant to ricin but nevertheless bind about as much lectin
to their surface as normal, sensitive cells. It is likely that in
these variants the block is in some later event in cell killing,
for example, in entry or an inability of the lectin once in the cytoplasm to affect protein synthesis.

     In several of the resistant BHK cell lines isolated by our-
selves and in the single clone of CHO cells described by Gottlieb
et al. (10) the number of ricin receptors is only 1% of normal. One
explanation of this result is that these cells lack certain surface
oligosaccharide sequences that function as receptors for ricin in
the first step (adsorption) in cell killing induced by this lectin.
Alternatively the receptors may contain altered carbohydrate units

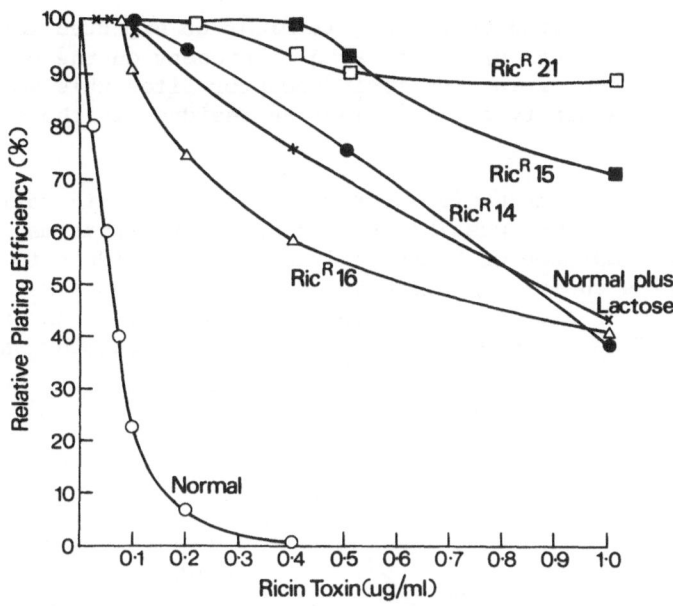

Figure 10. *Baby hamster kidney cells are killed by ricin at low concentrations. Control plates containing no ricin develop many colonies when about 10^3 single cells are incubated for several days. With ricin present however, fewer or no colonies grow. Variant cell lines picked as resistant colonies from a plate containing mutagenized cells and ricin show greatly decreased sensitivity on re-testing with the toxin.*

Table 1. *Resistance to ricin toxicity : possible mechanisms*

1. Insensitive ribosomes.

2. Endocytotic "blocks".

3. Loss of surface receptors:

    (a) Shortened carbohydrate chains;

    (b) Elongated carbohydrate chains (increased sialylation);

    (c) Altered carbohydrate chains (anomeric conversions);

4. Loss of exposed surface receptors ("crypticity").

5. Increased turnover of surface receptors.

that are not recognized by ricin. It might be expected for instance
that if a large amount of sialic acid is present on the peripheral
sugar sequence containing the ricin receptor sites this would lead
to a decreased affinity of the galactose residues in those sequences
for the lectin.

   Other possible mechanisms for resistance to ricin toxicity are
shown in Fig. 11. One possible biosynthetic step that may be involved is the addition of N-acetylglucosaminyl residues to growing
oligosaccharide chains. Failure to add these residues during biosynthesis, due to a defective N-acetylglucosaminyl transferase for
example, would mean premature termination of the carbohydrate chains.

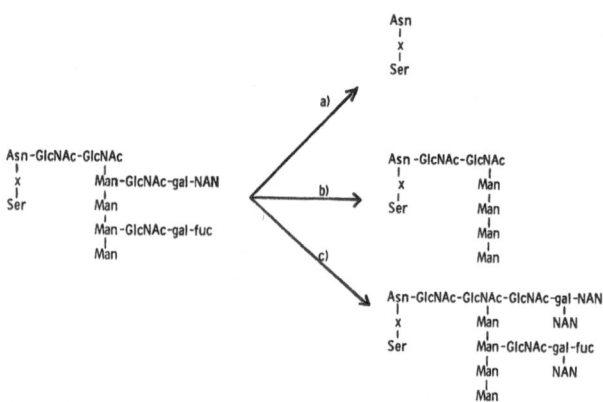

Figure 11. *Possible structural changes in surface carbohydrates lead-
          ing to loss of ricin binding and increased resistance to
          lectin cytotoxicity.*

   (a) *Addition of the first sugar of the carbohydrate chain to
        the polypeptide is blocked, e.g. by lack of the appropr-
        iate N-acetylglucosaminyl transferase enzyme. Further
        elongation of the chain is not possible and the receptor
        sequence containing galactose is not present.*

   (b) *Another enzymic block is possible at a later stage of
        carbohydrate chain assembly. The unfinished carbohydrate
        chains do not bind ricin although they do bind Con A
        normally.*

   (c) *A second type of mechanism conferring resistance to ric-
        in. The sugar sequences recognized by ricin are heavily
        substituted with other sugars in this case by N-acetyl-
        neuraminic acid. In this way the receptor sequence may be
        so modified that it is no longer recognized as a ricin
        receptor or alternatively that ricin binding to galactose
        is inhibited.*

Even if the resistant cells contain normal levels of galactose and sialic acid adding enzymes as well as the necessary intermediates, these sugar residues could not be added since the necessary acceptor sequence for galactosyl transfer, i.e. glycoproteins carryig terminal residues of N-acetylglucosamine would be lacking. We have found one BHK cell variant, Ric^R14, of this type. Thus, the level of N-acetylglucosaminyl transferase activity assayed using a neuraminidase, β-galactosidase and βN-acetylglucosaminidase treated glycoprotein as acceptor substance is only about 10% that of normal cells. By contrast the activities of sialyl or galactosyl transferases assayed using appropriate acceptor glycoproteins are similar to those of normal cells.

The effects of the enzymic lesion on cell surface receptor activity are quite profound. Thus, in addition to the binding of considerably less ricin these cells also bind less *Phaseolus vulgaris* phytohaemagglutinin than do normal BHK cells and are resistant to the weak cytotoxicity of this lectin towards BHK cells. This result is expected since the binding sites for ricin and the common bean lectin are very similar. Both involve galactose or N-acetylgalactosamine residues. It is reasonable therefore that the loss of ricin receptors from the surface of resistant cells means the simultaneous loss of *P. vulgaris* receptors.

By contrast, loss of ricin receptors by the mechanism described for the cell like Ric^R14 exposes terminal sugars that are recognized by other lectins. Thus, concanavalin A which recognizes mannose residues in glycoproteins binds more effectively to the ricin resistant cells than to normal, sensitive cells. Similarly, the ricin resistant cells are more sensitive to the weak cytotoxicity of concanavalin A than are normal BHK cells. Therefore response of the cells towards ricin or concanavalin A varies in a reciprocal manner simply by the manipulation at the cell surface of carbohydrate structures involved in receptor activity.

CONCLUSIONS

It is tempting to extrapolate from these results, obtained by using cells isolated by artificial selection techniques, to the surface changes occurring during *in vivo* selection for resistance to invading organisms and noxious agents. So far however, we do not understand the biochemical details of resistance or susceptibility to viral infections although it is clear that some cells are deficient in receptor sites for virus binding and that viral replication is blocked in these cells. It is hoped that the continued isolation of lectin-resistant cell lines and characterization of the biochemical events leading to their resistance may provide more clues concerning the mechanisms of specificity and resistance to disease. The replication of viruses in lectin-resistant cells lacking surface

carbohydrate receptors is also clearly of interest and may ultimately provide the necessary details of these very important early
events in disease.

One final point : the existence on cells of substances showing
receptor activity for virulent organisms such as the cholera organism and viruses is of course of great medical importance. There is
no reason to suppose however that these receptors have persisted at
the surface of mammalian cells simply to provide convenient entry
of these agents. Preservation of these receptors under strong selection conditions suggests that they have some advantage for cells.
Elucidation of their functions in normal cells remains a major challenge for the future.

REFERENCES

1.   BANG, F.B. (1972). Specificity of viruses for tissues and
          hosts. *In : Microbial pathogenicity in man and animals.*
          (SMITH, H. and PEARCE, J.H., Eds.). *22nd Symp. Soc. Gen.*
          Microbiol., 415 - 419.

2.   BENNET, V., O'KEEFE, E. and CUATRECASAS, P. (1975). Mechanism
          of action of cholera toxin and the mobile receptor theory
          of hormone-receptor-adenylate cyclase interactions. *Proc.*
          natn. Acad. Sci. U.S.A., 72, 33 - 37.

3.   BOULANGER, P.A., HOUDRET, N., SHARFMAN, A. and LEMAY, P. (1972).
          The role of sialic acid in adenovirus adsorption. *J. gen.*
          Virol., 16, 429 - 434.

4.   BUTTERS, T.D. and HUGHES, R.C. (1975). Solubilization and
          fractionation of glycoproteins and glycolipids of KB cell
          membranes. *Biochem. J.,* 140, 469 - 478.

5.   BUTTERS, T.D. and HUGHES, R.C. Unpublished results.

6.   CUATRECASAS, P. (1973). Interaction of *Vibrio cholerae* entero-
          toxin with cell membranes. *Biochemistry, N.Y.,* 12, 3547 -
          3558.

7.   CUATRECASAS, P. (1973). Gangliosides and membrane receptors
          for cholera toxin. *Biochemistry, N.Y.,* 12, 3558 - 3566.

8.   CUATRECASAS, P. (1974). Membrane receptors. *A. Rev. Biochem.,*
          43, 169 - 214.

9.   DALES, S. (1973). Early events in cell-animal virus interact-
          ions. *Bacteriol. Rev.,* 37, 103 - 135.

10. GOTTLIEB, G., SKINNER, A.M. and KORNFELD, S. (1974). Isola-
    tion of a clone of Chinese hamster ovary cells deficient
    in plant lectin binding sites. *Proc. natn. Acad. Sci.*,
    U.S.A., 71, 1078 - 1082.

11. GOTTSCHALK, A., BELYAVIN, G. and BIDDLE, F. (1973). Glyco-
    proteins as influenza virus haemagglutinin inhibitors
    and as cellular virus receptors. *In : Glycoproteins;
    their composition, structure and function.* (GOTTSCHALK,
    A., Ed.). Part B, 1082 - 1096. Elsevier Publishing Co.,
    Amsterdam, London and New York.

12. HOLLAND, J.J. (1964). Viruses in animals and in cell culture.
    In : Microbial behaviour, 'in vivo' and 'in vitro'.
    (SMITH, H. and TAYLOR, J., Eds.). *14th Symp. Soc. gen.
    Microbiol.*, 257 - 271.

13. HUGHES, R.C. (1973). Glycoproteins as components of cellular
    membranes. *Prog. Biophys. mol. Biol.*, 26, 189 - 268.

14. HUGHES, R.C. and MAUTNER, V. (1972). Interaction of adenovirus
    with host cell membranes. *In : Membrane mediated informa-
    tion.* (KENT, P., Ed.). Vol. I, 104 - 125. Medical and
    Technical Publishing Co. Ltd., Lancaster.

15. HYMAN, R., LOCORBIERE, M., STAVEREK, S. and NICOLSON, G.L.
    (1974). Derivation of lymphama variants with reduced
    sensitivity to plant lectins. *J. Nat. Cancer Inst.*, 52,
    963 - 969.

16. MEAGER, A., BUTTERS, T.D., MAUTNER, V. and HUGHES, R.C. (1975).
    Interactions of KB cell glycoproteins with an adenovirus
    capsid protein. *Eur. J. Biochem.* (In press).

17. MEAGER, A., UNGKITCHANUKIT, A., NAIRN, R. and HUGHES, R.C.
    (1975). Ricin resistance in baby hamster kidney cells.
    Nature, Lond., 257, 137 - 139.

18. NICOLSON, G.L. (1974). The interaction of lectins with animal
    cells. *Int. Rev. Cytol.*, 39, 89 - 190.

19. NORRBY, E. (1969). The structural and functional diversity of
    adenovirus capsid components. *J. gen. Virol.*, 5, 221 -
    236.

20. PAPPENHEIMER, A.M. and GILL, D.M. (1973). Diphtheria. *Science,
    N.Y.*, 182, 353 - 358.

21. PEPPER, D.S. (1964). The sialic acids of horse serum with
     special reference to their virus inhibitory properties.
     Biochim. biophys. Acta, 156, 317 - 326.

22. PETERSON, O.H. (1974). Cell membrane permeability change :
     an important step in hormone action. *Experimenta*, 30,
     1105 - 1107.

23. PHILIPSON, L., LONBERG-HOLM, K. and PETERSON, U. (1969). Virus-
     receptor interaction in an adenovirus system. *J. Virol.*,
     2, 1064 - 1075.

24. RAPIN, A.M.C. and KALCKAR, H.H. (1971). The relation of bact-
     eriophage attachment to lipopolysaccharide structure.
     In : Microbiol Toxins. Vol. IV. *Bacterial Endotoxins.*
     (WEINBAUM, G., KADIS, S. and AJL, S.J., Eds.), 267 - 307.
     Academic Press, New York and London.

25. REFSNES, K., OLSNES, S. and PIHL, A. (1974). On the toxic
     proteins ricin and abrin. Studies of their binding to
     and entry into Ehrlich ascites cells. *J. biol. Chem.*,
     249, 2557 - 3562.

26. SCHLESINGER, R.W. (1969). Adenoviruses : the nature of the
     virion and of controlling factors in productive or ab-
     ortive infection and tumorigenesis. *Adv. Virus Res.*, 14,
     1 - 61.

27. SIMPSON, L.L. and RAPPORT, M.M. (1971). Ganglioside inactiv-
     ation of botulinum toxin. *J. Neurochem.*, 18, 1341 - 1343.

28. SPRINGER, G.F., SCHWICH, H.G. and FLETCHER, M.A. (1969). The
     relationship of the influenza virus inhibitory activity
     of glycoproteins to their molecular size and sialic acid
     content. *Proc. natn. Acad. Sci.*, *U.S.A.*, 64, 634 - 641.

29. STECK, T.L. (1974). The organization of proteins in the human
     red blood cell membrane. *J. Cell Biol.*, 62, 1 - 19.

30. SUTTAJIT, M. and WINZLER, R.J. (1971). Effect of modification
     of N-acetyl neuraminic acid on the binding of glycoprot-
     eins to influenza virus and in susceptibility to cleavage
     by neuraminidase. *J. biol. Chem.*, 246, 3398 - 3404.

31. VAN HEYNINGEN, S. (1974). Cholera Toxin : interaction of sub-
     units with ganglioside GMI. *Science*, *N.Y.*, 183 , 656 -
     657.

32. VAN HEYNINGEN, S.E. and MELLANBY, J. (1968). The effect of cerebroside and other lipids on the fixation of tetanus toxin by gangliosides. *J. gen. Microbiol.*, 52, 447 - 454.

33. VAN HEYNINGEN, W.E., CARPENTER, W.C., PIERCE, N.F. and GREENOUGH, W.B. (1971). Deactivation of cholera toxin by ganglioside. *J. infect. Dis.*, 124, 415 - 419.

34. WINZLER, R.J. (1970). Carbohydrates in cell surfaces. *Int. Rev. Cytol.*, 23, 77 - 125.

CONTRIBUTIONS

HANCHEY, P.   Ultrastructural demonstrations of silver methenamine reactive sites in diseased plants.

MANSFIELD, J.W.   Role of lysosomes in controlling hypersensitive cell death in French beans.

MUSSELL, H.   Do plant cell walls contain recognition sites for phytopathogens ?

SCHEFFER, R.P.   Host receptor or sensitive sites for toxins from *Helminthosporium victoriae*, *H. maydis* and *Periconia circinata*.

SUMMARY OF POINTS FROM CONTRIBUTIONS AND DISCUSSIONS
BY
P. HANCHEY

Chairman and Discussion Leader

The existence of recognition sites for infectious organisms or their toxic products and the location and nature of such sites within the cell were discussed as bases of specificity by four speakers following the lecture by Hughes. Scheffer examined the binding abilities of three selective toxins in experiments similar to those done by Strobel with helminthosporoside from *Helminthosporium sacchari*. The only significant site for action of HM-T toxin from *H. maydis* race T (a maize pathogen), is in the mitochondrion. Claims that the toxin inhibits ATP-ase activity in microsomes were not confirmed. Reports that the toxin affects mitochondria from susceptible but not from resistant plants, causing swelling and uncoupled oxidative phosphorylation, were confirmed. These effects were reversible if toxin-affected mitochondria were washed before much damage had been caused. Evidently, HM-T toxin does not bind firmly to the mitochondrial site, although it is active at extremely low concentrations.

In similar experiments with victorin from *H. victoriae*, membrane preparations from oats were treated with toxin, solubilized with detergent, and passed through Sephadex columns to separate bound from free toxin. Toxin was detected in the protein fractions from both resistant and susceptible plants; further, the eluate from control preparations (containing toxin and detergent but without membrane) also contained as much "bound" toxin as did the preparations with membranes (Damann, 1974). Equilibrium dialysis indicated that membrane preparations from resistant and susceptible plants bound small but equal amounts of toxin. These data indicate that victorin is not firmly bound to a specific component of susceptible cells, although the permeability properties of susceptible but not of resistant cells are changed significantly. Similar experiments with a host-selective toxin from *Periconia circinata* (a pathogen of grain sorghum) gave similar results. These three toxins do not appear to bind firmly to active sites in sensitive cells. Thus, their relationships to sensitive cells differ from the relationship reported by Strobel for toxin from *H. sacchari*.

To determine the nature of some chemical changes in oat roots treated with victorin, Hanchey used the silver methenamine reagent to demonstrate reducing group activity. Frequently, this stain is used with periodic acid which oxidizes adjacent glycol and α-amino alcohol groups (such as found in glycoproteins). The aldehydic groups which result reduce the silver in the silver methenamine reagent to free silver as in the Tollen's test, and the resulting deposits are electron dense. However, since some reducing groups are present in tissue sections and reduce silver in the absence of periodic acid, the oxidative step was usually omitted. Thin sections from oat roots exposed to victorin for thirty minutes, then fixed in glutaraldehyde and mounted on gold grids for staining, showed silver deposition in plasma membranes, cell walls, ribosomes and nuclei. Healthy oat roots showed little staining in these regions without prior oxidation, staining being primarily confined to cell walls. Oxidation of sections from victorin-treated roots prior to staining greatly enhanced the density of the deposits. Staining in the cytoplasmic side of cell walls of healthy roots could be almost entirely eliminated by washing with sodium thiosulphate but in toxin-treated roots staining was reduced but not eliminated. Pre-treatment of sections with iodoacetic acid reduced both silver metal and silver salt formation in toxin-treated roots. These results indicate that early effects of victorin include the following : 1. increased exposure of oxidizable groups, as shown by the effects of periodic acid, 2. increased numbers or exposure of anionic sites, perhaps resulting from permeability changes, as shown by sodium thiosulphate washes, 3. increased numbers of reducing groups, including sulfhydryl groups, as shown by pre-treatments with iodoacetic acid.

Mussell reported on the specificity of purified endopolygalacturonase (endo-PG) from *Verticillium albo-atrum* for cotton cell

walls. Both the pathogen and the isolated enzyme cause interveinal
necrosis on cotton. Endo-PG prepared from *Fusarium oxysporum* f. sp.
lycopersici, race 1, elicited a silvering reaction which was probably related to cell wall separation and light reflection. The two
enzymes have a similar molecular weight but their pH optima and isoelectric points differ. Incubation of cell walls from several plant
species with purified endo PG from *Verticillium* or *Fusarium* resulted
in release of peroxidase. More peroxidase was released from cotton
cell walls by the *Verticillium* enzyme.

Indoleacetic acid (IAA) oxidase was also released by the *Verticillium* enzyme from walls of one susceptible cotton variety but
not from two resistant varieties. The *Fusarium* enzyme did not release IAA oxidase. Mussell suggested that the release of IAA oxidase from host walls by endo PG may be important in the regulation of
hormone balance in infected tissues.

Mansfield proposed a mechanism to account for specificity in
bean anthracnose caused by *Colletotrichum lindemuthianum*. The resistant response in this disease is characterized by a rapid hypersensitive necrosis soon after penetration. Histochemical studies
showed release of β-glycerophosphatase from lysosome-like particles
in response to an incompatible race. In contrast, infection by a
compatible race had no effect on particulate staining. Mansfield
suggested that the lysosomal membrane may act as a site of recognition of infectious agents. Infection by a compatible race leads
to stabilization of lysosomal membranes, delayed host cell death
and, consequently, insufficient phytoalexin production. Disruption
of lysosomal membranes in the incompatible interaction leads to rapid
release of hydrolases resulting in necrosis and subsequent early
accumulation of phytoalexins.

These reports illustrate the problems encountered in studies
on disease mechanisms and demonstrate the dangers in generalizing
from one system to another. A biological product may bind firmly
or competitively to an "active site" thus preventing a normal metabolic reaction, or it may destroy some cell component. The consequences of such reactions may lead to secondary reactions which may
then play a role in determining specificity.

CELL SURFACES AND THE BIOCHEMISTRY OF VARIETAL SPECIFIC PATHOGENESIS

PETER ALBERSHEIM

Department of Chemistry, University of Colorado

Boulder, Colorado, U.S.A.

This paper was presented under the following headings.

I. GENERAL RESISTANCE

A. Introduction

B. The Role of the Cell Wall and Wall-Degrading Enzymes in Infective Processes - An Example of General Resistance

II. VARIETAL SPECIFICITY IN HOST-PATHOGEN SYSTEMS

A. An Hypothesis to Account for Varietal Specificity in Gene-for-Gene Host-Pathogen Systems

B. Examples which Demonstrate that Cell Surface Recognition Phenomena are Mediated through the Interaction of Carbohydrate-Containing Macromolecules and Proteins

 1. *Agglutinins of sexual mating types in yeast.*

 2. *A species specific aggregation factor of sponge cells.*

 3. *The determinants of A, B and O red blood cells.*

 4. *The structure of bacteriophage receptors.*

 5. *A specific surface component is required for virulence of a plant pathogenic bacterium.*

 6. *Lectins and the host specificity of nitrogen-fixing bacteria.*

103

7.   *Species specific antigens on the surface of bacteria and fungi.*

8.   *The mitogenic effect of lectins.*

9.   *The specificity of colicins is determined by cell surface receptors.*

10.   *The specificity of several eukaryotic toxins is determined by cell surface receptors.*

11.   *A toxin receptor in the plasma membrane of a plant.*

12.   *The role of carbohydrates in hepatic-plasma membrane recognition of circulating glycoproteins in animals.*

III. ELICITORS OF PHYTOALEXIN PRODUCTION

    A.   Introduction

    B.   An Elicitor Isolated from a Fungus which causes Stem and Root Rot in Soybeans

    C.   An Elicitor Isolated from the Fungus which causes Anthracnose in Beans

    D.   Are Elicitors the "Antigenic" Determinants of Gene-for-Gene Pathogens ?

IV. FURTHER CONSIDERATION OF THE HYPOTHESIS AND HOW THE GENE-FOR-GENE RELATIONSHIP MAY HAVE EVOLVED

The substance of the above is contained in the following reference to which the reader is referred.

ALBERSHEIM, PETER and ANDERSON-PROUTY, ANNE, J. (1975). Carbohydrates, proteins, cell surfaces, and the biochemistry of pathogenesis. *A. Rev. Pl. Physiol.*, 26, 31 - 52.

CONTRIBUTIONS

BRAMBL, R.   A possible role of cell surface glycoproteins in the haustorium-host cell interaction.

CALLOW, J.A.   Plant lectins, a general view.

CALLOW, J.A.   The role of plant lectins in recognition and defense.

SOLHEIM, B.   The possible role of lectins in the infection of
   legumes by *Rhizobium* spp.

SUMMARY OF POINTS FROM CONTRIBUTIONS AND DISCUSSIONS
BY
R. P. SCHEFFER

Chairman and Discussion Leader

   Albersheim's report on glucans as highly active inducers
("elicitors") of phytoalexins was of special interest. His talk
was speculative and provocative, and stimulated much discussion.
First, there was confusion over use of the term "specificity";
this was clarified, following a question by Durbin. Albersheim had
used the word with two different meanings; one was in relation to
the ability of specific compounds to induce the biosynthesis of phytoalexins, and the other was in relation to host specificity of a
micro-organism (the main theme of the Institute). The elicitors
described by Albersheim have little or no genotype specificity for
induction of phytoalexins. For example, the glucan elicitor from
all races of *Phytophthora megasperma* var. *sojae* induced phytoalexins
in all genotypes of soybean, whether resistant or susceptible to the
fungus. How then do we explain the ability of a race to infect certain cultivars and not others ? How does this elicitor lead to accumulation of glyceollin in response to one race of the fungus, but
not to others ? These and related questions were posed by Keen,
Deverall, Kuć and others. Albersheim stated that the work is at
an early stage, and that there are no satisfactory answers to such
questions at this time.

   Lack of sufficient data did not dampen further discussion of
glucans and other substances as special elicitors of phytoalexins.
Day was concerned that fungal components other than glucans might
act as elicitors, and that soybean cells themselves may contain elicitors. Albersheim stated that the only active elicitor found in
fungal cell walls was the glucan; a mannan-containing glycoprotein
may be active, but not as a phytoalexin elicitor (this was not explained further). Attempts to obtain elicitors from soybean cells
were not successful. Hadwiger pointed out that a peptide (monili-
colin from *Monilinia fructicola*) was reported to induce a phytoalexin in one non-host species, but did not induce the compound in
other non-hosts or in host tissue. Even if not highly active, such
inducers might be present in sufficient concentrations to induce
phytoalexin synthesis. Albersheim replied that much effort had
gone into the search for active elicitors, and that none other than
glucans had been found. In response to queries from Mansfield,
Albersheim stated that he has no data on the ability of glucans from
P. megasperma var. *sojae* to induce phaseollin in *Phaseolus vulgaris*,

but that an elicitor from *Colletotrichum lindemuthianum* elicited
glyceollin in soybeans and phaseollin in *P. vulgaris*. DeVay suggested that elicitors must be secreted by haustoria or growing fungal tips in order to be involved in host-pathogen interactions. There
are no data on this but Albersheim suggested an experimental approach.

Questions by Sequeira and Keen were concerned with lectins and
how they fit into the elicitor hypothesis. Albersheim stated that
the glucan elicitors must have receptor molecules, and that lectins
are a rational possibility that has not been examined. Membrane-
bound lectins, which may be overlooked by use of the usual extraction procedures, should be considered. Sequeira asked for comments
in the work of Bohlool and Schmidt (Science, 185, 269 - 271, 1974),
which indicated that only the species of *Rhizobium* capable of inducing nodules were agglutinated by soybean lectins. Sequeira then
suggested that agglutinated cells should be less active and less
effective in nodulation than are free cells; we might suggest that
the non-nodulating bacteria would be the ones that are agglutinated.
Albersheim responded that the nodules of legumes contain live bacteria that are not multiplying. Perhaps the binding to lectin prevents mitosis; lectins are known to limit multiplication of animal
cells. Thus, lectins may be involved in maintaining the bacterioid
state rather than in the infection process.

Many of the participants were left with the impression that
Albersheim's elicitor hypothesis has potential significance, but
there were doubts that it will lead to an understanding of host-
specificity in plant pathogens. Perhaps some relationship to host-
specificity will be evident when more data are available.

Plant lectins were discussed further in two brief papers by
Callow. First, he gave a general review of lectin chemistry, distribution, cellular localization, and genetic control of synthesis.
Callow proposed that we define plant lectins as those proteins or
glycoproteins of plant origin that have the ability to bind specifically to carbohydrates or to molecules containing carbohydrates.
This binding may cause haemagglutination or other dramatic effects;
however, such effects should not be included in a general definition,
because they may have no bearing on the physiological functions of
lectins. Callow then discussed the possible role of lectins in recognition and defense in plants. Lectins are involved in certain
complementary molecular interactions because of their ability to
bind specifically to carbohydrate-containing molecules. Therefore,
lectins are rational candidates for roles in many recognition phenomena, including host-pathogen interactions. Callow reviewed recent
evidence on the possible roles of plant lectins or lectin-type compounds in the following phenomena : 1. gametegamete interactions in
algae, 2. pollen-style incompatibility, 3. sexual agglutination
in yeasts, 4. morphogenetic aggregation in slime-molds, 5. recognition of *Rhizobium* cells by legume roots, 6. defense against path-

ogenic micro-organisms,   7. binding of toxins.

Solheim described briefly his experiments and observations on the possible role of lectins in infection of legumes by species of *Rhizobium*. The work was based on a suggestion by Hamblin and Kent (Nature, New Biol., 245, 28 - 30, 1973) that lectin in roots is capable of binding the bacterial cells at sites suitable for infection. Further background data from Bohlool and Schmidt (Science, 185, 269 - 271, 1974) showed that partly purified lectin from soybean binds to effective but not to ineffective cells of *Rhizobium* (with some exceptions). They suggested that the ability of lectin to bind bacterial cells may account for the host-specificity of the nitrogen-fixing species of *Rhizobium*. Solheim examined the binding of cells of *Rhizobium* to roots of *Vicia hirsuta*, using scanning electron microscopy. Bacterial cells pretreated with lectin from *V. hirsuta* were bound to the roots much more quickly than were untreated cells. There appears to be an interaction between lectin from *V. hirsuta* and slime from compatible but not from incompatible cells. It was then suggested that binding of *Rhizobium* spp. to legume roots that contain lectin should be compared with binding to legumes that contain no known lectins. A model of the early events in the infection process was also suggested (Solheim and Raa, J. gen. Microbiol., 77, 241 - 247, 1973).

Finally, Brambl outlined a proposed study of glycoproteins and other potential determinants of specificity in plant infections. He was especially interested in learning whether or not such substances are present on the surface of fungal haustoria, and how such compounds may be involved in gene-for-gene systems and in hypersensitive reactions.

MODELS OF INTERACTION BETWEEN HIGHER PLANTS AND BACTERIA[*]

KLAUS RUDOLPH

*Institut für Pflanzenpathologie und Pflanzenschutz
der Universität
Göttingen, Germany*

INTRODUCTION

Three models of interaction between higher plants and bacteria
have been distinguished, compatibility, incompatibility and plants
infected by saprophytic bacteria (25).

Only the compatible or susceptible combination leads to the
development of typical disease symptoms and high bacterial populations within the plant tissue. During incompatible reactions visible necrosis may appear, but not always; the bacteria do multiply
within the host tissue but usually their concentration does not reach
more than 1×10^{-3} that of the susceptible combination. Incompatibility results mainly from two different combinations; virulent
bacteria in a resistant host plant, or phytopathogenic bacteria in
a non-host plant. Resistance of the host plant may also result from
age or from environmental effects (23).

Compatible and incompatible combinations can also be differentiated by the numbers of bacteria necessary to induce a visible
response. Ercolani (11) concluded that during compatible combinations bacteria act independently, that is, one bacterial cell can
cause a lesion whereas incompatible combinations are best described
by a hypothesis of combined action.

[*] A reference in the text such as "Patel and Walker (in 26)" means
that the Patel and Walker reference may be found in reference
26 at the end of this paper.

In the third model, saprophytic or, more precisely, non-phyto-
pathogenic bacteria, essentially do not multiply within tissues of
any plant. However, the bacteria can sometimes survive for considerable periods within plant tissues.

In the following our present knowledge of these three models
in regard to specificity will be reviewed. This means mainly a comparison of compatible and incompatible combinations because the model
of saprophytic bacteria in plants does not justify much consideration.
In addition to specificity the term susceptibility will be used too,
because specificity means that only one or a restricted number of
host plant species are susceptible to a particular bacterial species.

I shall not deal in detail with rhizobia, with tumours induced
by *Agrobacterium* spp. or *Pseudomonas savastanoi*, or with diseases
caused by viruses, mycoplasmas and rickettsiae. If we consider the
large number of bacterial species which have been described and that
for most cultivated plants not more than five pathogenic species of
bacteria are known, it is obvious that the susceptible reaction of
a plant to a bacterial species is very rare. It may also be worth-
while to speculate why only five genera are known to cause plant diseases, of which two, *Xanthomonas* and *Agrobacterium*, were coined for
plant pathology alone.

As with fungal diseases, specificity beyond the species level
has been reported. Races or strains with different patterns of host
specificity have been described for many phytopathogenic bacteria
such as *Corynebacterium michiganense*, *Pseudomonas glycinea*, *P. phaseolicola*, *P. solanacearum*, *P. syringae*, *Xanthomonas malvacearum*, *X.
oryzae*, *X. pruni* and *X. vesicatoria*. The physiological or biochemical mechanisms underlying differentiation of races are probably more
subtle than are those for species and will not be discussed here.

Host specificity of a pathogenic bacterium has been explained
by inability of the bacterium to induce the hypersensitive reaction
(HR) (26). If this be sufficient, it seems necessary to analyse the
HR in order to understand the nature of susceptibility. Therefore,
a few words about this phenomenon are justified. HR occurs when
high concentrations of phytopathogenic bacteria are infiltrated into
tissue of a non-host plant and is characterized by a rapid necrotic
reaction (25). HR does not occur when saprophytic bacteria are used.
Therefore, this technique has been used to differentiate phytopathogenic from saprophytic bacteria.

If the statement "susceptibility = inhibition of the hypersensitive reaction" is correct, it would imply that hypersensitivity is
the basic resistance mechanism towards bacterial pathogens. Several
points, however, indicate that this may not always be so.

FACTORS REFUTING THE GENERAL IMPORTANCE OF HR AS A RESISTANCE
MECHANISM

The conditions under which HR is induced are very artificial
because the concentrations of bacteria required are much higher than
those reached by bacteria within a non-host plant under natural conditions. Temperature is also decisive. Thus, many *Xanthomonas* spp.
will only induce HR consistently at high temperatures (42). But at
lower temperatures without visible HR xanthomonads will not multiply
extensively in a non-host plant. As Klement and Goodman (26) have
pointed out the HR may nevertheless play a decisive role in these
instances because only a few cells within a tissue may collapse,
that is, react in a hypersensitive manner. Recently, Turner and
Novacky (51), using Evans blue staining, detected single dead plant
cells, distributed randomly throughout the tissue, when low concentrations of *Pseudomonas pisi* were used. This model is probably near-
er to the situation which occurs in nature. It is questionable
whether the appearance of a few dead cells within an otherwise apparently healthy plant tissue can be called a HR. In any case, the
mechanism by which bacteria are inhibited may be different in a tissue with confluent necrosis and in a tissue with a few dead cells.
In the first case further multiplication of bacteria may be inhibited
by the desiccation of the tissue. Also, collapse of the majority of
the cells means that the quantity of generally biocidal and probably
unspecific compounds liberated during necrosis may be much larger
than when only a few cells die. Although resistance is also induced
in tissue with a few dead cells, this reaction type can less easily
be described as an effect of a non-specific necrosis. Other mechanisms may have to be considered.

Disease resistant cultivars have been developed which do not
show the typical HR even when high concentrations of bacteria are inoculated into tissues. This happens in bush bean varieties developed
from crossings with the primitive bean P.I. 150 414. This reaction
towards *P. phaseolicola* was designated as tolerant by Patel and
Walker (in 26). The multiplication of *X. oryzae* at the site of inoculation was similar in a susceptible and a resistant cultivar.
However, 1.0 cm away from the inoculation point, a lower rate of
multiplication was observed in the resistant variety (36). It must
be concluded from these and similar cases with other varieties that
mechanisms other than HR may play a role in resistance towards pathogenic bacteria.

Saprophytic bacteria which multiply little if at all within
plant tissue do not induce the HR. Moustafa and Whittenbury (in 23)
concluded that phytopathogenic pseudomonads had in common several
features which were not observed with saprophytic pseudomonads. In
addition to the capacity to induce the HR, they listed differences
in Mn-tolerance, H_2O_2 formation, arginine-dihydrolase activity,

oxidase and phenoloxidase activity. It may be argued, therefore, that saprophytic bacteria are not adapted to physiological conditions in plant tissue,so that a specific response of the plant is not necessary for their growth to be prevented. Obviously HR is not involved in this inhibition which occurs very early in infection.

The HR can be inhibited without making plant tissue susceptible. Thus, the HR is inhibited by different treatments such as heat killed bacteria (23), low concentrations of HR inducing bacteria (Novacky *et al.* , in 51), ethanol or ammonium sulphate precipitates from bacterial sonicates (Sleesman *et al.,* in 23), proteinaceous bacterial fractions (Sequeira *et al.,* in 23), cytokinins (32), plant extracts (48), calcium and other salts (Cook and Stall, in 22). But, in all these cases the protected tissue still prevents multiplication of the bacteria or does not allow a susceptible reaction to develop.

According to Ercolani (11) the relationship between inoculated cells of heterologous bacteria during growth *in vivo* is described best by the hypothesis of combined action. This means that a single bacterium or bacterial numbers below a certain threshold do not induce HR. These small numbers of bacteria are obviously restricted from multiplication by mechanisms other than the HR.

Bacteria can multiply in leaves in which the HR develops. This was reported for *X. phaseoli* in pepper plants which were maintained in the dark or treated with chlorophenyldimethylurea to inhibit photosynthesis (41). Under these conditions the photosynthetic capacity of the host appeared to influence resistance.

To conclude this section, the primary importance of the HR in resistance has not yet been established. It can be argued that the HR is a consequence rather than the cause of host resistance to infection as has been postulated for systems involving fungi (13). Therefore, we should continue to look for other mechanisms of resistance towards bacteria, in order to explain the difference between compatible and incompatible reactions.

OTHER MECHANISMS OF RESISTANCE

Nutrition

The restriction of the HR inducing capability to phytopathogenic bacteria indicates the fundamental importance of this activity. It may be considered in light of the main effect of the HR which is to increase membrane permeability leading to electrolyte leakage and loss of turgor (5). Alteration of structural membrane protein may be responsible for this effect (21). Although the inducing mechanism has not been elucidated for leaf spot diseases, several

reports ascribe the HR inducing activity to a heat and pronase labile bacterial fraction,(Sleesman *et al.*, Gardner and Kado, in 11).
Polymethylgalacturonase activity was also reported in this context
(19); so have components in the surface layers of bacteria (11).
In soft rots, caused by *Erwinia* spp., pectinases, especially *trans*-
eliminases seem to be mainly responsible for the increased permeability. If we consider,that with few exceptions, *(Pseudomonas tolaasi*
and mycoplasmas) bacteria propagate within intercellular spaces, it
seems obvious that nutrients from cells must be available to account
for the high concentrations of bacteria within plant tissues. The
HR inducing effect of bacteria would thus describe their capacity
to release nutrients from the cells into intercellular spaces. This
capacity has been called a pathogenicity factor by Ercolani (11)
and differentiates phytopathogenic from saprophytic bacteria. However, only in susceptible combinations are bacteria able to regulate
the increased cell permeability so that it is not detrimental to
them. An increase in permeability of the leaf tissue has also been
observed in susceptible reactions (Williams and Keen,and Burkowicz
and Goodman, in 53).

In bean leaf tissue inoculated with *P. phaseolicola*, permeability
increases were higher in resistant tissue during the early stages of
the reaction, whereas permeability increases started with rather low
values in the susceptible tissue and later increased steadily (58).
Obviously, it is the speed of this permeability increase which is
decisive for the final reaction. The extreme speed of the HR, induced by bacteria is well established (25). How bacteria retard HR
in susceptible reactions is largely unknown. For *P. phaseolicola*
on bean leaves it was suggested that the bacterial toxin is at least
partly responsible for this effect. If healthy leaves are treated
with purified toxin the permeability of the tissue is decreased and
concomitantly the osmotic value in the tissue increases (57). This
may explain why the toxin of *P. phaseolicola* can break the resistance
of bean leaves towards *P. phaseolicola* and *X. phaseoli* var. *fuscans*
and thus elicit a susceptible reaction (38). The mechanism of the
permeability decrease induced by the toxin is not yet known. An
analysis of lipids especially those of chloroplasts revealed an increase of the content of unsaturated as compared to saturated fatty
acids (56).

The influence of nutrition on final reactions can be considered
qualitatively and quantitatively. The qualitative aspect has been
formulated as the balance hypothesis of parasitism and the nutrition-
inhibition hypothesis. But these hypotheses have never been substantiated because the inability of auxotrophic mutants to multiply
(Garber and Schaeffer, in 26) certainly does not explain resistance
towards bacterial pathogens which do not have specific nutritional
requirements if we exclude mycoplasmas and rickettsiae. Many investigations have, however, reported on the influence of host nutrition on bacterial diseases but the causes have not been analysed and

show only more or less modifying effects. The quality of nutrition
may be more important in its influence on synthesis of specific compounds by the pathogen such as toxins or enzymes. Thus, out of several media only one synthetic medium induced toxin production by *P.
phaseolicola* (37). Toxins of other phytopathogenic bacteria also
are only produced on specific media whereas growth of the bacteria
does not show this dependence.

Quantitative aspects of nutrition may be even more important.
As in other diseases, bacterial infection sites accumulate or retain
organic and inorganic substances (43). Also, failure to obtain typical disease symptoms on primary leaves of bean after inoculation
with *P. phaseolicola* is due to depletion of nutrients in these leaves
by translocation into trifoliate leaves since this effect can be prevented by pruning the developing shoot and leaves above the primary
leaves (Rudolph, unpublished).

Since bacteria multiply within intercellular spaces, it is the
composition of the intercellular fluid which is decisive for their
nutrition. A method for obtaining intercellular fluid was first
described by Söding (in 45) and developed by Klement (in 26). It
has been reported that organic and inorganic constituents in intercellular fluid increased dramatically during the susceptible reaction.
Thus, carbohydrates, amino acids, protein, Mg^{2+} and K^+ increased 10-
fold (44). This would be expected in view of the high bacterial
concentrations reached in susceptible reactions. In several host
parasite combinations maximal bacterial concentrations have been
estimated as $5 \times 10^8 - 5 \times 10^9$ bacteria/cm^2 leaf (Ercolani and
Crosse, in 11, 21, 53). Assuming an intercellular space of 0.01 ml/
cm^2 leaf this gives concentrations of $5 \times 10^{10} - 5 \times 10^{11}$ bacterial
cells/ml which are at least 10 times higher than those which occur
in vitro in intercellular fluid from healthy leaves after inoculation
with phytopathogenic or saprophytic bacteria (Rudolph, unpublished).
Besides the increased leakage of cell constituents breakdown of host
cellulose has also to be considered as a source of nutrients. However, the much lower multiplication of bacteria in some incompatible
combinations as well as the inability of saprophytic bacteria to
multiply cannot be explained by nutritional effects alone.

Inhibiting Compounds

The existence of bacteriostatic or bactericidal compounds in
plants was reported in 1914 by Wagner (in 45). Nevertheless, their
importance has remained an open question, although workers have obtained direct and indirect evidence for their presence in plant tissues.

Direct evidence. Phytoalexins have for a long time served to
explain resistance in fungal diseases but not in bacterial diseases.

Thus, Cruickshank (in 22) reported that pisatin in reasonable concentrations was not toxic to plant pathogenic bacteria, and its
formation in the plant was not induced by bacteria according to
Stholasuta *et al.* (in 23). Although phaseollin is induced in bean
by an incompatible race of *P. phaseolicola*, phaseollin has no effect
on bacterial growth *in vitro*. Recently induction of hydroxyphaseollin (=glyceollin) and related isoflavanoids has been reported in
the HR of soybeans to *P. glycinea*. Two of these compounds inhibited
development of colonies of *P. glycinea* in bioassays (22). After HR
to *P. phaseolicola* and *P. mors-prunorum* had developed in bean leaves
three bacterial growth inhibitors were isolated one of which was
coumestrol (O'Brien and Wood, in 22). Several antibacterial compounds were detected by thin layer chromatography of extracts from
bean leaves infected with *P. phaseolicola, P. syringae,* avirulent
P. syringae and *P. tabaci*. The size of inhibition zones, however,
was not correlated with the resistance reaction. These compounds
also appeared in susceptible reactions (7).

Before a definite conclusion can be drawn on the role of phytoalexins in resistance in a particular bacterial disease, evidence
for their effective concentration at the place of interaction between host and parasite has to be obtained.

For resistance of pea tissue towards pseudomonads Hildebrand
(18) suggested a decisive role for preformed substances such as
homoserine which was tolerated only by the pea pathogen *P. pisi*.
Tannin-like materials have been implicated in resistance to *X. pelargonii* (52).

Indirect evidence. This has been obtained for a number of
anti-bacterial compounds in many systems (26). Constitutive
systems mainly based on preformed substances have been distinguished
from induced formation of anti-bacterial compounds (23). Correlations have been reported between resistance and retarded growth of
bacteria in extracts from potato (55), corn (Hartman *et al.,* in 23),
alfalfa (4), pepper (Stall and Cook, in 25) and rice (34) among
others. Hildebrand and Schroth (in 17) obtained indirect evidence
for a bacterial inhibitor which was active when bean leaves were
comminuted.

With intercellular fluid from bean leaves of different bean
cultivars there is a good correlation between inhibition of growth
of bacteria and resistance (39). The activity of a bacteriostatic
principle in the fluid was enhanced by treating leaves with water
from a spray gun and subsequent incubation at high temperatures.
Bacteriostatic compounds were not detectable when plants were kept
at 100% relative humidity in the dark after treatment. When larger
growth chambers were used to obtain more material the earlier results were not always reproducible probably because use of larger
amounts of plant material enhanced the possibility of damage to

cells so that nutrients from them would mix with intercellular
fluid and thus the effect of antibacterial compounds would be nullified. It was easier to demonstrate correlations between the resistant tissue and anti-bacterial activity of intercellular fluid
by concentrating the fluid by lyophilization.

As in fungal diseases, phenolic and glycosidic compounds have
also been implicated in anti-bacterial effects (34). Thus, a correlation between increased peroxidase activity and resistance has
been proposed for *P. phaseolicola* (40), *P. tabaci* by Lovrekovich
et al. (in 23) and for *X. phaseoli* var. *sojensis* (35). Lovrekovich
et al. (in 23) concluded that higher reducing capacity of host tissue is characteristic of rots of potato tuber caused by *E. carotovora*. Also, Moustafa and Whittenbury (in 53) reported that *P. mors-
prunorum*, *P. syringae*, *P. tabaci* and *P. phaseolicola* counteracted
phenolic compounds in host plants whereas non-pathogenic species
were inhibited. However, increased phenolic metabolism in diseased
tissue should not be regarded as evidence of increased resistance
without proof.

Further indirect evidence for the action of anti-bacterial
compounds in resistance tissue has been obtained by 'protection'
experiments (in 11, 23).

Last but not least should be mentioned the vast number of antibiotics and saponins which have been isolated from higher plants
(31, 50). Their role in resistance in bacterial diseases is unknown.

Other Mechanisms

Bacterial diseases have a strong requirement for a high moisture
content of infected tissues. Cessation of bacterial growth during
typical HR may be explained by desiccation, and water-soaked, trans-
lucent, greasy looking spots are typical of many bacterial diseases.
Also, the promotion of naturally occurring bacterioses by high humidity is well known. Mechanisms by which bacteria induce the water
soaking of lesion is not fully understood, although Keen and Williams
(in 22) isolated a lipomucopolysaccharide from *P. lachrymans* which
induces water-soaking in cucumber leaves. Failure of bacteria to
induce a water-soaked lesion will probably favour a resistant reaction of the host plant. Goodman (in 22) showed that in the incompatible combination tobacco/*P. pisi* the bacterial population increased during the 48 hours of the experiment when plants were kept
in a humidity chamber although membrane damage occurred. Similarly,
Young (54) reported that pathogens in heterologous relationships
with beans *(P. lachrymans* and *P. syringae)* multiplied at a rate equal
to that of the homologous *P. phaseolicola* when leaves were saturated
with water after inoculation. Also non-pathogenic bacteria increased

in numbers under these conditions. The effect of water congestion in increasing bacterial multiplication has been ascribed by Sasser *et al.* (in 25) to a decrease of the repressive effect of a high osmotic potential in the intercellular fluid.

A resistance mechanism common in fungal diseases, the establishment of a mechanical barrier, is not so common in bacterial diseases. However, a sealing off of *P. syringae* cankers in plum trees by phellogen activity has been proposed (12) and limitation of spread of *E. carotovora* in potato tissue has been explained by a mechanical barrier (14).

INTERACTION BETWEEN HIGHER PLANTS AND BACTERIA

Saprophytic Bacteria

The third model of interaction between higher plants and bacteria is that of saprophytic bacteria occurring in and on plants. This subject was reviewed recently (17).

In the tissue. It has been demonstrated that saprophytic bacteria occur within apparently healthy plant tissue such as tomato fruit, turnips, stems of *Phaseolus vulgaris* and roots of various plants (27). Multiplication of saprophytic bacteria inoculated into leaves was studied by Söding (45). Usually the bacteria died after three to four days but sometimes, especially in old, detached leaves there was some multiplication. Also phytopathogenic bacteria in a saprophytic phase have been reported in healthy plant tissue such as soft-rotting bacteria in vegetables (Meneley and Stanghellini, in 17), *P. syringae* in soybean plants (28) and *E. amylovora* in pears (49).

On plant surfaces (epiphytes). Nearly all plant surfaces are colonized by bacteria as reported by Berthold (in 45) and more recently by Leben (in 17). Epiphytic bacteria may be antagonists of phytopathogenic bacteria and fungi and in this way add to the barriers to be overcome by pathogens during infection. Also, phytopathogenic bacteria can occur on apparently healthy plant surfaces in a resident phase, for example, *P. syringae* and *P. mors-prunorum* on fruit trees (6, 9, in 17). Analysis of interactions between saprophytic bacteria and higher plants is difficult because usually they do not involve pronounced physiological or biochemical changes.

Mixed Inoculations

If bacteria which induce different reaction types are inoculated together, interesting effects can be observed some of which are described below.

Protection. Susceptibility of plants towards bacterial pathogens can be decreased by simultaneous or earlier inoculation with
other bacterial species or with bacterial extracts. This effect
is called 'protection' since an alteration of the reaction of the
host plant is probably more important than direct effects of a bacterium or extract.

Protection of plant tissue can be induced by avirulent mutants,
incompatible bacteria, weak pathogens, or extracts from virulent
bacteria. Inoculation or treatment 24 hours before inoculation with
the pathogen is in general more effective than simultaneous treatment. For example Averre and Kelman (in 23) reported decreased
virulence of *P. solanacearum* by mixed inoculation with avirulent
cells. Lovrekovich and Farkas (in 26) induced protection against
P. tabaci by heat-killed bacteria. Apple and pear tissue can be
protected towards *E. amylovora* by cell free sonicates of *E. amylovora* (30). In bean and tomato leaves, pre- or simultaneous inoculated with incompatible *X. phaseoli, X. vesicatoria, X. campestris*
and *P. fluorescens,* decreased multiplication of compatible pathogens
(20).

The biological or chemical mechanisms causing protection are
unknown. Hsu and Dickey (20) suggested that incompatible pathogens
induce plant responses which inhibit the growth of the compatible
pathogens. Antagonism of *E. herbicola* towards *E. amylovora* has been
related to hydroquinone formation from arbutin by *E. herbicola* (3).

Synergistic effects. Synergistic effects between bacteria have
been observed, usually as the shifting of the heterologous combination towards compatibility by simultaneous or earlier inoculation
with the homologous pathogen such as heterologous pseudomonads on
bean by *P. phaseolicola* and a *Corynebacterium* sp. on carnation by
P. caryophylli. It was suggested that compatible combinations lead
to an increased release of nutrients from the host which favours
development of incompatible bacteria (Brathwaite and Dickey, in
23).

In a few cases enhancement of disease by weak or non-pathogens
has been described. *P. syringae* and a species of *Achromobacter* enhanced infection by *P. phaseolicola* of bean, probably contributing
to breakdown of cell walls (29).

                            FACTORS OF VIRULENCE

The understanding of the specific susceptibility of plants toward a few bacterial species may also be improved by analysing the
pathogenicity and virulence factors of the pathogen. Pathogenicity
factors have been defined by Ercolani (in 11) as "factors which
control general metabolic activities common to all pathogenic bacteria

in plant tissue, whereas virulence factors govern the specific activities resulting in the induction of a progressive disease in true
host plants". By this definition virulence factors are important
for understanding specificity.

Toxins. One case is known of a host-specific toxin produced
by a phytopathogenic bacterium which can be regarded as a virulence
factor. Polysaccharide fractions of the ooze matrix of *E. amylovora*
caused wilting only of cuttings of rosaceous plants and rate of
wilting was most rapid with the more susceptible hosts (15). Therefore, it has been suggested that this toxin be used to screen for
resistance towards *E. amylovora*. Chemically it consists largely of
a high molecular weight polysaccharide with galactose as the main
sugar residue. More detailed investigations will show whether the
reported correlation between host specificity of the pathogen and
its toxin always applies. If it does elucidation of the mechanism
involved would be very interesting.

However, certain non-host specific toxins may be important in
establishing compatible host-parasite relationships. Thus, a
susceptible reaction can be induced in resistant bean leaves by application of the *P. phaseolicola* toxin (38), and a non-toxigenic
mutant of *P. phaseolicola* with altered pathogenicity has been reported (33). There is the further point that the optimum temperatures for toxin production by *P. phaseolicola* and for disease are
similar.

It has been argued that the toxin of *P. phaseolicola* is unimportant during pathogenesis since it is not produced on resistant
cultivars, so that its formation on susceptible cultivars might be
regarded as of secondary importance. This argument is only conclusive on the assumption that only one compound or one mode of action
decides whether a compatible relationship will result. However, this
does not seem to be the case in most host-parasite combinations.
If, on the other hand, we assume that several successive or concomitant interactions are necessary to produce the compatible reaction, the toxin may well play a decisive role during pathogenesis
as was indicated by the described experiments.

The importance of toxins in inducing typical disease symptoms
in bacterial wilts has been emphasized as follows : *Corynebacterium*
spp. (47), *X. campestris* (in 47), *P. solanacearum* (in 22).

In the case of *P. syringae* pathogenic on peach a membrane-
damaging toxin, syringomycin, has been described by Backman and
DeVay (in 7). Here, rapid development of a necrotic reaction cannot be regarded as a HR but is typical of the compatible combination.
However, *P. syringae* isolates attacking other host plants did not
show this relationship, as shown by Otta and English, and Delgado
(in 7). When Delgado investigated the capacity of isolates of

P. syringae pathogenic to bean to produce syringomycin the correlation between pathogenicity to bean and the production of syringomycin, appeared to be negative because with increasing production
of syringomycin pathogenicity decreased. An explanation for
this phenomenon might be that the antibiotic properties of syringomycin are more important than its phytotoxicity, thus enabling
weakly pathogenic strains of *P. syringae* to survive on leaf surfaces
in competition with other epiphytic bacteria.

   Other factors. Although considerable information has accumulated on other pathogenicity factors of bacteria, it has not
contributed much to our understanding of specificity. Such factors
as enzymes, growth substances and flagellar motility can only be
related to a general capacity to infect plants, that is, to induce
leakage of nutrients from host cells, to degrade large molecular
weight plant substances and to counter oxidative processes in the
host tissue. It remains unknown as to why only in very few combinations can bacteria use these factors without simultaneously triggering plant defence reactions. Counteraction of the HR during the
compatible interaction has been attributed by Ercolani (in 11) to
the so called virulence factor and the nature or mode of action of
factors which counteract the HR or other resistance mechanisms has
been the subject of speculation, some of which are described below.

   DeVay (see this book) has postulated that a common antigen is
a prerequisite for a compatible relationship and that an exchange
of substances might have less disruptive effects on cell metabolism
if these substances provide common information on synthesis. For
several plant-bacteria combinations DeVay and co-workers have demonstrated the existence of common antigens although their role during pathogenesis was not determined. Other laboratories have been
less successful in demonstrating such common antigens (1).

   Pathogenicity of *P. lachrymans* and *P. phaseolicola* has been
correlated with specific capsular antigens (16). Similarly it has
been suggested that the specificity of rhizobia for their host might
be a function of the specific polysaccharides in the bacterial coat
(8) and recently, plant lectins have been proposed as recognition
sites for these polysaccharides (Solheim and Paxton, see this book).

   It was suggested by Ercolani (10) that during the compatible
interaction a virulence factor is responsible for the stable attachment of bacteria to multiplication sites in the host. Experimental
evidence for this attachment is still lacking. Only for *A. tumefaciens* does bacterial attachment to a specific wound-site seem to
be essential for development of typical disease symptoms, that is
tumour initiation, as shown by Lippincott and Lippincott (in 23).
It seems more likely that during the incompatible combination agglutination of bacteria occurs on cell walls and this may involve
lectin-like material (Sequeira, see this book). In view of the high

concentrations of bacteria which build up in susceptible tissue the term 'multiplication site' should be regarded as the optimal milieu for the bacterial multiplication which results from the interaction between host and parasite.

It can be concluded that the host specificity of phytopathogenic bacteria is the result of interactions between two living organisms. This is supported by the simple observations that specific susceptibility of the plants is lost when the plant tissue is killed by heat or frost injury, or when its tissue is homogenized. Even the intercellular fluid obtained from plants during or after resistance reactions does not always inhibit growth of incompatible bacteria. Detached leaves, tissue cultures (24) or plants at 100% humidity (54) also lose their specific susceptibility and can be colonized by bacteria which cannot infect the normal living plant tissue. Also the breaking of resistance by compatible bacteria (46) or by a bacterial toxin (38) indicates that reactions of the plant have to be countered by bacteria in order to allow the development of a susceptible reaction. Obviously, higher plants have developed during evolution diverse and complicated mechanisms to counteract actively pathogenic organisms. Most investigations show that simple models or simple causal analyses are not justified. We have to assume that interactions between higher plants and bacteria are complex. Their full elucidation probably requires more sophisticated techniques than those so far used.

REFERENCES

1.   CAROLL, R.B., LUKEZIC, F.L. and LEVINE, R.G. (1972). Absence of a common antigen relationship between *Corynebacterium insidiosum* and *Medicago sativa* as a factor in disease development. *Phytopathology*, 62, 1351 - 1360.

2.   CHAKRAVARTI, B.P., LEBEN, C. and DAFT, G.C. (1972). Numbers and antagonistic properties of bacteria from buds of field-grown soybean plants. *Can. J. Microbiol.*, 18, 696 - 698.

3.   CHATTERJEE, A.K., GIBBINS, L.N. and CARPENTER, J.A. (1969). Some observations on the physiology of *Erwinia herbicola* and its possible implications as a factor antagonistic to *Erwinia amylovora* in the "fire-blight" syndrome. *Can. J. Microbiol.*, 15, 640 - 642.

4.   CHO, Y.S., WILCOXSON, R.D. and FROSHEISER, F.I. (1973). Differences in anatomy, plant-extracts, and movement of bacteria in plants of bacterial wilt resistant and susceptible varieties of alfalfa. *Phytopathology*, 63, 760 - 765.

5.   COOK, A.A. and STALL, R.E. (1967). The effect of *Xanthomonas vesicatoria* on permeability in resistant and susceptible pepper leaves. *Phytopathology*, 57, 807.

6.   CROSSE, J.E. (1966). Epidemiological relations of the *Pseudomonas* pathogens of deciduous fruit trees. *A. Rev. Phytopath.*, 4, 291 - 310.

7.   DELGADO, M.A. (1974). Zur Pathogenese und Epidemiologie der durch *Pseudomonas syringae* van Hall verursachten Braunfleckenkrankheit der Buschbohne (*Phaseolus vulgaris* L.). Dissertation. Göttingen.

8.   DIXON, R.O.D. (1969). Rhizobia, with particular reference to relationship with host plants. *A. Rev. Microbiol.*, 23, 137 - 158.

9.   DOWLER, W.M. and WEAVER, D.J. (1975). Isolation and characterization of fluorescent pseudomonads from apparently healthy peach trees. *Phytopathology*, 65, 233 - 236.

10. ERCOLANI, G.L. (1970). Bacterial canker of tomato. III. The effect of auxotrophic mutation on the virulence of *Corynebacterium michiganense* (E.F. Sm.) Jens. *Phytopath. medit.*, 9, 145 - 150.

11. ERCOLANI, G.L. (1973). Two hypotheses on the aetiology of response of plants to phytopathogenic bacteria. *J. gen. Microbiol.*, 75, 83 - 95.

12. ERIKSON, D. (1945). Certain aspects of resistance of plum trees to bacterial canker. Part II. On the nature of bacterial invasion of *Prunus* spp. by *Pseudomonas morsprunorum* Wormald. *Ann. app. Biol.*, 32, 112 - 117.

13. ÉRSEK, T., BARNA, B. and KIRÁLY, Z. (1973). Hypersensitivity and the resistance of potato tuber tissues to *Phytophthora infestans*. *Acta phytopath. Acad. Sci. hung.*, 8, 3 - 12.

14. FOX, R.T.V., MANNERS, J.G. and MYERS, A. (1972). Ultrastructure of tissue disintegration and host reactions in potato tubers by *Erwinia carotovora* var. *atroseptica*. *Potato Res.*, 15, 130 - 145.

15. GOODMAN, R.N., HUANG, S. and HUANG, P. Y. (1974). A host specific phytotoxic polysaccharide from apple tissue infected by *Erwinia amylovora*. *Science, N.Y.*, 183, 1081 - 1082.

16. GROGAN, R.G., LUCAS, L.T. and KIMBLE, K.A. (1965). The cor-
     relation of pathogenicity in *Pseudomonas lachrymans* and
     P. phaseolicola with specific capsular antigens. *Phyto-
     pathology*, 55, 1060.

17. HAYWARD, A.C. (1974). Latent infections by bacteria. *A. Rev.
     Phytopath.*, 12, 87 - 97.

18. HILDEBRAND, D.C. (1973). Tolerance of homoserine by *Pseudo-
     monas pisi* and implications of homoserine in plant resi-
     stance. *Phytopathology*, 63, 301 - 302.

19. HOPPER, D.G., VENERE, R.J., BRINKERHOFF, L.A. and GHOLSON, R.K.
     (1975). Necrosis induction in cotton. *Phytopathology*,
     65, 206 - 213.

20. HSU, S.T. and DICKEY, R.S. (1972). Interaction between *Xantho-
     monas phaseoli*, *Xanthomonas vesicatoria*, *Xanthomonas
     campestris* and *Pseudomonas fluorescens* in bean and tomato
     leaves. *Phytopathology*, 62, 1120 - 1125.

21. HUANG, J.S., HUANG, P.Y. and GOODMAN, R.N. (1974). Ultra-
     structural changes in tobacco thylakoid membrane protein
     caused by a bacterially induced hypersensitive reaction.
     Physiol. Pl. Path., 4, 93 - 97.

22. KEEN, N.T. and KENNEDY, B.W. (1974). Hydroxyphaseollin and re-
     lated isoflavanoids in the hypersensitive resistance re-
     action of soybeans to *Pseudomonas glycinea*. *Physiol. Pl.
     Path.*, 4, 173 - 185.

23. KELMAN, A. and SEQUEIRA, L. (1972). Resistance in plants to
     bacteria. *Proc. R. Soc.* B, 181, 247 - 266.

24. KENNEDY, R.W., MEW, T.W. and OLSON, L. (1971). Reaction of
     soybean tissue culture to pathogenic and saprophytic bac-
     teria. *Proc. 3rd int. Conf. Pl. pathogenic Bacteria*,
     Wageningen, 201 - 202.

25. KLEMENT, Z. (1971). Development of the hypersensitivity re-
     action induced by plant pathogenic bacteria. *Proc. 3rd
     int. Conf. Pl. pathogenic Bacteria, Wageningen*, 157 - 164.

26. KLEMENT, Z. and GOODMAN, R.N. (1967). The hypersensitive re-
     action to infection by bacterial plant pathogens. *A. Rev.
     Phytopath.*, 5, 17 - 44.

27. LANGE, R.T. (1966). Bacterial symbiosis with plants. *In :
     Symbiosis*. (HENRY, S.M., Ed.), Vol. I, 99 - 170. Academic
     Press, New York, London.

28. LEBEN, C. and MILLER, T.D. (1973). A pathogenic pseudomonad
     from healthy field-grown soybean plants. *Phytopathology,*
     63, 1464 - 1467.

29. MAINO, A.L., SCHROTH, M.N. and VITANZA, V.B. (1974). Synergy
     between *Achromobacter* sp. and *Pseudomonas phaseolicola*
     resulting in increased disease. *Phytopathology,* 64, 277 -
     283.

30. McINTYRE, J.L., KUČ, J. and WILLIAMS, E.B. (1973). Protection
     of pear against fire blight by bacteria and bacterial
     sonicates. *Phytopathology,* 63, 872 - 877.

31. NICKELL, L.G. (1959). Antimicrobial activity of vascular
     plants. *Econ. Bot.,* 13, 281 - 318.

32. NOVACKY, A. (1972). Suppression of the bacterially induced
     hypersensitive reaction by cytokinins. *Physiol. Pl.
     Path.,* 2, 101 - 104.

33. PATIL, S.S., HAYWARD, A.C. and EMMONS, R. (1974). An ultra-
     violet-induced nontoxigenic mutant of *Pseudomonas pha-
     seolicola* of altered pathogenicity. *Phytopathology,* 64,
     590 - 595.

34. PURUSHOTHAMAN, D. and PRASAD, N.N. (1971). Multiplication of
     bacterial leaf blight pathogen in leaf extracts of three
     rice cultivars. *Annamalai Univ. agric. Res. Annu.,* 3,
     90 - 92.

35. RAMA, R., URS, N.V. and DUNLEAVY, J.M. (1974). Bactericidal
     activity of horseradish peroxidase on *Xanthomonas pha-
     seoli* var.*sojensis. Phytopathology,* 64, 542 - 545.

36. REDDY, A.P.K. and KAUFFMAN, H.E. (1973). Multiplication and
     movement of *Xanthomonas oryzae* in susceptible and re-
     sistant hosts. *Pl. Dis. Reptr.,* 57, 784 - 787.

37. RUDOLPH, K. (1969). Ein phytotoxisches Polysaccharid von
     Pseudomonas phaseolicola. Naturwissenschaften, 56,
     569 - 570.

38. RUDOLPH, K. (1972). The halo-blight toxin of *Pseudomonas
     phaseolicola* : Influence on host parasite relationships
     and counter effect of metabolites. *In : Phytotoxins in
     plant diseases.* (WOOD, R.K.S., BALLIO, A. and GRANITI,
     A., Eds.), 373 - 375. Academic Press, London and New
     York.

39. RUDOLPH, K. and CINAR, O. (1971). Bacteriostatic compounds from bean leaves. *Acta phytopath. Acad. Sci. hung.*, 6, 105 - 113.

40. RUDOLPH, K. and STAHMANN, M.A. (1964). Interactions of peroxidases and catalases between *Phaseolus vulgaris* and *Pseudomonas phaseolicola* (halo blight of bean). *Nature, Lond.*, 204, 474 - 475.

41. SASSER, M. (1974). Evidence against the involvement of hydrogen peroxidase in bacterial leaf spot of pepper. *Phytopathology*, 64, 793 - 796.

42. SCHROTH, M.N. and HILDEBRAND, D.C. (1967). Host reaction as a method of differentiating bacterial plant pathogens.*Proc. 2nd int. Conf. Pl. pathogenic Bacteria, Lisbon.* Contr.no.23.

43. SHAW, M. and SAMBORSKI, D.J. (1956). The physiology of the host-parasite relations. I. The accumulation of radioactive substances at infections of facultative and obligate parasites including tobacco mosaic virus. *Can. J. Bot.*, 34, 389 - 405.

44. SINCLAIR, M.G., SASSER, J.M. and GULYA, T.J. (1970). Pepper leaf intercellular fluid composition after inoculation with *Xanthomonas vesicatoria*. *Phytopathology*, 60, 1314.

45. SÖDING, H. (1959). Über das Verhalten von Bakterien in lebenden Blättern. *Arch. Mikrobiol.*, 34, 103 - 131.

46. STALL, R.E., BARTZ, J.A. and COOK, A.A. (1974). Decreased hypersensitivity to xanthomonads in pepper after inoculations with virulent cells of *Xanthomonas vesicatoria*. *Phytopathology*, 64, 731 - 735.

47. STROBEL, G.A. (1974). Phytotoxins produced by plant parasites. *A. Rev. Pl. Physiol.*, 25, 541 - 566.

48. SÜLE, S., COLENO, A. and LE NORMAND, M. (1973). Inhibition de la réaction hypersensible par des extraits de feuilles. *Acta phytopath. Acad. Sci. hung.*, 8, 71 - 75.

49. THOMPSON, S.V., SCHROTH, M.N., MOLLER, W.J. and REIL, W.O. (1975). Occurrence of fire blight of pears in relation to weather and epiphytic populations of *Erwinia amylovora*. *Phytopathology*, 65, 353 - 358.

50. TSCHESCHE, R. and WULFF, G. (1965). Über die antimikrobielle Wirksamkeit von Saponinen. *Z. Naturf.*, 206, 543 - 546.

51. TURNER, J.G. and NOVACKY, A. (1974). The quantitative relation between plant and bacterial cells involved in the hypersensitive reaction. *Phytopathology*, 64, 885 - 890.

52. WAINWRIGHT, S.H. and NELSON, P.E. (1972). Histopathology of *Pelargonium* species infected with *Xanthomonas pelargonii*. *Phytopathology*, 62, 1337 - 1347.

53. YOUNG, J.M. (1974). Development of bacterial populations *in vivo* in relation to plant pathogenicity. *N.Z. Jl. agric. Res.*, 17, 105 - 113.

54. YOUNG, J.M. (1974a). Effect of water on bacterial multiplication in plant tissue. *N.Z. Jl. agric. Res.*, 17, 115 - 119.

55. ZALEWSKI, J.C. and SEQUEIRA, L. (1973). Inhibition of bacterial growth by extracts from potato tissues. *Phytopathology*, 63, 942 - 944.

56. ZELLER, W. (1972). Zur Wirkung des Toxins von *Pseudomonas phaseolicola* (Burkh.) Dowson im Hinblick auf Permeabilitätsveränderungen. Dissertation, Göttingen.

57. ZELLER, W. and RUDOLPH, K. (1972). Einfluss des Toxins von *Pseudomonas phaseolicola* auf die Permeabilität und den osmotischen Wert von Mangoldblättern *(Beta vulgaris* L.). *Z. Pflanzenphysiol.*, 67, 183 - 187.

58. ZELLER, W., RUDOLPH, K. and FUCHS, W.H. (1973). Permeabilitätsveränderungen bei resistenten und anfälligen Buschbohnensorten nach Inokulation mit *Pseudomonas phaseolicola* (Burkh.) Dowson. *Phytopath. Z.*, 77, 363 - 372.

CONTRIBUTIONS

BRETHAUER, T.S. Evidence implicating specific recognition phenomena in the establishment of *Rhizobium japonicum* in the roots of soybean *(Glycine max)*.

NOVACKY, A. Transmembrane potentials in bacterial hypersensitivity and disease.

SOLHEIM, B. A model of the recognition-reaction between *Rhizobium trifolii* and *Trifolium repens*.

WELVAERT, W. Internal presence of bacteria in artichoke.

SUMMARY OF POINTS FROM CONTRIBUTIONS AND DISCUSSIONS
BY
A. NOVACKY

Chairman and Discussion Leader

Wood opened the discussion with the reminder that the Klement table on hypersensitive reactions (HR) or typical disease symptoms in different combinations of host plants and bacteria may apply mainly to leaf spotting pseudomonads. There are several species of plant-pathogenic bacteria which do not cause HR in leaves. He then responded to Rudolph's remark that confluent HR necrosis - HR induced by the introduction of bacteria into leaves - is "unnatural" by pointing out that the confluent necrosis develops if only a small proportion of leaf cells (about 20%) are killed. Since the number of bacteria introduced bear a reasonable relation to the number of leaf cells killed (about 1:1), we can accept HR as a natural phenomenon. Rudolph explained that in nature high concentrations of bacteria rarely invade the intercellular space of non-host tissue. Although inhibition of bacterial multiplication in HR can be explained by desiccation, this is not convincing when there are few bacterial cells in the inoculum. It has not been shown whether the few dead plant cells are the cause of the resistant reaction or a secondary response. He emphasized that the question is not so much the number of dead plant cells that can be called natural but what is the mechanism of resistance involved when only a few bacterial cells invade a non-host tissue.

Rudolph commented on the terms "resistant" and "avirulent". He suggested avoiding these terms in non-host-pathogen interactions and that "resistant" should be reserved for host-pathogen combinations where susceptibility is an alternative. When susceptibility has never been observed (and therefore is not an alternative) a different terminology should be used.

Brethauer then discussed his work in collaboration with Paxton on the interaction of soybean lectin with extracellular material from *Rhizobium japonicum*. In Ouchterlony double diffusion plates a precipitin reaction occurs between an aqueous extract (20 mg protein/ml) of defatted soybean flour and the supernatant from cultures of *Rhizobium japonicum* (strains USDA 311 and 110 and Nitragin 61A89). A similar preparation of *R. leguminosarum* (ATCC 10314 and Nitragin 128050) failed to react with this soybean extract. Incorporation of galactose in the agar plates (> 0.9 mg/ml) prevented the formation of precipitin bands while glucose or mannose (up to 90 mg/ml) did not affect the reaction. The discrepancy between the lack of serological and chemical specificity of exopolysaccharides and species specificity reported here was questioned by Albersheim. Brethauer suggested that a great deal of information may be contained in parts of molecules which are not antigens. Until shown otherwise,

it is reasonable to assume that plant lectins may recognize differences in these regions of the polysaccharide sufficient to account
for the specificity observed in nodulation.

Solheim then presented a model of the recognition reaction between *Rhizobium trifolii* and *Trifolium repens*. In filtrates from
cultures of *R. trifolii* and in root media of *T. repens* inoculated
with this bacterium he found at least two factors able to deform
root hairs of *T. repens*. The deforming substances in culture filtrates are heat-labile whereas those in inoculated root medium are
stable. The heat-stable, root hair-deforming component is formed
when culture filtrate is mixed with non-inoculated root medium.
The culture filtrate of the bacterium does not have any effect on
the infection process whereas root exudates from *T. repens* speed up
the infection significantly. Solheim discussed the role of these
substances in the binding of rhizobia to clover roots.

Attention then turned to Welvaert who presented his observations
on resident bacterial flora in apparently healthy artichoke tissues.

Finally, the electrophysiological study of membranes in tissues
infected with bacteria reported by Novacky was discussed. Transmembrane potentials (PD) on cotyledons of cotton cultivars, susceptible and resistant to bacterial blight, were measured during
development of disease and HR symptoms induced by an isolate of
Xanthomonas malvacearum and also during HR development induced by
Pseudomonas pisi, a non-pathogen of cotton. Using KCN to separate
passive and active components of PD he found that in all three combinations the passive (diffusion) component was decreased. However,
the active component (electrogenic pumps) was different. In tissues
developing bacterial blight symptoms, it increased, bringing the
total PD to the level of control tissue. A similar change occurred
during HR development in a cotton cultivar resistant to this isolate
of *X. malvacearum*. This, however, had only a short duration. No
increase in the active component of PD was found in cotton tissues
undergoing HR caused by *P. pisi*.

The discussion then focused on various aspects of electrophysiological work: position of the microelectrode tip (Gay), types of
cells measured (Heath), length of the measurement (Scheffer), measurements of tissue segments versus intact tissue (Sivak), and symptom
development in detached leaf pieces versus whole plant (Mansfield).
Novacky explained the problems of PD measurements. After impalement
of cell with microelectrodes the fine tip is positioned mostly in
the vacuole. Data presented represent measurements of spongy parenchyma cells since intercellular spaces of this tissue are infiltrated with bacteria. A successful measurement may last from 30
minutes to a few hours. Technically it would be very difficult to
measure cells of intact tissue, although such measurements have been

performed in roots by Mertz in Higginbotham's laboratory. Symptoms of HR or bacterial blight developed in a similar fashion in both attached and excised cotyledons.

REACTIONS OF CYTOPLASM AND ORGANELLES IN RELATION TO HOST-PARASITE

SPECIFICITY *

WILLIAM R. BUSHNELL

Cereal Rust Laboratory, Agricultural Research Service
U.S. Department of Agriculture
University of Minnesota, St. Paul, Minnesota, U.S.A.

INTRODUCTION

Cytoplasm and organelles of host and parasite are probably involved in most of the host-parasite interactions that determine specificity. Space does not permit a comprehensive treatment of this topic. Instead, two aspects of the involvement of cytoplasm and organelles in host-parasite specificity will be considered. First, the degree of compatibility between cytoplasm from unlike organisms will be discussed in an attempt to establish if host-parasite specificity might be related to cytoplasmic specificity. Second, several ways in which host cytoplasm responds to the approach and invagination by parasites will be described and evaluated with respect to the possible role of each response in specificity.

CYTOPLASMIC COMPATIBILITY

Although cytoplasm of hosts does not ordinarily come into direct contact with cytoplasm of parasites, an understanding of host-parasite specificity depends, in part, on whether or not the cytoplasm from one organism is compatible with the cytoplasm from a second, alien organism. Is cytoplasmic incompatibility generally

* Cooperative investigations, U.S. Department of Agriculture, Agricultural Research Service, and the University of Minnesota, Department of Plant Pathology. Paper no. 1602, Misc. Journal Series, Minnesota Agricultural Experiment Station.

present between unlike organisms, and is it involved in development of incompatibility between host and parasite ? Experimental mixing of cytoplasm from a host and parasite, which might answer these questions, has apparently not been done. However, cytoplasm of certain closely related isolates of fungi is mixed through fusion of hyphae or plasmodia, and cytoplasm of unrelated species of higher plants and animals has been mixed experimentally using the technique of protoplast fusion. The consequences of the mixing of cytoplasm in these systems are suggestive of possible cytoplasmic relation- ships in host-parasite systems.

Somatic Fusion in Fungi

Generally, fusion of hyphae or plasmodia in the fungi occurs only between closely related fungi, such as two strains within a single species. Cytoplasm of the two strains mixes after fusion and remains alive only if the two organisms are vegetatively (somatically) compatible. If the two organisms are incompatible, as often happens when the strains originate from distant geograph- ical areas, death of the mixed cytoplasm results. Somatic incompatibility sometimes relates to sexual incompatibility in that the same genes condition both phenomena in some fungi (15). Further- more, somatic incompatibility can preclude the sexual process in fungi such as the Hymenomycetes and the Ustilaginales in which a stable vegetative dikaryon must be formed by fusion of two haploid hyphae before karyogamy becomes possible (16).

The genetics and physiology of somatic incompatibility in the fungi have been studied extensively in two slime molds, *Physarum polycephalum* (9, 10) and *Didymium iridis* (11, 23, 24), and in two ascomycetes, *Neurospora crassa* (34, 36, 37) and *Podospora anserina* (2, 3). Each of these species is known to have several genetic loci that condition somatic compatibility. At each locus there are usually two allelic genes. If the same gene is not present at that locus in each of the two strains of a fungus, the two strains will be incompatible. In slime molds, genes at some loci condition the fusion of plasmodia, whereas genes at other loci condition compatibility after fusion. In *P. anserina,* several nonallelic genes also condition compatibility in that semi- incompatibility results when two strains of this fungus combine to pair up two of the genes in certain combinations, and complete incompatibility results when two such gene combinations are present (15).

The development of incompatibility is readily observed after fusion of slime molds, in which lethal reactions may occur within one to five minutes after fusion (24) or only after several hours (9). The part of the combined plasmodium after fusion that contains both types of protoplasm dies, and the lethal reaction may spread

so that one member of the pair (the "sensitive" one) dies, leaving
the other plasmodium (the "killer") as a survivor. In less intense
reactions, the killer digests nuclei of the sensitive strain, so
that nuclei of the killer predominate in the remaining plasmodium
(9, 24).

The biochemical events leading to such incompatibility reactions in the fungi are not well understood, although an initial
disruptive effect on membranes is often implicated. Hyphae or
plasmodia dying from incompatible reactions look much like hyphae
or plasmodia treated with agents that disrupt membranes such as
deoxycholate (36), other surfactants, or lysolecithin (9). With
P. anserina, Blaich and Esser (3) speculated that the membranes
of lysosome-like vesicles within the cytoplasm break down, releasing
aminopeptidase which a cytological test had shown to increase in
activity during lethal interactions but which did not increase as
measured in mycelial homogenates. With N. crassa, Wilson (36)
isolated two incompatibility factors that caused a lethal reaction
when injected into hyphae. The factors were proteins; one was
soluble and the other was obtained in a light membrane fraction.
Williams and Wilson (35) postulated that the incompatibility
factors might be structural components of membranes, so that differences resulting from incorporation of different factors into membranes might lead to structural instability when membranes fuse.

For poorly understood reasons, somatic incompatibility in fungi
has evolved to isolate effectively strains within species from one
another (10). It is under strict genetic control and has been
demonstrated only between closely related strains in which fusion
can occur spontaneously.

Protoplast Fusion in Higher Plants

Recent success in several laboratories in fusing protoplasts of
diverse species of higher plants provides good examples of what
happens when whole cytoplasm from two unrelated species is allowed
to intermix. Generally, plant protoplasts will fuse after treatment
with agents that promote agglutination of the protoplasts. The most
effective treatment has been polyethylene glycol (mol. wt. 1 500 -
6 000) applied at a concentration of about 25% in the presence of
calcium ions. Successful interspecific fusions include those made
between species within a single genus such as Torenia baillonii
with T. fournieri (28) and those made between species from unrelated
genera such as soybean with rapeseed (20), pea, barley, or corn (19).

Under appropriate inducing procedures, fusion rates can be
high, involving up to 35% of surviving protoplasts in a suspension
containing two species according to Kao et al. (19). With similar
techniques organelles or nuclei isolated from one species have been

introduced into protoplasts of a second species, as, for example, algal chloroplasts into carrot protoplasts (4) or nuclei of *Petunia hybrida* into protoplasts of *Nicotiana glauca* or *Zea mays* (29).

Generally, there are no signs of incompatibility in such fusion experiments. Fused protoplasts can live for several days and sometimes divide one or more times (19).

These fusion experiments indicate that the mixing of cytoplasm from two higher plants alien to one another, or the introduction of organelles from one species into another does not produce incompatibility reactions, at least under the experimental procedures used. The fusion experiments also indicate that plasma membranes can fuse without generating incompatibility, although these membranes may have been altered by the crude pectinase and cellulase enzyme preparations used in the isolation of the protoplasts from leaves or suspension cultures or by the agents used to promote agglutination. Similar results have been obtained with somatic fusion between diverse animal species (14, 31). Thus cytoplasm of diverse unrelated species of higher plants and animals seems to be compatible when mixed regardless of the presence of sexual incompatibility between the species. Cytoplasm of higher plants and animals is apparently protected by walls and other means so that incompatibility systems against alien cytoplasm have not evolved.

Based on this apparent general cytoplasmic compatibility between species, and in place of direct experimental evidence, we can speculate that cytoplasm of a host and a parasite, such as that of a higher plant and a fungus, will be generally compatible. If this is true, any mechanism of host-parasite specificity would be superimposed on this general compatibility and could not involve any general cytoplasmic incompatibility factors. The experimental mixing of cytoplasm from host and parasite, in this case, might help establish if determinants of specific incompatibility are located in cytoplasm.

If, as seems less likely, there is general cytoplasmic incompatibility, the experimental mixing of cytoplasm might prove useful in establishing the nature of the basic compatibility that exists in highly biotrophic host-parasite associations; i.e., it might help determine if this compatibility results from evolution toward specific cytoplasmic compatibility or from compartmentalization that keeps a general cytoplasmic incompatibility from being expressed.

CYTOPLASMIC RESPONSES OF HOSTS TO PARASITES

Most responses of host cytoplasm to attack by parasites are general ones which apparently do not have a primary role in the determination of specificity, but which may have a secondary role associated with the expression of specificity. Some of these responses will be described here, mainly as they have been observed in my laboratory in barley cells under attack by the obligate parasite, *Erysiphe graminis* f. sp. *hordei*. So far as possible, each response will be related to any possible role it may have in specific host-parasite interactions.

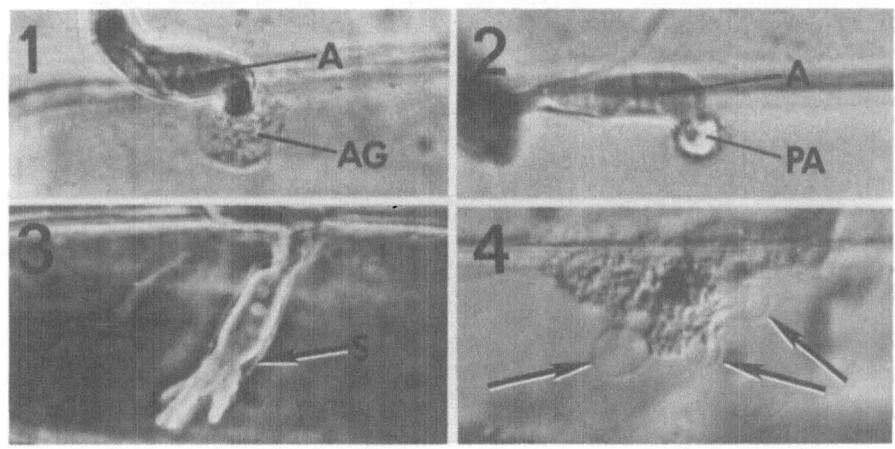

Figure 1. *An aggregate (AG) of cytoplasm in a barley epidermal cell beneath the tip of an appressorium (A) of* Erysiphe graminis *f. sp.* hordei, *11 hours after inoculation. Bright field microscopy, x 1000 (Adapted from Bushnell)(6).*

Figure 2. *Papilla (PA) beneath the tip of an appressorium (A) as in Fig. 1, but 64 hours after inoculation. Bright field microscopy, x 1000 (Adapted from Bushnell and Bergquist)(8).*

Figure 3. *Haustorium of* E. graminis *f. sp.* hordei, *partly formed in a barley epidermal cell. Finger-like lobes have formed on apex of the haustorium; lobes are yet to be formed near haustorial neck. Haustorial sac (arrow) visible on both sides of main haustorial body, visibly separated from haustorial wall. Fixed in 1% OsO$_4$. Negative phase contrast microscopy, x 905.*

Figure 4. *Secondary vacuoles (arrows) near a cytoplasmic aggregate beneath an appressorium (out of focus) of* E. graminis *f. sp.* hordei *in an epidermal cell of barley. Differential interference contrast microscopy, x 1000.*

Cytoplasmic Aggregates

About two hours before an appressorium of *E. graminis* produces a haustorium in a host cell, cytoplasm of the host aggregates abruptly at the site of attack as shown in Fig. 1 (8). In one time-lapse study, an aggregate developed to full size in just 6.5 minutes (Bushnell and Zeyen, unpublished data). Aggregates remain for three to four hours and then disperse. They are always present when any haustoria are produced by the fungus or when any papillae are deposited by the host.

The cytoplasm contains abundant organelles which move rapidly within the aggregate in living specimens. In an investigation by electron microscopy, Bushnell and Zeyen (unpublished data) found that mitochondria predominated among the organelles in the aggregate. Dictyosomes and rough endoplasmic reticulum were abundant and were suspected of being involved in the synthesis of materials that are later deposited in papillae (see below).

Bushnell and Bergquist (8) determined the frequency and timing of cytoplasmic aggregates in six compatible combinations of barley variety and mildew race and three combinations that were incompatible by virtue of single corresponding genes in host and parasite. Aggregates were produced by 93 - 100% of the appressoria observed in all the host-parasite combinations except for a compatible one which had 71% ('Hanna'-Race 9). Thus, the aggregate appeared to be a general response to mildew attack, unrelated to specific incompatibility.

Aggregates sometimes extended to cells adjacent to the one under attack, suggesting that chemical substances were involved in the induction of this response. However, glass microneedles can also induce aggregates as has been described by Russell and Halliwell (30).

In the study of Bushnell and Bergquist (8), the aggregate seemed to be a precondition for deposition of papillae by the host. Although only about half of the aggregates deposited papillae, aggregates were always present when papillae were deposited. Other than its probable role in papillae deposition, the aggregate had no apparent effect on the course of host-parasite interaction.

In some other cases, aggregates have been reported to be associated with specific hypersensitive death of host cells. For example, Kitazawa *et al.* (21), working with *Phytophthora infestans* on potato, found that cytoplasm gathered more consistently and in greater amounts at penetration sites in incompatible than in compatible host-parasite combinations. Thus, the aggregate of cytoplasm can be more pronounced in some hypersensitive hosts than in

susceptible hosts, even though the aggregate is a general response
to parasites and other stimuli.

Papillae

The papilla is also a general response to pathogens that may
relate to some types of incompatibility.   Generally, the papilla
has been thought of as a barrier that is produced by a host as a
fungus tries to enter a host cell (7). Some authors (12, 22) have
suggested that toxins in the host cell might stop fungus growth when
papillae are produced. Smith's (32) observations of fixed specimens
suggested that the papilla would stop the fungus from entering the
interior of cells only if it was laid down rapidly enough.

In the study of Bushnell and Bergquist (8), papillae were
deposited in response to *E. graminis* about two hours after the

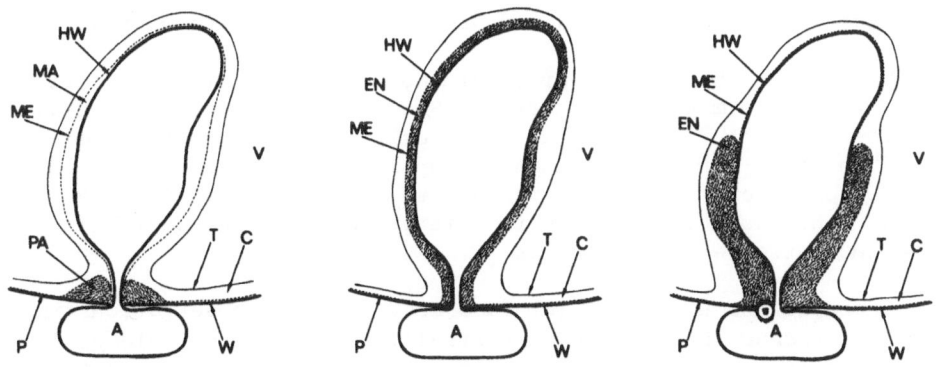

Figure 5. *Diagram of fungal haustorium. Host plasma membrane (P)
    is continuous with extrahaustorial membrane (ME). Other compo-
    nents : host wall (W), host cytoplasm (C), host tonoplast (T),
    host vacuole (V), appressorium or haustorial mother cell (A),
    haustorial wall (HW), papilla (PA), extrahaustorial matrix (MA).
    The haustorial sac includes ME, C and T.*
Figure 6. *Encasement (EN) which originated by action of extra-
    haustorial membrane. Encasement deposited along entire length
    of the membrane. Designation as in Fig. 5.*
Figure 7. *Encasement (EN) which originated from extension of de-
    posits in papilla region. Encasement is deposited by plasma
    membrane at upper edge of the cup-shaped encasement. The extra-
    haustorial membrane does not deposit encasement materials but
    instead remains close to the haustorial wall. Portions of the
    extrahaustorial membrane became trapped between the encasement
    and the haustorial wall. Encasement may later extend to en-
    close the entire haustorium. Designations as in Fig. 5.*

aggregate first appeared. Each papilla (Fig. 2, 5) was a mound-
shaped deposit on the inner surface of the host cell wall, outside
the plasma membrane of the host. By transmission electron microscopy,
the papilla appeared to be deposited in two stages: a first layer of
osmiophilic substances, probably lipid materials, followed by a
second layer of more homogeneous material that stained lightly with
osmium (Zeyen and Bushnell, unpublished data). The plasma membrane
was highly irregular and convoluted, especially when the first layer
was being deposited. Other ultrastructural studies have also
indicated that papillae are deposited in distinct layers (27, 33).

        Bushnell and Bergquist (8) first saw papillae 1.3 hours on the
average after they first saw the aggregate of cytoplasm. This was
about 30 minutes before haustoria were first seen, an insignificant
time difference in view of the difficulties encountered in seeing
young papillae and haustoria in the midst of active aggregates and
the intervals (45 minutes) between observations. Their timing data
did indicate that papillae and haustoria were produced at nearly the
same time and that small differences in timing might, therefore,
determine whether a papilla would prevent haustorium formation.

        The formation of the papilla did not correlate with the failure
of the fungus to make a haustorium. Some haustoria were formed when
papillae were not (32% of observed cases); other haustoria were
formed in spite of the formation of papillae (35% of observed cases).
Only 20% of observed appressoria induced papillae while failing to
produce a haustorium, and only in this group could the papillae have
prevented entrance of the fungus.

        Bushnell and Bergquist (8) made most of their observations in
epidermal tissue which had been partially isolated from coleoptiles
because of the excellent optical properties of such tissue in the
living condition. In leaves, they found a closer correlation
between the presence of the papilla and the absence of the haustorium
than they had in coleoptile tissue. About two-thirds of appressoria
induced a papilla and failed to produce a haustorium. In both
coleoptiles and leaves there was no relationship between papillae
and specific compatibility, as approximately one-half to two-thirds
of observed appressoria induced papillae in both the compatible and
incompatible host-parasite combinations described earlier.
Incompatibility was expressed two or more hours after any papillae
had formed either as hypersensitive death of host cells, or by
retarded fungus development.

        The papilla may, however, relate to species specificity as
results elsewhere have suggested (12). Eighty to ninety per cent of
observed appressoria induced a papilla and produced no haustorium
when *E. graminis* f. sp. *tritici* was placed on leaves of barley (8),
a finding in need of further investigation.

In summary, the papillae occurred with a large part of the
mildew-parasite population in all host-parasite combinations; its
occurrence did not relate to gene-for-gene specificity; and its
occurrence may relate to some cases of species specificity. It is
not clear, however, from this study or others, whether the papilla
actually does stop fungus growth (as concluded by Bushnell and
Bergquist (8)), or is only a secondary structure that forms when
the fungus stops growing for some other reason.

To decide between these alternatives, we are trying to determine more precisely when papillae form in relation to the initiation
of haustoria, but so far we have been unable to see the penetration
peg and young haustorium well enough in living specimens to do this.
Aist and Israel (1) are trying to establish the role of papillae
in host-parasite systems by experimental manipulation with micro-
tools, chemicals, and varied environments. For the present, the
role of the papilla in excluding fungi is uncertain.

Invaginated Cytoplasm

As *E. graminis* enters host epidermal cells, it apparently
invaginates the host cytoplasm that borders the cell, so that a
thin layer of cytoplasm encloses the young haustorium as it extends
into the vacuole of the host (Fig. 3). The layer of cytoplasm acts
as an osmotic sac. It expands on hypotonic media and contracts on
hypertonic media (6). The layer retains the ability to expand
and contract around the central body of the haustorium after the
haustorium forms finger-like branches. The mature haustorium may
extend obliquely across the host cell so that its fingertips come
into contact with parietal host cytoplasm, but most of the
haustorium (with its cytoplasmic sac) remains within the host
vacuole. Occasionally, organelles can be seen in the sac, but they
move so slowly that their movement cannot be detected by eye.

The diagram in Fig. 5 represents the haustorium of *E.graminis*
and most other haustorial fungi. As shown, the invaginated layer of
cytoplasm (the sac) is bounded on the outside by the tonoplast of
the host and on the inside, by a derivative of the plasma membrane
of the host. This membrane (the extrahaustorial membrane) stains
differently from the plasma membrane, is more resistant to disruption (5, 7), and looks different in freeze-etch preparations both in
mildews (F. Gil, personal communication) and in rust fungi (5).
Depending on the type of fungus involved, this membrane sometimes
contains vesicles, invaginations, or other specialized configurations
that appear to be involved in secretory activity across the membrane
(5, 13).

A matrix is invariably found between the layer of cytoplasm
and the haustorial wall and has been designated variously as the

extrahaustorial matrix, sheath, or encapsulation (5, 7, 13). This matrix varies in thickness, but tends to be thicker in powdery mildews than in other diseases. The composition and possible function of the matrix are not known, although it possibly contains secretory products of either the fungus, the host, or both. The matrix is the principal interface between haustorium and host cell. As such, it is a likely site for chemical determinants that may control specific host-parasite interactions, particularly since gene-for-gene specificity is not usually expressed with haustorial parasites until haustoria are produced (7).

Encasements

Sometimes entire haustoria become encased in papilla-like deposits, particularly in the rusts (5). These encasements* can occur in two ways. As shown in Fig. 6, the invaginated plasma membrane (the extrahaustorial membrane) can retain an ability to produce encasement materials and does so all around the haustorium (5, 17). Or, as is more common with haustorial parasites, and as shown in Fig. 7, the encasement apparently develops as an extension of papilla-like deposits on the host cell wall, forming a cup that eventually closes over the top of the haustorium. As a result, membranes and cytoplasm of the host tend to get trapped between the haustorial wall and the encasement. Heath (18) has developed this concept of encasement development more clearly than anyone else. Apparently with many haustorial parasites the invaginated cytoplasm loses the ability to make encasement. The host retains an ability to make encasement in the cytoplasm around the base of the haustorium and can encase haustoria through the cup configuration in some circumstances.

The role of the encasement, like that of the papilla, is not clear. Encasements are found around aborted or dysfunctional haustoria (7). They can occur at low frequencies in compatible host-parasite combinations (17), but they are more frequent in some incompatible combinations (18). We do not know if the encasement causes the haustorium to become dysfunctional, or, as seems more likely, the encasement develops after the haustorium becomes dysfunctional.

Secondary Vacuoles

The cytoplasmic sac that encloses the functional haustorium

* The term "encasement" is used here following Bracker and Little-field (5). "Sheath", as used earlier (7), is avoided here because some authors have used it for "extrahaustorial matrix", others for "encasement".

bears a striking resemblance to the large secondary vacuoles that
are sometimes seen in both healthy and diseased plant cells. These
secondary vacuoles occur within the primary vacuole of plant cells.
They usually consist of a thin layer of cytoplasm which encloses a
vacuole of its own (Fig. 4). Mahlberg and co-workers (25, 26)
have described secondary vacuoles extensively at both the light and
electron microscope levels. Mahlberg (25) postulated that secondary
vacuoles are formed as shown in Fig. 8. In this scheme, the plasma
membrane lifts away from a small portion of the cell wall and
invaginates the adjacent cytoplasm. The invagination enlarges until
it extends into the primary vacuole of the cell. Connections to
the wall are pinched off (as in pinocytosis) so that the secondary
vacuole becomes fully enclosed. The structure usually remains in
contact with parietal cytoplasm or transvacuolar cytoplasmic strands,
and can migrate from one part of the cell to another. Like the
haustorial sac, the outer membrane of the secondary vacuole is derived from the tonoplast, the inner membrane from the plasma membrane.
Mahlberg *et al.* (26) have found small invaginations at the ultrastructural level which they believe to be the beginnings of the secondary vacuole corresponding to the early stages of Fig. 8. Neither
the cause nor the function of secondary vacuoles has been determined.

Figure 8. *Mahlberg's (25) interpretation of the development of
secondary vacuoles. Invagination of the plasma membrane (left)
along the cell wall (shaded) forms a small sac that enlarges
through the cytoplasm (dotted) and subsequently protrudes into
the primary vacuole (V). The plasma membrane probably fuses
at the orifice (dashed lines) to form detached secondary vacuoles which then can move along peripheral cytoplasm and along
transvacuolar strands. Drawing and most of legend from
Mahlberg (25).*

Secondary vacuoles were seen occasionally in studies of haustorium formation by *E. graminis* in epidermal cells of barley (6) (the structures were termed vesicles). The vacuoles undulated rapidly, and remained in any one place only a few minutes. Recently (Bushnell, unpublished) secondary vacuoles of various sizes have been seen in healthy epidermal cells of barley in which they migrated slowly from one place to another. When a haustorium was forming, the secondary vacuoles would sometimes migrate to the young haustorium and remain there for a few minutes, sometimes in clusters. No clear evidence was obtained that the structures were newly formed at sites of infection.

Nevertheless, the secondary vacuole is a structure that plant cells can produce in the absence of an invading fungus that resembles the cytoplasmic sac around haustoria. Thus, the sac may not be the result of physical invagination by the intruding haustorium as has generally been assumed in the past. Instead, the haustorial sac may be produced by the host in the same way that secondary vacuoles are produced. Perhaps the secondary vacuole is a way in which plant cells compartmentalize unwanted materials and the host cell uses this device to enclose the haustorium.

ACKNOWLEDGEMENT

The author thanks Dr. P. Mahlberg and the editor and publishers of the American Journal of Botany for permission to publish Fig. 8, and thanks Dr. S. Akai, Dr. S. Ouchi and the Phytopathological Society of Japan for permission to publish Fig. 1.

REFERENCES

1. AIST, J.R. and ISRAEL, H.W. (1975). Papillae and penetration: cytological aspects of host responses to fungal attack. *Proc. Am. Phytopath. Soc.*, 1, 132.

2. BEGUERET, J. and BERNET, J. (1973). Proteolytic enzymes and protoplasmic incompatibility in *Podospora anserina*. *Nature New Biol.*, 243, 94 - 96.

3. BLAICH, R. and ESSER, K. (1971). The incompatibility relationships between geographical races of *Podospora anserina*. V. Biochemical characterization of heterogenic incompatibility on cellular level. *Molec. gen. Genetics*, 111, 265 - 272.

4. BONNETT, H.T. and ERIKSSON, T. (1974). Transfer of algal chloroplasts into protoplasts of higher plants. *Planta*, 120, 71 - 79.

5. BRACKER, C.E. and LITTLEFIELD, L.J. (1973). Structural concepts of host-pathogen interfaces. *In : Fungal pathogenicity and the plant's response.* (BYRDE, R.J.W. and CUTTING, C.V., Eds.), 159 - 318. Academic Press, London, New York.

6. BUSHNELL, W.R. (1971). The haustorium of *Erysiphe graminis* : An experimental study by light microscopy. *In : Morphological and biochemical events in plant-parasite interaction.* (AKAI, S. and OUCHI, S., Eds.), 229 - 254. The Phytopathological Society of Japan, Tokyo.

7. BUSHNELL, W.R. (1972). Physiology of fungal haustoria. *A. Rev. Phytopath.,* 10, 151 - 176.

8. BUSHNELL, W.R. and BERGQUIST, S.E. (1975). Aggregation of host cytoplasm and the formation of papillae and haustoria in powdery mildew of barley. *Phytopathology,* 65, 310 - 318.

9. CARLILE, M.J. (1972). The lethal interaction following plasmodial fusion between two strains of the myxomycete *Physarum polycephalum. J. gen. Microbiol.,* 71, 581 - 590.

10. CARLILE, M.J. (1973). Cell fusion and somatic incompatibility in myxomycetes. *Ber. dt. bot. Ges.,* 86, 123 - 139.

11. CLARK, J. and COLLINS, O.R. (1973). Further studies on the genetics of plasmodial incompatibility in a Honduran isolate of *Didymium iridis. Mycologia,* 65, 507 - 518.

12. CORNER, E.J.H. (1935). Observations on resistance to powdery mildews. *New Phytol.,* 34, 180 - 200.

13. EHRLICH, M.A. and EHRLICH, H.G. (1971). Fine structure of the host-parasite interfaces in mycoparasitism. *A. Rev. Phytopath.,* 9, 155 - 184.

14. EPHRUSSI, B. (1972). *Hybridization of somatic cells.* Princeton University Press, Princeton, New Jersey, 175 pp.

15. ESSER, K. and KUENEN, R. (1967). *Genetics of fungi.* Translated by E. STEINER. Springer Verlag, Berlin, 500 pp.

16. FINCHAM, J.R.S. and DAY, P.R. (1965). *Fungal genetics.* Second edition. Blackwell Scientific Publications, Oxford, 326 pp.

17. HARDWICK, N.V., GREENWOOD, A.D. and WOOD, R.K.S. (1971). The fine structure of the haustorium of *Uromyces appendiculatus* in *Phaseolus vulgaris. Can. J. Bot.,* 49, 383 - 390.

18. HEATH, M.C. (1971). Haustorial sheath formation in cowpea
     leaves immune to rust infection. *Phytopathology*, 61, 383 -
     388.

19. KAO, K.N., CONSTABEL, F., MICHAYLUK, M.R. and GAMBORG, O.L.
     (1974). Plant protoplast fusion and growth of intergen-
     eric hybrid cells. *Planta*, 120, 215 - 227.

20. KARTHA, K.K., GAMBORG, O.L., CONSTABEL, F. and KAO, K.N. (1974).
     Fusion of rapeseed and soybean protoplasts and subsequent
     division of heterokaryocytes. *Can. J. Bot.*, 52, 2435 -
     2436.

21. KITAZAWA, K., INAGAKI, H. and TOMIYAMA, K. (1973). Cinephoto-
     micrographic observations on the dynamic responses of
     protoplasm of a potato plant cell to infection by *Phytoph-
     thora infestans*. *Phytopath. Z.*, 76, 80 - 86.

22. LIN, M.R. and EDWARDS, H.H. (1974). Primary penetration process
     in powdery mildewed barley related to host cell age, cell
     type, and occurrence of basic staining material. *New
     Phytol.*, 73, 131 - 137.

23. LING, H. and LING, M. (1974). Genetic control of somatic cell
     fusion in a myxomycete. *Heredity*, 32, 95 - 104.

24. LING, H. and UPADHYAYA, K.C. (1974). Cytoplasmic incompatibil-
     ity studies in the myxomycete *Didymium iridis* : Recovery
     and nuclear survival in heterokaryons. *Am. J. Bot.*, 61,
     598 - 603.

25. MAHLBERG, P. (1972). Further observations on the phenomenon
     of secondary vacuolation in living cells. *Am. J. Bot.*,
     59, 172 - 179.

26. MAHLBERG, P.G., TURNER, F.R., WALKINSHAW, C. and VENKETESWARAN,
     S. (1974). Ultrastructural studies on plasma membrane
     related secondary vacuoles in cultured cells. *Am. J. Bot.*,
     61, 730 - 738.

27. MCKEEN, W.E. and RIMMER, S.R. (1973). Initial penetration pro-
     cess in powdery mildew infection of susceptible barley
     leaves. *Phytopathology*, 63, 1049 - 1053.

28. POTRYKUS, I. (1971). Intra and interspecific fusion of proto-
     plasts from petals of *Torenia baillonii* and *Torenia
     fournieri*. *Nature New Biol.*, 231, 57 - 58.

29. POTRYKUS, I. and HOFFMANN, F. (1973). Transplantation of nuclei
     into protoplasts of higher plants. *Z. PflPhysiol.*, 69,
     287 - 289.

30. RUSSELL, T.E. and HALLIWELL, R.S. (1974). Response of cultured
       cells of systemic and local lesion tobacco hosts to micro-
       injection with TMV. *Phytopathology*, 64, 1520 - 1526.

31. SELL, E.K. and KROOTH, R.S. (1972). Tabulation of somatic cell
       hybrids formed between lines of cultured cells. *J. cell.
       Physiol.*, 80, 453 - 461.

32. SMITH, G. (1900). The haustoria of the Erysipheae. *Bot. Gaz.*,
       29, 153 - 184.

33. STANBRIDGE, B., GAY, J.L. and WOOD, R.K.S. (1971). Gross and
       fine structural changes in *Erysiphe graminis* and barley
       before and during infection. *In : Ecology of leaf surface
       micro-organisms*. (PREECE, T.F. and DICKINSON, C.H., Eds.),
       367 - 379. Academic Press, New York and London.

34. WILLIAMS, C.A. and WILSON, J.F. (1966). Cytoplasmic incompat-
       ibility reactions in *Neurospora crassa*. *Ann. N.Y. Acad.
       Sci.*, 129, 853 - 863.

35. WILLIAMS, C.A. and WILSON, J.F. (1968). *Neurospora Newsl.*, 13,
       12.

36. WILSON, J. (1974). *Neurospora Newsl.*, 21, 6.

37. WILSON, J.F., GARNJOBST, L. and TATUM, E.L. (1961). Heterokary-
       on incompatibility in *Neurospora crassa* . Micro-injection
       studies. *Am. J. Bot.*, 48, 299 - 305.

CONTRIBUTIONS

GIL, F. Specific differentiation on the host plasmalemma after
    infection by *Erysiphe pisi*.

HEATH, M.C.   Ultrastructural responses of hosts and non-hosts to
    rust fungi.

WELVAERT, W.   Reaction of some organelles to mycoplasma infection.

YODER, O.C.   Control of specificity by nuclear and cytoplasmic
    genomes.

SUMMARY OF POINTS FROM CONTRIBUTIONS AND DISCUSSIONS
BY
J. L. GAY

Chairman and Discussion Leader

The discussion centred on the host response in forming papillae
and on the role of haustoria. Johnson enquired whether variation
in papilla formation was due to fungus or host variability. Bushnell's evidence was that leaf epidermal cells were heterogeneous and
that those near stomates responded less frequently. Coleoptile cells
of a particular variety also varied but cells of one coleoptile were
more uniform in forming papillae. The nature of papilla material
was questioned and Albersheim pointed out that many erroneous reports
of callose in papillae arose from alkali solubility and aniline blue
staining which were not specific for callose. Heath reported that
papillae could be digested by β,1-3 glucanase but results of periodic
acid - silver staining indicated that other substances were present.
Bushnell also pointed out that they have membranous inclusions.

A lively discussion followed questions on the function of haustoria. Were they vestigial organs as Gaümann had suggested ?
(Sivak). What were the relative contributions of haustoria and
mycelia to the metabolism of the pathogen ? (Daly). Scheffer thought
it inconceivable that powdery mildews could obtain nutrients other
than through haustoria but the question remained open. What was
the hard evidence of their role in transfer of substances from host
to pathogen ? (Wood). Radio-tracer studies indicated movement from
host to haustorium and strong evidence for their importance was
provided in incompatible infections by powdery mildews where the
fungus never continued development if haustoria failed to form
(Ellingboe). A similar conclusion had been reached from haustorial
excision experiments (Bushnell). In crown rust of oat also, micro-
autoradiography indicated the active role of haustoria. Assimilates
were taken up by infection hyphae but pyrimidine nucleoside was
hardly absorbed unless haustoria were present and then it also spread
into hyphae (Tani). Dekhuijzen stated that haustoria show respiratory activity and, therefore, were likely to be functional. However, there was evidence of direct passage of lithium into germ
tubes of powdery mildews when the host was supplied with lithium
salts which move in the free space (Bushnell and Durbin). Gay raised
the possibility that haustoria may be nutritionally functional only
until their boundaries became differentiated and Bushnell pointed
out that function may also be impaired by widening of the extrahaustorial matrix as haustoria age or when the host or hyphal cell
dies. When questioned by Dekhuijzen on the possibility that secondary cytoplasts move to the extra-haustorial boundary where their
remaining organelles were picked up and finally digested by enzymes
in the sheath matrix, Bushnell considered this unlikely because in

continuous observations, secondary cytoplasts rarely made contact
with haustoria. Schönbeck remarked that haustorium-like bodies in
vesicular - arbuscular mycorrhizae ultimately became divided and
their contents digested leaving the fungal wall. Bushnell then
observed that if host cells could be trained to digest haustoria we
might have a new way of controlling plant diseases.

Four short contributions followed, each giving particular examples of effects on organelles. Gil showed that in peas infected
by *Erysiphe pisi* the region of the host plasmalemma adjacent to the
haustorium became specifically modified. Conclusions were drawn
from light and electron microscopy of haustoria in host cells and
after isolation of the haustoria. The haustorial region of the
plasmalemma retains its semi-permeability but is less permeable than
elsewhere. Freeze fracture replicas show that the two regions differ
in molecular structure but most outstanding is the difference in
stability. The contact region withstands mechanical stresses, osmotic gradients and detergent treatments which disrupt the unmod-
ified part. Stability and low permeability are most probably due
to impregnation with polysaccharides and treatment with commercial
cellulolytic and pectolytic enzymes eliminates both. Further modification occurs at the junction of the stabilized and unstabilized
regions where a narrow band of osmiophilic material seals the membrane firmly to the fungal wall. The biological significance of
the changes can only be tentatively assessed. The seal retains
fluid in the zone between host and pathogen through which nutrients
presumably pass. The impregnating polysaccharides and the reduced
permeability of the membrane are also likely to be significant in
the physiology of disease. It is postulated that the modifications
to the contact region of the host plasmalemma are essential to the
continued function of the infected cell and that if they are not
formed the host cell may die (hypersensitivity) leading to restriction of the pathogen and consequent specific disease resistance.

During discussion Day pointed out that the membrane differentiation that occurs in haustorial development suggests effects of
the pathogen that could be interpreted as induction of susceptibility
in the host.

To elucidate the determination of specificity Heath described
resistant reactions in three host-pathogen combinations. By light
microscopy, each reacted similarly with one or two haustoria and
rapid necrosis of the invaded cell but the reactions could be distinguished by electron microscopy as follows.

1.   Non-host interaction - *Uromyces phaseoli* var. *typica* (bean rust)
on cowpea. Haustoria developed and remained ultrastructurally normal
despite subsequent disorganization of the invaded cell. Therefore,
cell death was not accompanied by production of substances immediately toxic to the fungus and lack of fungal development was probably
due to starvation.

2.   Varietal response - *U. phaseoli* var. *vignae* (cowpea rust) on
cowpea cv. Dixie Cream. Haustoria were abnormal and subsequently
they and invaded cells died simultaneously. Infection hyphae appeared normal but seldom initiated a second haustorium. Thus cell
death was probably accompanied by toxin formation which contributed
to resistance.

3.   Varietal response - *U. phaseoli* var. *vignae* on cowpea cv. 'Queen
Anne'. In 60% of infection sites, the haustorium and invaded cell
disorganized simultaneously suggesting that a fungal toxin contributed to resistance. At other sites, neither haustorium nor host
cell died but the haustorium became encased in callose-containing
material. Before encasement was complete, vacuolation and increase
of lipids in haustoria suggests restriction of passage from the host.
The substantial accumulation of phospholipid-like material on the
intervening host plasmalemma may have caused this. Thus, in this
cultivar there were two types of mechanism responsible for host resistance.

These experiments show that each rust-host combination has unique
features and extrapolations should be minimized until the detailed
mechanisms in each are known.

In discussion Kuć commented on multiple mechanisms for resistance to a single pathogen. He warned that elimination of one mechanism without destroying resistance is often incorrectly interpreted
to mean that the mechanism is unimportant or not concerned with resistance. Even in single gene resistance it may be incorrect to
assume only one mechanism for resistance.

Welvaert spoke on the reaction of organelles to infection by
mycoplasmas. Mycoplasma-like organisms (MLO) were seen in ultra-
thin sections of degenerated flowers of *Hydrangea macrophylla* by
electron microscopy. They were transmitted through *Cuscuta subinclusa*
to *Vinca rosea* where they caused green flowering. Infected and the
adjacent phloem cells of these plants showed typical symptoms. Some
were collapsed and in others abnormal formation of callose thickened
the sieve plates. In the surrounding cells, the chloroplasts gradually degenerated. They became filled with starch granules and finally ruptured liberating their contents. Mitochondria were not affected. The effects are characteristic of MLO infections. Abnormal
callose deposits and deterioration of chloroplasts associated with
mycoplasma infection have been reported by other investigators.

In the discussion which followed, van den Heuvel asked if an
incompatible reaction could be induced in plants by a mycoplasma.
Welvaert replied that so far none has been clearly recognized. However in incompatibility studies it is important to distinguish between *Spiroplasma* where the isolated organism has proved to be the

causal organism and *Acholeplasma* types where there is doubt about pathogenicity, and also the normally saprophytic types, e.g. *A. laidlawii* which cause no symptoms but can be re-isolated from infected plants.

Evidence of specific effects on the plasmalemma and mitochondria in diseases caused by *Helminthosporium* spp. was given by Yoder. In most plant diseases, specificity is controlled by nuclear genes in the host. For example, in leaf spot of maize caused by *Helminthosporium carbonum* race 1, specificity depends on one nuclear gene. The most convincing evidence for cytoplasmic genetic control of specificity has been obtained from southern corn leaf blight caused by *H. maydis* race T where susceptibility is due to Texas and other male-sterile cytoplasms. Both pathogens produce host-specific toxins which only affect plants susceptible to the fungi. Thus the question of location of the gene(s) product(s) for fungus-host specificity can be considered in terms of the sites of action of the two host specific toxins. Evidence suggests that effects of the *H. carbonum* toxin are confined to the plasma membrane whereas *H. maydis* toxin only affects mitochondria of susceptible hosts.

Cells exposed to the *H. carbonum* toxin immediately take up and retain larger amounts of certain solutes(nitrate, sodium, chloride, 3-0-methylglucose and leucine) without altering with respect to others (nitrite, potassium, calcium, phosphate, sulphate and glutamic acid). Also, the electrical potential of the plasma membrane changes within five minutes. However, there is no detectable effect on the physical or biochemical properties of the membrane systems of mitochondria or chloroplasts and the toxin does not affect intracellular compartmentalization.

H. maydis toxin disrupts mitochondria isolated from susceptible but not from resistant cells and has no effect on chloroplasts obtained from susceptible or resistant corn. Evidence that the toxin also affects the plasmalemma is not convincing because attempts to repeat experiments have failed and the evidence has mostly been obtained from work with impure toxin. Whereas impure toxin affects both mitochondria and plasma membranes, only mitochondria are damaged by short exposures to partially purified toxin. Thus the molecular bases for cellular specificity in the two diseases seem to parallel the mechanisms of genetic control.

In discussion Bushnell mentioned work at the University of Minnesota in which mitochondria of callus cultures adapted to grow in solutions of T-toxin have proved to be insensitive. Watrud and Koeppe at Illinois have shown that specificity to *H. maydis* resides in the outer but not the inner mitochondrial membrane (Paxton). Possible interactions between nuclear and cytoplasmic factors were suggested by Johnson who cited work by Barratt and Flavell at the

Plant Breeding Institute, Cambridge, who showed that nuclear genes which restore male fertility in plants carrying Texas cytoplasm also reduce the sensitivity of mitochondria to race T toxin.

SPECIFIC INTERACTIONS INVOLVING HORMONAL AND OTHER CHANGES *

J. M. DALY

Laboratory of Agricultural Biochemistry, University of Nebraska Lincoln, Nebraska, U.S.A.

INTRODUCTION

In their advice to participants at this Institute, the organizers asked that presentations be critical reviews and provocative speculation. The critical review of hormones in plant disease recently has been accomplished by Sequiera (34) and Van Andel and Fuchs (41) while certain other aspects of hormones in diseased plants were covered in a recent symposium to be published (10). Consequently, it seems more appropriate for me to emphasize the speculative, although I am not sure whether this colloquial effort will be considered provocative, provocation or mere idiosyncrasy. An attempt will be made to develop the notion that growth regulators, including natural hormones, are potential candidates as causes of specificity in plant disease. By necessity, the attempt is colloquial because critical evidence is lacking; yet our experience with rust diseases has caused us at least to acknowledge the idea.

Hormonal involvement in plant disease almost instinctively is coupled conceptually with the development of the extensive growth distortions typical of a number of classical diseases such as crown gall. Usually, the investigative focus is centered on the nature and quantities of hormones which are involved in symptom expression and with the mechanisms by which hormonal imbalance is achieved.

* Published with the approval of the Director, Nebraska Agricultural Experiment Station, as Paper No. 4058, Journal Series.

Rust and mildew diseases frequently are characterized by massive growth distortions, for example white pine and cedar apple rust diseases, while mild, almost unobtrusive, growth irregularities are nearly universal in diseases caused by biotrophs. Because distortions are particularly obvious in late stages of infection, there is a tendency to view them merely as secondary outcomes of pathogenesis, not as primary determinants in the establishment of a parasite. In the case of cereal rusts and mildews, this view appears reasonable because obvious growth distortion is rare at any stage of infection. As pointed out a number of years ago (9), the extent of distortion is as much a function of tissue responsiveness to hormones as it is to the presence of a growth regulator. Generally, the mature leaf tissue of cereals is unresponsive to application of hormones. Consequently, failure to see a growth response during infection by cereal rusts, or in any disease, does not rule out a fundamental role for hormones in pathogenesis. A compatible or incompatible relationship still could be mediated by an effect of hormones on metabolism, but without growth.

HORMONES AND VIRULENCE

It seems increasingly clear that plant hormones do not fit the classical definition of circulating, regulatory molecules with a target site distant from the site of synthesis. Rather, they may be considered as endogenous regulators of cellular activity whose local concentration, and interaction with each other, are responsible for differentiation and development through altered metabolism. It is not necessary to document here the manifold influence of hormones on metabolism of healthy plants, but it is appropriate to point out that such hormonal effects on enzyme and respiratory activity, isozyme patterns, nucleic acid metabolism etc., are duplicated in the literature on host-parasite interactions.

Logically, there is no reason therefore to exclude indigenous hormones or parasite-produced growth regulators from the potential equipment for virulence possessed by pathogens. Pathologists by habit, largely, are more comfortable when virulence is explained through toxins or through degradative enzymes, rather than growth regulators. The demonstration that fusicoccin has growth regulatory properties (21, 23), as does helminthosporol (19, 38), may heighten appreciation of the potential of endogenous hormones in determining host-parasite specificity through metabolic regulation. In both cases, these compounds are toxic to hosts, but cause growth effects on non-host species. Perhaps of greater interest are the reports by Evans and co-workers that the host-specific toxins of *Helmintho-sporium victoriae* and *H. maydis*, race T cause rapid growth response in oats and corn respectively (15). Similarly, fusicoccin can substitute for indole acetic acid in the rapid response of cereal coleoptiles (6).

Unfortunately, the observations of Safter and Evans (31) with oats and victorin must be interpreted with caution. Both resistant and susceptible cultivars responded to toxin preparations by added increments of growth in the first few minutes, but subsequent inhibition of elongation occurred in susceptible tissues and was followed by contraction in size. The obvious conclusion is that victorin had affected membrane reactions leading to growth (31), but this was accompanied by a toxic effect in susceptible cultivars. Until completely pure preparations of victorin can be documented satisfactorily, however, it remains possible that two distinct components are involved: a non-specific impurity affecting growth in both resistant and susceptible tissues and victorin specifically causing toxicity in susceptible tissue. The results on ion leakage with crude and partially purified corn blight toxin presented by Yoder during this meeting are of unusual interest in this regard. Despite reservations in specific instances, it is not surprising that toxins affect growth if their toxicity depends on disturbance of membrane functions. The central role of membranes in biochemistry and the accumulating evidence for hormonal interaction (18, 23, 27, 39) with membranes ultimately may be related to the action of toxins in plant disease.

HORMONES IN DISEASES CAUSED BY BIOTROPHIC ORGANISMS

The growth distortions caused by biotrophic organisms has led to the identification of IAA (9) and gibberellins (2) which are present in amounts sufficient to cause the extra growth. In our early work with safflower rust, increases in IAA were found just prior to the increased elongation characteristic of infected hypocotyls. This extra growth of the host was associated with mycelial development and ceased at sporulation. We ascribed to IAA a functional role in eliciting a condition in the host conducive to parasite development; that is, hormonal imbalance caused metabolic changes leading to higher respiratory activity and increased availability of substrates which the parasite could utilize (12). The same sort of argument was developed by Bushnell and Allen (4) for the similarity between the green islands of rust and mildew infections and green islands caused by kinetin. In addition, it was reported that kinetin prevented the change to a susceptible infection type in excised leaves (32). In both cases, tissues appear to be maintained in a "juvenile" condition and to serve as "sinks" for movement of substrates. Among several possible mechanisms (22), the massive movement of carbon to infected leaves in bean rust disease (22) might be under cytokinin control. Király (20) and Dekhuijzen and Staples (14) presented evidence for the formation of additional "cytokinin-like" substances during successful infections by rust fungi.

The phrase "during successful infections" is deliberately chosen because, until recently, my view of the role of hormones has
been as outlined above. Hormones were assumed to be of importance
in compatible interactions *after* the primary determination of compatibility or incompatibility had been made at some earlier stage
during the infection process (9, 12). The potential of hormonal
changes as causative factors in determining host-parasite specificity was brought to focus by our experience with the *Sr6* allele for
resistance of wheat to stem rust fungi. After approximately five
years of effort, we found that our results were not readily explicable by current theories of induced, active mechanisms determining
incompatibility (resistance). The results were left unpublished
until, during the course of pursuing a failure to associate peroxidase with resistance, we attempted to induce peroxidase and resistance in the normally susceptible, near-isogenic line of wheat
carrying the recessive *sr6* allele (13). Not unexpectedly, ethylene
treatment did cause an increase in peroxidase activity but it failed
to affect the normally compatible reaction of the line with race 56
of *Puccinia graminis* f. sp. *tritici*.

The unexpected finding in these studies was the reversal of
the incompatible reaction of *Sr6* plants at $20°C$ by ethylene (13).
There was, however, no similar ethylene effect on the incompatible
reaction controlled by the *Sr11* allele (11). At that point, our
negative findings between phenolic content and resistance, and those
of the group at Winnipeg with hydroxyputrescine derivations of cin-
namic acids and resistance, caused us to offer in 1970 an alternative
to induced resistance mechanisms (11, 13); later this was done in
much more detail (8).

INDUCED SUSCEPTIBILITY

As Dr. Wheeler pointed out after Dr. Day's lecture, and as I
stated in 1972 (8), the concept of induced susceptibility to explain
host-parasite specificity did not originate with me. Others have
expressed similar ideas, but I believe that my reasons (8, 13) for
suggesting an alternative to induced resistance are different in
substance from others that have been advanced. It should also be
clear that I am not wedded to the concept of induced susceptibility
as the only explanation of host-parasite interactions in rust diseases (8).

Apart from sheer frustration in attempting to find a suitable
lead for investigating the metabolic events causing resistance,
there were some biological phenomena which appeared amenable with
the idea that the induction in the host of reactions leading to
susceptibility could be an alternative to induced resistance. For
stem rust of wheat, at least, the infection phenotypes (37) represent a continuum of fungus development in host tissue, proceeding

first from the immune reaction (no visible symptom) through 0;
(white flecks but no spores), to type 1 (necrosis with some sporulation). The other resistant phenotype has even more sporulation
in the center of a green island with a necrotic ring at the outside
(infection type 2). The so-called compatible reactions (infection
types 3 and 4) are distinguished largely by the amount of sporu-
ation, pustule size and the degree of chlorosis surrounding the infected site. Although the divisions among resistant and susceptible
reaction classes are somewhat arbitrary, the infection types appear
to be true genotypes. Thus, infection of 'Marquis' wheat by race
2 gives infection type 2 not an immune reaction.

For the above, and other reasons (8), it seemed appropriate to
consider, as one alternative, that the growth of the rust fungi was
dependent on a temporally rigid sequence of changes induced in the
host to accommodate development of a parasite. The demand by the
pathogen on the host for the supply of necessary substrates is certainly not small and it represents a significant change from normal
metabolism as reflected, for example, in translocation patterns
(20, 22). We suggested that compatibility required an active switching of normal metabolism in several stages. If a metabolic switch
was not turned on, the parasite failed to develop beyond a certain
point. Each switch might be triggered by changes in endogenous
hormones such as IAA, gibberellins or cytokinins, and certainly
ethylene.

It is also possible, however, to consider the induction of
susceptibility as the turning off a normal sequence; that is, cells
may have to degenerate to a degree before a parasite is successful
in colonization. I wish to stress, however, that this does not mean
that a specific sequence imparting "resistance" has been turned off.
I use the term resistance in the sense of a repressive or inhibitory
action against the invading micro-organism in contrast to the situation where a parasite simply does not have the proper environment
for development. The action of host-specific toxins may be used as
an example. Generally, these impair cells at a site removed from
the parasite, and, as far as is known, antifungal compounds play no
role in determining their action or their specificity.

Responses causing compatibility do not necessarily require new
synthesis of enzymes, as suggested for induction of incompatibility
(17). The enzymic gene products normally present for a particular
function may be regulated through hormonal or regulatory (allosteric) control to divert metabolism to accommodate the parasite. A
resistance mechanism in the usual sense also may not be necessary;
the host simply would not respond to accommodate the parasite. An
allele for "resistance" might produce an enzymic gene product which
functions in the metabolism of an uninfected plant but is not capable of activation during infection. Failure to activate at any

of several necessary and sequential steps in parasite development
would lead to several classes of infection types.

For some of my colleagues more familiar with genetics than I
am (see Dr. Ellingboe's remarks), the gene-for-gene hypothesis
developed by Flor (16) poses some problems in considering the induction of susceptibility since one of its outcomes, the quadratic
check (30), indicates that the only unique genetic relationship is
represented by the quadrant for resistance. With all deference to
those colleagues, certain aspects of the data upon which the gene-
for-gene hypothesis rests have always disturbed me. In order to
arrive at reasonable genetic ratios, it was necessary to group several phenotypes of different reaction classes, for example, by combining infection types immune, 0 , 1 and 2 in the stem rust reaction
scheme. I agree with Dr. Day's comment that there was a practical
necessity for this and that the gene-for-gene relationship is a
useful concept for which we owe Flor a large debt. The point I would
emphasize is that we are now trying to establish molecular models.
Differences among infection types in a single reaction class must
be explained. The existence of several infection types for resistance seems to me to imply modifying genes either for different amounts or for different kinds of gene products as a minimal explanation, if specificity is determined by a major gene which results
in restriction of the pathogen. Is it possible, then, to talk about
gene-for-gene control of disease resistance ?

A serious obstacle to the proper molecular interpretation of
the gene-for-gene hypothesis is the so-called "intermediate" reaction that Flor was forced to ignore. The mesothetic reaction of
stem rust of wheat is a case in point. It is very difficult to explain infections in which individual pustules range from 0; to
type 4 on the same leaf of certain varieties. It is my understanding
that spores from any of the individual pustules will produce the
same range in subsequent inoculations of the same varieties. If
compatibility, rather than resistance, is induced, the mesothetic
reaction might involve competition, as first suggested by Shaw and
Hawkins (35), among infection centers for some essential process
which is activated in a limited way, for example, an energy dependent transfer of available nutrients.

SOME COMMENTS ON INDUCED RESISTANCE

In one discussion at this meeting, the question was asked,
rather emphatically, if induced susceptibility was worth considering
seriously because 1. there was no documentation on it and 2. it
seems to ignore a substantial volume of literature on induced resistance which I had not bothered to read. General and *a priori*
acceptance of the concept of active resistance has pretty well shaped the nature of research on specificity. Documentation on induced

susceptibility *is* scanty but growing. At this meeting we have
heard from Dr. Ouchi and Dr. Manners on the possible role of compatibility in the "induced accessibility" of mildew fungi to hosts
they normally would not attack. As pointed out earlier (8),
Tomiyama's data (40) clearly show that infection with a compatible
race of *Phytophthora infestans* will cause potato to respond in a
compatible manner to a normally incompatible race and these results
have been confirmed by Kuč in his lecture. Also, there is no question that host-specific toxins alter cells and the work of Comstock
and Scheffer (7) indicates that a non-pathogen will behave as a
pathogen if the proper host-specific toxin is supplied.

I must confess to not reading all the literature that I should
read and also volunteer the fact that much of it which I do read, I
tend to ignore. Some of the biochemical and biological evidence
referred to by Dr. Keen I have read critically, and I am not convinced that the interpretations offered are the only ones consistent with the data (8). I know of no unequivocal evidence, for
example, that resistance entails the activation of protein biosynthesis in the host. All that is known is that some enzymes of
aromatic biosynthesis, particularly phenylalanine ammonia lyase,
show greater activity. The best isotopic experiments designed to
show protein synthesis do not, in fact, rule out inhibition of the
degradation of enzymes occurring as a part of normal protein turnover. The data on the biosynthesis of antifungal compounds during incompatible disease reactions are limited to a handful of
papers. If examined carefully, these data do not rule out 1. accumulation of antifungal compounds by inhibition of further transformation or 2. formation by mixing of separate cellular compartme-
nts of pre-existing substrates and enzymes for an inter-conversion.
This may be only a result of non-specific injury. The necessary
biochemical evidence undoubtedly will be forthcoming, but it is not
yet of the quality that permits one to embrace, as proven, the concept of the induction of biochemical pathways for specific resistance. As discussed elsewhere (8), there remain substantial questions about the significance of chemicals which accumulate during
incompatible reactions. Although such comments were made several
years ago, I have seen no data subsequently which answer all of the
pertinent, open questions.

Two types of biological experiments mentioned by Dr. Keen are
more difficult to analyze. The conversion of a resistant tissue to
a condition of susceptibility by inhibitors of protein or nucleic
acid synthesis is not as straight-forward as he implies. Zucker
and El-Zayat (42) obtained extra growth of a contaminating *Pseudomonas* species on potato discs in the presence of cycloheximide but
how this can be related to natural resistance is not known. Such
inhibitors metabolically may create a culture medium on which any
self-respecting saprophyte would be expected to grow. Cycloheximide
and similar inhibitors may be affecting much more than the biosyn-

thesis of enzymes specially associated with resistance. In current thinking about diseases caused by biotrophs, Dr. Tani's results presented at this meeting are even more difficult to interpret. It is interesting, however, that only blastocidin S, out of all the inhibitors tested, caused resistant oats to become susceptible to the rust pathogen. Unfortunately, the genetic control of this disease reaction is not known and it would be interesting to see whether blastocidin S shows a specificity in causing susceptibility as does ethylene (11).

Resistance induced by pre-inoculation with an avirulent pathogen, "challenge inoculations" such as described by Dr. Johnson in his lecture, are subject to similar uncertainty because it is impossible, at present, to decide whether the observed response is due to the same mechanism as occurs in natural resistance. It may be a secondary phenomenon. As far as I am aware, induction of resistance requires an appreciable time between inoculation of the incompatible and compatible pathogens, usually 48 hours. I do not know of a case where a mixture of incompatible and compatible pathogens applied to a host simultaneously has resulted in the significant suppression of the compatible race. Although one can argue that the resistance is very localized and hence will not be exerted at the sites of compatible infections there is a pitfall in this logic when applied to rust diseases. Since the immune reaction in rusts is localized in a small number (perhaps only one or two) of host cells, as documented by Dr. Heath, there is no reason to expect that cells away from the affected areas would be immune at 48 hours. If resistance induced by pre-inoculation is to be good evidence for a natural mechanism for specificity through active resistance, it should reflect the specificity inherent in a given infection type. That is the resistance induced by infection type 2 should lead to an infection type 2 following challenge with a compatible race, and should not lead to an immune or infection type 1 reaction. As far as I know, such results have not been reported but perhaps only because no one has tried the experiments.

THE ROLE OF HYPERSENSITIVITY AND THE TIMING OF EVENTS DETERMINING SPECIFICITY

In remarks on the opening day of the meeting, Professor Brian indicated that the induction of compatibility was a possibility for explaining specificity, but not a very likely probability. He stressed, for example, the strong correlation of the "hypersensitive response" of nearly all higher plants with the development of incompatibility. Dr. Rohringer will summarize some results in which RNA appears to induce a hypersensitive response specifically with certain dominant alleles for resistance to stem rust fungi.

On the other hand, Mayama *et al.* (25) found nearly identical hypersensitive responses to race 56 of *Puccinia graminis* f. sp. *tritici* at 20 - 21°C and at 25 - 26°C in a line of wheat containing the *Sr6* allele for incompatibility, yet at 25 - 26°C the line is completely compatible with race 56, while at 20 - 21°C it is incompatible. Detachment of leaves of other near-isogenic lines and of 'Khapli' wheat increased the hypersensitive response and also increased the susceptibility of the leaves (24). At present, therefore, we can only conclude, with Brown *et al.* (3, 28), that the hypersensitive response is not a cause of incompatibility. Unlike the varieties used previously (3, 28), with our lines of wheat (25) there were no indications of incompatibility at any stage of the infection of *Sr6* lines at 26°C, yet the hypersensitive response could not be distinguished either in its time course or relative intensity (i.e. number of cells involved) from the hypersensitive response of an incompatible reaction at 20°C.

There appear to be several reasons why so much emphasis has been placed on the hypersensitive response in diseases, especially in those caused by biotrophs. The collapse and death of a cell is dramatic and it is natural to assume that death is in some way related to failure of a biotroph to develop. It is worth pointing out that, almost without exception, parasite development is described as "inhibited" in incompatible responses, and frequently the parasite is described as dead. As pointed out elsewhere (8, 25), in some diseases growth can be resumed by transfer (5) of a mass of incompatible tissue, *including* necrotic cells, into compatible tissue. Thus it is difficult to see why the presence of dead cells is a determinant of incompatibility, unless one postulates that a compound produced by dying cells is responsible. If so, its diffusion is extremely limited because available evidence indicates that incompatibility is localized. Why such a compound should be ineffective when the same tissue is transplanted is unexplained. The validity of the transplant argument is flawed by the fact that the percentage of successful transplants from incompatible tissue is not high but it should be remembered that such experiments deal with mycelium, not reproductive or survival structures such as spores. One would expect a normal decrease in viability of mycelium with time.

A second reason for the wide popularity of the hypersensitive response as an explanation for specificity is the fact that it is one of the earliest events known in incompatible reactions, first appearing in rust diseases between 12 and 24 hours after deposition of spores and increasing in intensity with time (25, 36). There is, apparently, a general tenet that unless a reaction occurs within a few hours after penetration it should not be considered of importance in determining compatibility or incompatibility. This is true certainly for diseases with short development times, but cereal rust diseases require several days for symptom expression. The

underpinnings of this tenet are, by circuitous reasoning, the hypersensitive response and, in addition, earlier work which appeared
to indicate that metabolic events, particularly changes in oxidative
metabolism, could be observed earlier and were of greater intensity
with incompatible than with compatible reactions. From the work of
Samborski and Shaw (33) this appeared to be true for cereal rusts,
but the data are not completely convincing largely because of in-
nate differences in the varieties being compared and differences
in the infection intensity and environmental conditions when comparisons were attempted. In later work Antonelli and Daly (1) found
parallel increases in rates of respiration in compatible and incompatible near-isogenic lines until the fifth day after inoculation, at which time the compatible tissue showed much greater rates
with the onset of sporulation. At the time of these experiments
there was a tendency on my part to dismiss such late events during
infection as factors crucial for host-parasitic specificity.

More recent studies on the development of rust fungi in compatible and incompatible hosts has caused a re-evaluation of that
dismissal. The data of Mayama *et al.* (24, 35) had suggested that
the hypersensitive response could occur in tissue which visually
was completely compatible in terms of time of appearance of symptoms and in the number and types of infection sites. It was of
interest to attempt to use glucosamine as a quantitative measure
of fungal growth in the presence and absence of a pronounced hypersensitive response during the early stages of infection of primary leaves of wheat. The techniques developed (26) were sensitive
enough to detect increases in glucosamine due to the growth of
germ tubes on the leaf surfaces in samples equivalent to a single
leaf. Surprisingly, no further increase in glucosamine was found
in either compatible or incompatible wheat at 20 - 21°C for at
least 60 additional hours. Clear differences between compatible
and incompatible tissues could not be established until just prior
to sporulation (26). Subsequently, appreciable increases in glucosamine occurred in the compatible reaction and, interestingly,
there were increments, though smaller, in incompatible tissues through the tenth day when experiments were terminated. Using a different technique, Mr. Pearce of University College, London has
found continued increase in glucosamine during incompatible reactions for as long as 14 days after inoculation (personal communication).

The use of glucosamine as an estimation of fungal growth is
not without reservation (26), but our results are not dissimilar
to those reported by Ride and Drysdale (29) for facultative parasites. Failure to detect increases in glucosamine may be due to
lack of glucosamine in young mycelium and/or a very slow rate of
fungal development in the early stages of infection. However, the
results are reasonably consistent with histological observations
of the same wheat lines infected with the same race, by Skipp and

Samborski (36). They detected no significant differences in hyphal lengths or colony size at 60 hours after deposition of spores. Only at 84 hours were differences obvious and these were not particularly large. As with glucosamine analysis, there was evidence for continued growth of the fungus in incompatible tissue. This is also indicated by the data of Brown and colleagues (3, 28) who could not distinguish differences in colony diameter until at least 48 hours after penetration in a number of resistant and susceptible varieties. Subsequently, increases in colony diameter resembled a geometric progression in both resistant and susceptible hosts, with much larger successive increments in the latter through sporulation.

As indicated above, these results have caused us to re-examine previous conclusions about the significance of the late respiratory rise as well as some of the semantics associated with the phenomenon of compatibility or incompatibility. How does one establish that there is an "inhibited" rate of fungal development in a host ? The use of the word implies a mechanism when, in fact, all that is known is that the rate is slower in incompatible lines than in compatible hosts. It is equally plausible that the growth rate is identical in the early stages of infection, perhaps through the third day (3, 26, 28, 36), but subsequently host metabolism is activated at a later stage in compatible tissues to permit more rapid growth and eventually sporulation to an extent determined by the nature and extent of the activation.   Incompatibility thus may not involve an inhibition of fungal development in the sense that there is an active, positive process or mechanism controlling the pathogen. Similar events but within a shorter time frame could take place in diseases other than cereal rusts. As stated previously, both an induced inhibition and an induced stimulation of a pathogen may be involved, depending on the genetic information being programmed. In either event, the known changes in hormones of rusted tissue may be determinants.

REFERENCES

1.   ANTONELLI, E. and DALY, J.M. (1966). Decarboxylation of indoleacetic acid by near-isogenic lines of wheat resistant or susceptible to *Puccinia graminis* f. sp. *tritici*. *Phytopathology*, 56, 610 - 618.

2.   BAILISS, K.W. and WILSON, I.M. (1967). Growth hormones and the creeping thistle rust. *Ann. Bot.*, N.S., 31, 195 - 211.

3.   BROWN, J.F., SHIPTON, W.A. and WHITE, N.N. (1966). The relationship between hypersensitive tissue and resistance in wheat seedlings infected with *Puccinia graminis tritici*. *Ann. appl. Biol.*, 58, 279 - 290.

4.   BUSHNELL, W.R. and ALLEN, P.J. (1962). Induction of disease
     symptoms in barley by powdery mildew. *Pl. Physiol.,
     Lancaster,* 37, 50 - 59.

5.   CHAKRAVARTI, B.P. (1966). Attempts to alter infection proc-
     esses and aggressiveness of *Puccinia graminis* var. *tritici.*
     Phytopathology, 56, 223 - 229.

6.   CLELAND, R. (1974). Fusicoccin as a tool for studying the
     mechanism of auxin action. *Pl. Physiol., Lancaster,* 53,
     A. Suppl. U.S. ISSN 0079 - 2241, 43.

7.   COMSTOCK, J.C. and SCHEFFER, R.P. (1973). Role of host-sel-
     ective toxin in colonization of corn leaves by *Helmintho-
     sporium carbonum.* *Phytopathology,* 63, 24 - 29.

8.   DALY, J.M. (1972). The use of near-isogenic lines in biochem-
     ical studies of the resistance of wheat to stem rust.
     Phytopathology, 62, 392 - 400.

9.   DALY, J.M. and INMAN, R.E. (1958). Changes in auxin levels
     in safflower hypocotyls infected with *Puccinia carthami.*
     Phytopathology, 48, 91 - 97.

10. DALY, J.M. and KNOCHE, H.W. (1975). Hormonal involvement in
     metabolism of host-parasite interactions. *In : Biochem-
     ical aspects of plant-parasite relationships.* (FRIEND, J.
     Ed.). Phytochemical Society Symposium Series, Vol. 13,
     Academic Press, New York, London. (In press).

11. DALY, J.M., LUDDEN, P. and SEEVERS, P. (1971). Biochemical
     comparisons of stem rust resistance controlled by the
     Sr6 and *Sr11* alleles. *Physiol. Pl. Path.,* 1, 397 - 407.

12. DALY, J.M. and SAYRE, R.M. (1957). Relations between growth
     and respiratory metabolism in safflower infected with
     Puccinia carthami. *Phytopathology,* 47, 163 - 168.

13. DALY, J.M., SEEVERS, P.M. and LUDDEN, P. (1970). Studies on
     wheat stem rust resistance controlled at the *Sr6* locus.
     III. Ethylene and disease resistance. *Phytopathology,*
     60, 1648 - 1652.

14. DEKHUIJZEN, H.M. and STAPLES, R.C. (1968). Mobilization fact-
     ors in uredospores and bean leaves infected with bean
     rust fungus. *Contr. Boyce Thompson Inst. Pl. Res.,* 24,
     39 - 51.

15. EVANS, M.L. (1974). Rapid responses to plant hormones. *A.
     Rev. Pl. Physiol.,* 25, 195 - 223.

16.   FLOR, H.H. (1956).   The complementary genetic systems in flax
        and flax rust. *Adv. Genet.*, 8, 29 - 54.

17.   HADWIGER, L.A. and SCHWOCHAU, M.E. (1969).   Host resistance
        responses - an induction hypothesis. *Phytopathology*, 59,
        223 - 227.

18.   HARDIN, J.W., CHERRY, J.H., MOORE, D.J. and LEMBI, C.A. (1972).
        Enhancement of RNA polymerase by a factor released by
        auxin from plasma membrane. *Proc. natn. Acad. Sci.*,
        U.S.A., 69, 3146 - 3150.

19.   HASHIMOTO, J. and TAMURA, S. (1967).   Physiological activities
        of helminthosporol and helminthosporic acid. II. Effects
        on excised plant parts. *Pl. Cell Physiol.*, *Tokyo*, 8,
        35 - 45.

20.   KIRÁLY, Z., EL HAMMADY, M. and POZSAR, B.I. (1967).   Increased
        cytokinin activity of rust-infected bean and broad bean
        leaves. *Phytopathology*, 57, 93 - 94.

21.   LADO, P., CALDOGNO, R., PENNACCHIONI, A. and MARRÉ, E. (1973).
        Mechanism of the growth-promoting action of fusicoccin.
        Planta, 110, 311 - 320.

22.   LIVNE, A. and DALY, J.M. (1966).   Translocation in healthy and
        rust-affected beans. *Phytopathology*, 56, 170 - 175.

23.   MARRÉ, E., LADO, P., FERRONI, A. and BALLARIN DENTI, A. (1974).
        Transmembrane potential increase induced by auxin, benzy-
        ladenine and fusicoccin. Correlation with proton extr-
        usion and cell enlargement. *Pl. Sci. Lett.*, 2, 257 -
        265.

24.   MAYAMA, S., DALY, J.M. and REHFELD, D.W. (1975).   The effect
        of excision on the hypersensitive response and the dev-
        elopment of rust fungi in resistant and susceptible
        wheats. *Phytopathology* (In press).

25.   MAYAMA, S., DALY, J.M., REHFELD, D.W. and DALY, C.R. (1975).
        Hypersensitive response of near-isogenic wheat carrying
        the temperature sensitive *Sr6* allele for resistance to
        stem rust. *Physiol. Pl. Path.* (In press).

26.   MAYAMA, S., REHFELD, D.W. and DALY, J.M. (1975).   A comparison
        of the development of *Puccinia graminis tritici* in res-
        istant and susceptible wheat based on glucosamine content.
        Physiol. Pl. Path. (In press).

27.   NEUMANN, P.M. (1971). Possible involvement of a glycerophos-
      phate compound in auxin induced growth. *Planta*, 99, 56 -
      62.

28.   OGLE, H.J. and BROWN, J.F. (1971). Quantitative studies of
      the post-penetration phase of infection by *Puccinia grami-
      nis tritici*. *Ann. appl. Biol.*, 67, 309 - 319.

29.   RIDE, J.P. and DRYSDALE, R.B. (1971). A chemical method for
      estimating *Fusarium oxysporum* f. *lycopersici* in infected
      tomato plants. *Physiol. Pl. Path.*, 1, 409 - 420.

30.   ROWELL, J.B., LOEGERING, W.Q. and POWERS, H.R. (1963). Genetic
      model for physiological studies of mechanisms governing
      development of infection type in wheat stem rust. *Phyto-
      pathology*, 53, 932 - 937.

31.   SAFTNER, R.A. and EVANS, M.L. (1974). Selective effects of
      victorin on growth and auxin response in Avena. *Pl. Phys-
      iol., Lancaster*, 53, 382 - 387.

32.   SAMBORSKI, D.J., FORSYTH, F.R. and PERSON, C. (1958). Metabolic
      changes in detached wheat leaves floated on benzimidazole
      and the effect of these changes on rust reaction. *Can. J.
      Bot.*, 36, 591 - 601.

33.   SAMBORSKI, D.J. and SHAW, M. (1956). The physiology of host-
      parasite relations. II. The effect of *Puccinia graminis
      tritici* on the respiration of the first leaf of resistant
      and susceptible species of wheat. *Can. J. Bot.*, 34, 601 -
      619.

34.   SEQUEIRA, L. (1973). Hormonal metabolism in diseased plants.
      A. Rev. Pl. Physiol., 24, 353 - 380.

35.   SHAW, M. and HAWKINS, A.R. (1958). A preliminary examination
      of the level of free endogenous indole acetic acid in
      rusted and mildewed cereal leaves and their ability to
      decarboxylate exogenously supplied radioactive indoleacetic
      acid. *Can. J. Bot.*, 36, 1 - 16.

36.   SKIPP, R.A. and SAMBORSKI, D.J. (1974). The effect of the *Sr6*
      gene for host resistance on histological events during
      development of stem rust in near isogenic wheat lines.
      Can. J. Bot., 52, 107 - 115.

37.   STAKMAN, E.C., STEWART, D.M. and LOEGERING, W.Q. (1962). Iden-
      tification of physiologic races of *Puccinia graminis* var.
      tritici. *Bull. U.S. Dep. Agric.*, E617, 53 pp.

38. TAMURA, S., SAKURAI, A., KAINUMA, K. and TAKAI, M. (1963).
     Isolation of helminthosporol as a natural plant growth
     regulator and its chemical structure. *Agr. Biol. Chem.*,
     27, 738 - 739.

39. TOMÉ, F. and BELLINI, E. (1974). Phenylalanine ammonia lyase
     activity in excised radish cotyledons : effects of water
     availability, fusicoccin and kinetin. *Pl. Sci. Lett.*,
     3, 413 - 418.

40. TOMIYAMA, K. (1966). Double infection by an incompatible race
     of *Phytophthora infestans* of a potato plant cell which
     has previously been infected by a compatible race. *Ann.
     phytopath. Soc. Japan*, 32, 181 - 185.

41. VAN ANDEL, O.M. and FUCHS, A. (1972). Interference with plant
     growth regulation by microbial metabolites. *In :
     Phytotoxins in plant diseases*. (WOOD, R.K.S., BALLIO,
     A. and GRANITI, A., Eds.), 227 - 247. Academic Press,
     London, New York.

42. ZUCKER, M. and EL-ZAYAL, M.M. (1968). The effect of cyclo-
     heximide on the resistance of potato tuber discs to in-
     vasion by a fluorescent *Pseudomonas* sp. *Phytopathology*,
     58, 339 - 344.

CONTRIBUTIONS

KERN, H. and NAEF-ROTH, S.   Formation of growth factors by *Taphrina*
     fungi.

MYERS, A. and WOOLSTON, J.A.   Relationship between leaf senescence
     and specific reaction to infection by biotrophs.

TANI, T.   Hormonal regulation of uredial formation in oat leaves
     infected by *Puccinia coronata* var. *avenae*.

SUMMARY OF POINTS FROM CONTRIBUTIONS AND DISCUSSIONS
BY
H. M. DEKHUIJZEN

Chairman and Discussion Leader

     In opening the discussion, Dekhuijzen pointed out that although
changes in auxin, ethylene, gibberellins, abscisic acid and cytokinins occur in compatible reactions involving obligate parasites,
mechanisms controlled by single genes remain unknown. The effect

of ethylene on induction of high peroxidase activity in resistant
and susceptible wheat lines carrying the *Sr6* genes infected with
Puccinia graminis tritici (race 56) is of interest but the results
show no causal relation between total peroxidase activity and resistance expressed as infection type or pustule density. Daly suggests that specificity may depend on induction of susceptibility
rather than of resistance and that hormonal changes might control
reactions leading to development of fungi in susceptible hosts. If
so, plants are resistant when fungi cannot induce these changes.
Although this hypothesis is attractive it is difficult to obtain
good evidence for it. Also, in crucifers zoosporangia of *Plasmodiophora brassicae* occur in resistant plants. Here, resistance is
expressed later in the life cycle when multinucleate plasmodia form
uni- or multinucleate daughter plasmodia which cannot penetrate
neighbouring host cells. It is likely that plasmodia induce host
cells to synthesize excessive auxins and cytokinins and that this
results in host cell division which is a prerequisite for the distribution of daughter plasmodia. It is possible that failure of a
host to produce hormones leads to the death of the pathogen and thus
causes a resistant reaction along lines proposed by Daly.

Tani described recent work on hormonal regulation of uredial
formation of uredosori in oat leaves infected by *Puccinia coronata*
var. *avenae*. Kinetin and N^6-benzyladenine at 10^{-4}M reduced numbers
of uredosori on leaves of susceptible oat var. Victoria 226-S (V2S)
but had no direct effect on fungal growth on epidermal strips. In
contrast, the time to formation of uredosori decreased from 8 to 6
days with 5×10^{-5}M abscisic acid applied two days after inoculation.
Chlorophyll breakdown was retarded by cytokinins and stimulated by
abscisic acid. Inhibition of formation of uredosori on detached
leaves by kinetin was nullified by later treatment with abscisic
acid and stimulation by abscisic acid was nullified by later treatment with kinetin. Breakdown of organelles in host mesophyll cells
by heat treatment before inoculation and decrease of protein synthesis to one twentieth by blasticidin S did not affect formation
of uredosori. The results suggest that senescence rather than juv-
enescence promotes formation of uredosori, a view supported by earlier work which showed that stem excision soon after infection remarkably accelerates rRNA synthesis of rust infected leaves.

Comments on this work included the suggestion that supra-optimal
kinetin concentrations may have been used and these changed the hormone balance and so decreased the formation of uredosori. Also,
kinetin may have changed the flow of nutrients to green islands at
infection sites.

Results of experiments on the relationship between leaf senescence and specific reactions to infection by biotrophs were summarized by Myers. DNA of barley leaves infected with

powdery mildew and of isolated chromatin was more rapidly extracted in 10% perchloric acid at 30°C than was DNA of uninfected leaves, an effect also found in powdery mildew of cucumbers and in barley yellow dwarf, but not in yellow rust. Leaves infected with biotrophic pathogens show many physiological changes associated with senescence such as DNA extractability which increases before aging is readily visible. DNA extractability differences disappeared when protein was stripped from DNA, but removal of histones did not affect the behaviour of chromatin to perchloric acid. Although there were no changes in DNA/histone ratio, the proportion of non-histone chromatin protein was increased by infection and senescence. Polyacrylamide gel electrophoresis showed loss of several bands of non-histone chromatin protein as the total increased. These proteins are concerned in control of gene expression so that the pathogen may induce changes in chromatin proteins which decrease protein synthesis and thus senescence may be necessary to prevent host resistance reactions from developing.

Growth regulating substances produced by *Taphrina* spp. were described by Kern and Naef-Roth. All ten species examined synthesize indole-3-acetic acid (IAA) and indole-3-lactic acid (ILA) from tryptophan with highest production (10 ppm) in cultures of *T. deformans*, *T. cerasi*, *T. betulina* and *T. amentorum*. Traces of tryptophan were formed only by *T. betulae* and *T. betulina* and tryptamine was absent in all cultures. Cell free culture filtrates had high cytokinin activity whereas cell homogenates and t-RNA-fractions were much less active. Zeatin and IPA {N^6 (γ,γ-dimethylallyl-amino) purine} were identified as exogenous cytokinins in concentrations of 10^{-3} ppm. In hydrolysates of t-RNA fractions only zeatin was demonstrated. Filtrates also contained a third unknown substance with high cytokinin activity and similar UV-absorption. 0.25 - 0.025 ppm zeatin reduced development of roots of duckweed *Spirodela oligorrhiza*, stimulated growth of stems and segments and changed the morphology of the latter to specific convex forms. The effect was strongly enhanced by a simultaneously applied high dose of IAA (1 000 x) whereas ILA had no effect. The different *Taphrina* species all produced the same hormones *in vitro* which suggests that deformation of organs by a particular species of *Taphrina* cannot be explained by production of specific growth hormones.

Discussion of this work commended that the growth hormones be analysed and characterized by gas-liquid chromatography mass spectrometry. It was emphasized that many bacteria and fungi including saprophytes produce hormones *in vitro* so that it may be difficult to decide on the origin of excess hormone in infected plants. This problem may be resolved by comparing hormone levels in healthy and infected plants, identifying hormones produced in culture and in infected and healthy plant, attempting to reproduce typical symptoms by applying hormones to plants and by obtaining data on synthesis of hormones by the pathogens *in vitro* and *in vivo*.

EFFECTS OF STRUCTURE OF ACTIVE COMPOUNDS ON BIOLOGICAL ACTIVITY AND SPECIFICITY

DANIEL H. RICH

School of Pharmacy, University of Wisconsin

Madison, Wisconsin, U.S.A.

It is probable that the chemical basis of resistance and specificity will be found in part from studies of the interactions of toxins with plants susceptible and resistant to those toxins. The phytotoxins with the most selective toxicity are the host-specific toxins; these are much more active in altering the metabolism of host plants susceptible to the toxin-producing micro-organism. Many host-specific toxins are unusually potent causing biological effects at concentrations less than 10^{-10} g/ml. The specificity and potency of these toxins suggest that they efficiently inhibit important biochemical processes in susceptible plants. Resistant plants either lack these pathways or possess mechanisms that protect against the action of the toxin. Elucidation of these resistance mechanisms will greatly increase our knowledge of specificity. Unfortunately the instability and chemical complexity of certain host-specific toxins have complicated mechanism of action studies because of uncertainties about chemical purity and concentration and by making structure determinations very difficult. No structure of a host-specific toxin has been established unambiguously although partial structural data have been obtained for many compounds and complete structures have been proposed for alternariolide (16) and helminthosporoside (25). Thus most of our concepts regarding the effects of structure on specificity must be derived from studies of other systems such as antibiotics and hormones. The objective of this paper is to describe in chemical terms the different ways in which the structure of a compound affects biological activity and to discuss criteria for interpreting effects on biological activity produced by chemically modifying the structure of a toxin.

Historically the structure of an antibiotic or phytotoxin has rarely provided much information about its mechanism of action.

Instead it has always been necessary to determine the mechanism of
action first before attempting to rationalize relationships between
structure and toxicity. Nevertheless, the structure of a toxin can
provide insight into its interaction with a biochemical system. For
example, puromycin (I) is a classical antimetabolite that inhibits
protein synthesis by acting as a non-functional analog structurally
similar to the terminal AMP residue of an aminoacyl t-RNA (II).
Notice that the structures are virtually superimposable except that
the oxygen atom in II has been replaced by a nitrogen atom. This
isosteric replacement produces an amide bond in I that is much more
stable than the ester bond in II, preventing further peptide chain
elongation. Puromycin would not be expected to show specificity
since amino acyl t-RNA II is needed for protein synthesis by all
species.

     Not all attempts to explain inhibition in terms of antimetabo-
lite theory have been successful,and these difficulties have led to
new concepts for rationalizing structure with activity. For example,
the close similarity between the structures of penicillin(III)and
D-Ala-D-Ala(IV)led to the suggestion that the antibiotic inhibited
the enzyme transpeptidase via an antimetabolite mechanism (26).
This hypothesis predicted that the 6-methyl penicillin analog(V)
which more closely resembles D-Ala-D-Ala than penicillin itself
should be a more active antibiotic. The synthesis and testing of V
showed that, contrary to the hypothesis, the analog was inactive and
indicated that penicillin did not closely resemble the D-Ala-D-Ala
substrate in the ground state.

     This result led to the suggestion that penicillin might be an
analog of D-Ala-D-Ala in the transition-state (12). It has been
pointed out that enzymes possess greatest affinity for substrates
in the transition-state. Stable analogs of the transition-state
would be expected to bind tightly to and inhibit the enzyme (13,
28). Lactones are thought to inhibit lysozyme by this transition-
state analog mechanism. These compounds bind to the enzyme 10^5 times
more tightly than to substrate (13, 28). Penicillin does resemble
more closely the transition-state form of the D-Ala-D-Ala substrate.
Furthermore this hypothesis predicts that 6-methylpenicillin(V)would
not be especially active, an observation in accord with the experimental results.

     Penicillin acylates a sulfhydryl group in the active site of
transpeptidase (26). Thus it can be called an active-site directed
irreversible inhibitor (2), or an affinity labelling reagent (23).
Several toxins inhibit by this mechanism including azaserine(VI)known
to alkylate an active-site sulfhydryl in formyl-glycinamideribotide :
L-glutamine ligase (3). Notice that azaserine(VI)closely resembles
the normal enzyme substrate, glutamine, and that the labile diazo-
group is located next to the carbonyl group that the enzyme normally

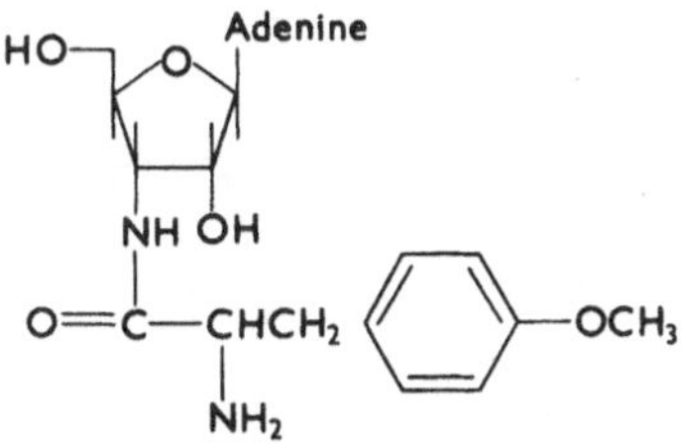

I

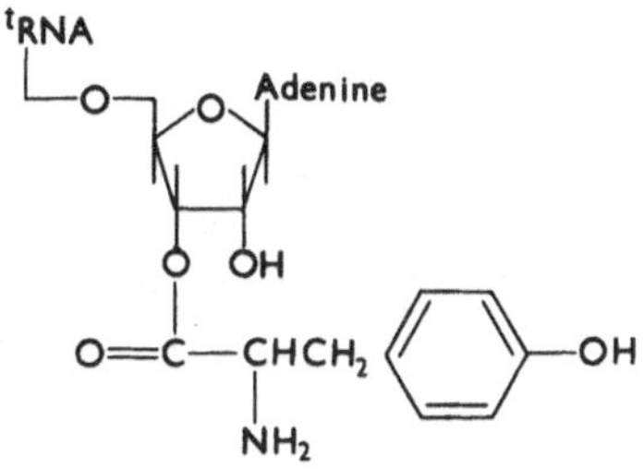

II

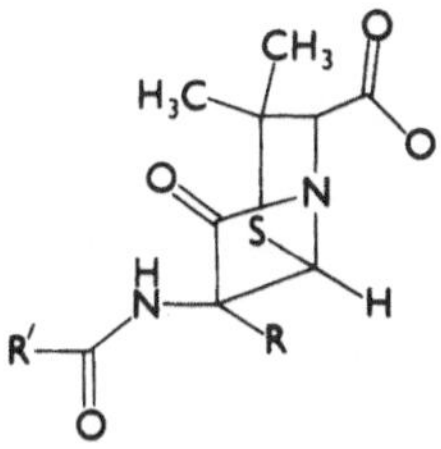

III   R = H

V   R = CH$_3$

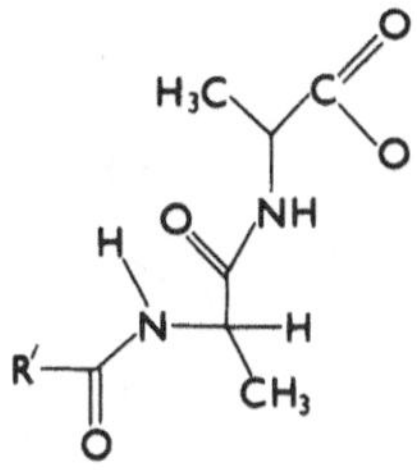

IV

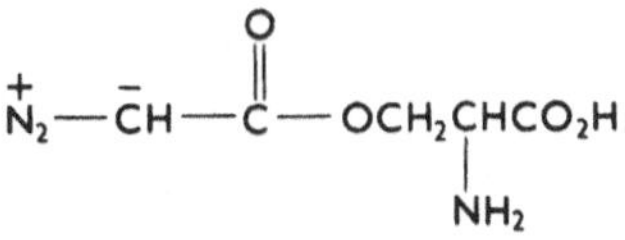

VI

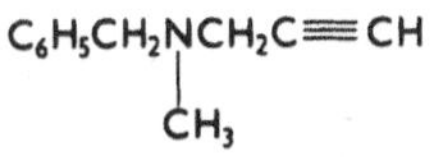

$C_6H_5CH_2NCH_2C \equiv CH$
$\quad\quad\quad |$
$\quad\quad\quad CH_3$

VII

$C_6H_5CH_2N{-}CH{=}C{=}CH_2$
$\quad\quad\quad\quad |$
$\quad\quad\quad\quad CH_3$

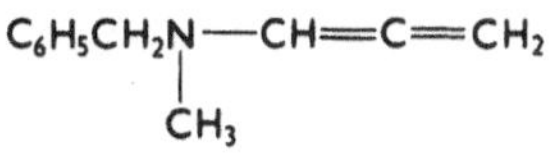

VIII

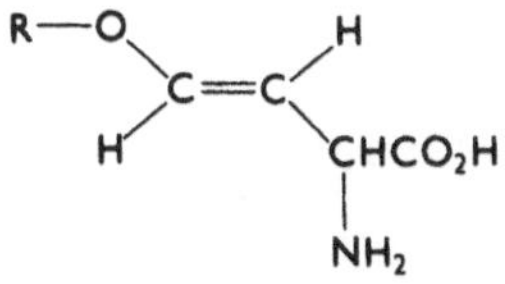

IX   R = HOCH$_2$CHCH
$\quad\quad\quad\quad\quad |$
$\quad\quad\quad\quad\quad NH_2$

XI   R = CH$_3$

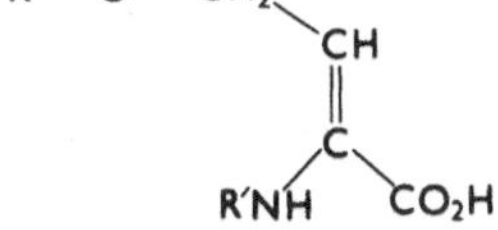

X

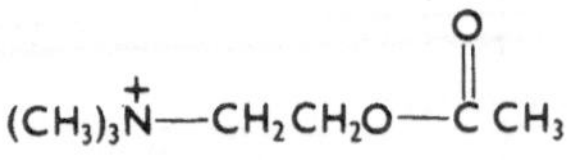

$$(CH_3)_3\overset{+}{N}-CH_2CH_2O-\overset{O}{\overset{\|}{C}}CH_3$$

XII

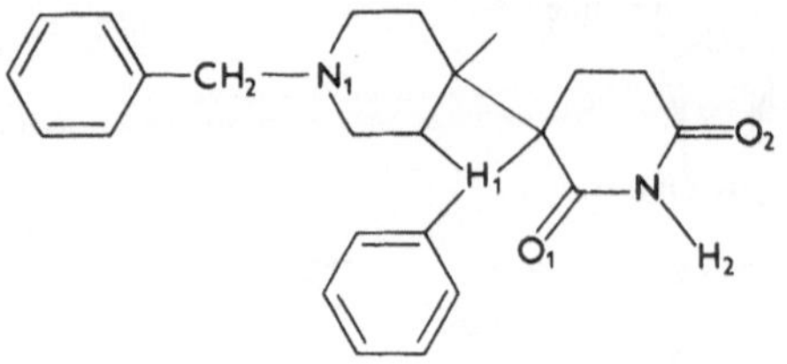

XIII

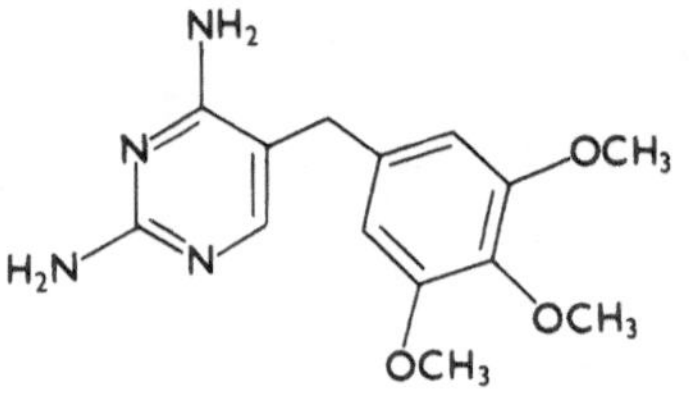

XV

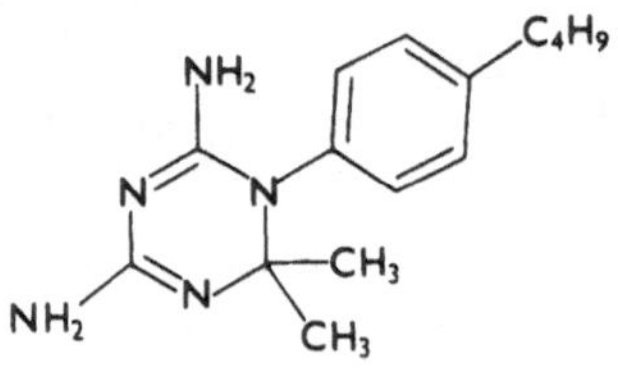

XVI

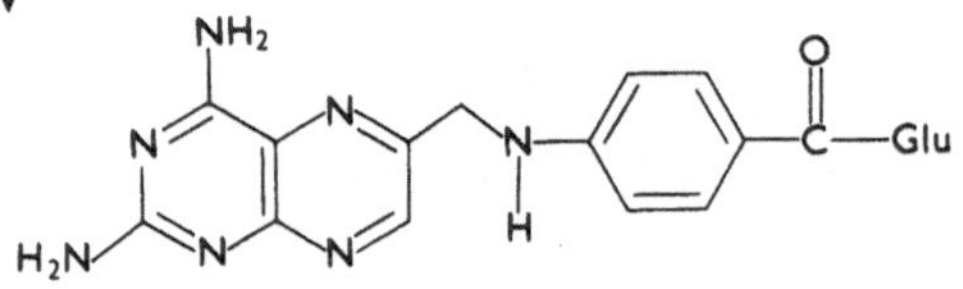

XIV

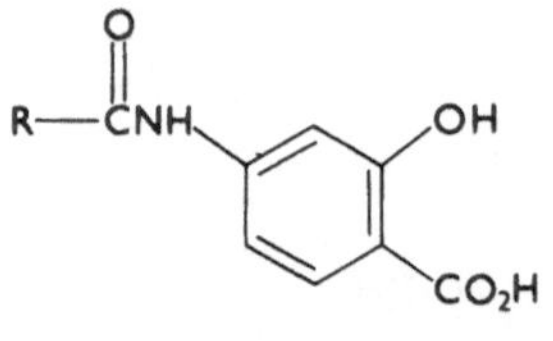

XVII   R = ICH$_2^-$

XVIII   R = C$_6$H$_5$O$^-$

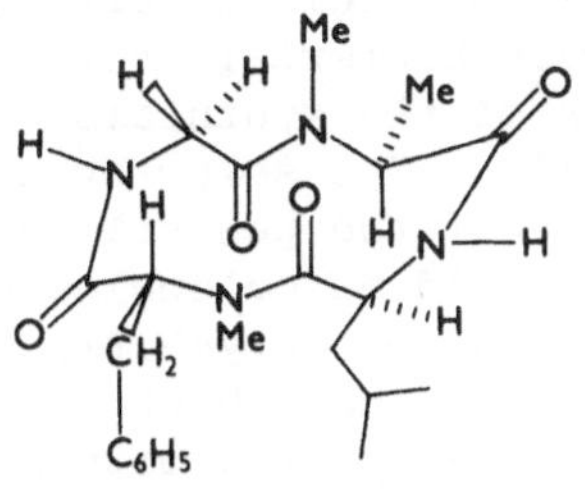

XXIII

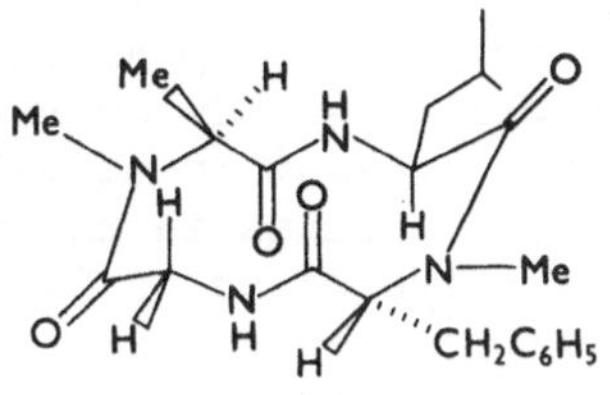

XXIV

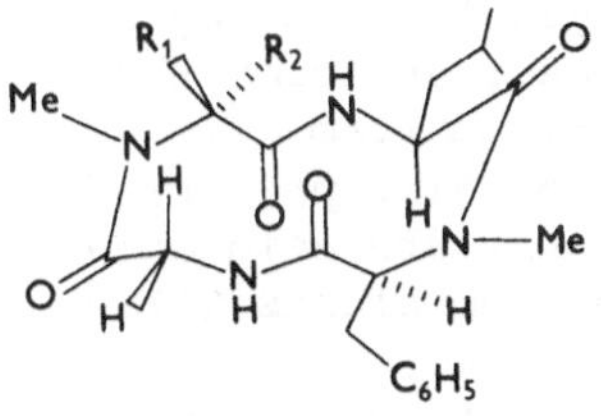

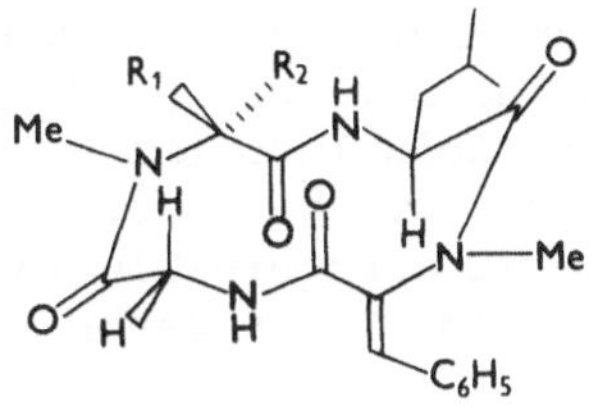

XXV $R_1 = CH_3$, $R_2 = H$

XXVI $R_1 = H$, $R_2 = CH_3$

XIX $R_1 = CH_3$, $R_2 = H$

XVII $R_1 = H$, $R_2 = CH_3$

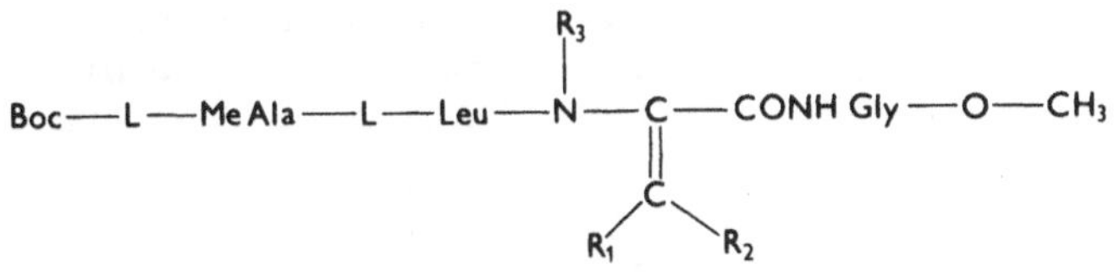

Boc—L—MeAla—L—Leu—N—C—CONH Gly—O—CH₃

XX $R_1 = C_6H_5$, $R_2 = R_3 = H$

XXI $R_1 = H$, $R_2 = C_6H_5$, $R_3 = H$

XXII $R_1 = C_6H_5$, $R_2 = H$, $R_3 = CH_3$

attacks. Thus it is clear why azaserine inhibits this glutamine
requiring enzyme. Because this ligase is required for purine biosynthesis in all species, azaserine is not selectively toxic. The
selective toxicity of penicillin toward bacterial cells and not
mammalian cells depends on inhibition of the synthesis of amino-
peptides not present in mammalian cells. The selectivity does not
depend on differences in the structure of the penicillin receptor.
A more selective class of inhibitors would be molecules that become alkylating agents only after the target enzyme transforms them
from an inactive to an active form. These have been called kcat
inhibitors (17). Paragyline(VII)irreversibly inhibits monoamine
oxidase by this mechanism. The enzyme transforms VII into the highly reactive allene(VIII)which then alkylates the enzyme.

Rhizobitoxin(IX)may inhibit β-cystathionase by a kcat mechanism
(18). The normal action of the enzyme transforms the β,γ-unsatu-
rated amino acid into the α,β-unsaturated form (X). This form alkylates the β-cystathionase. Rando (18) also predicted that amino
acid XI would irreversibly inhibit aspartate acid amino transferase.
The amino acid XI was later isolated from plants (6).

These examples illustrate that a sufficient number of inhibitors appear to act as anti-metabolites, active-site directed irreversible inhibitors, kcat inhibitors and transition-state analogs
to warrant comparing the structure of each newly characterized toxin
with the structures of known natural substrates, secondary messen-
gers,and hormones, to see if similarities can be found. Suggestions
for possible mechanisms of action may be found if the toxin is found
to resemble a known compound. Unfortunately any resemblance an
inhibitor may have with a natural substance may be subtle. Thus
x-ray crystallography studies of acetylcholine(XII)and its competitive inhibitor dexbenzetimide(XIII)(24) show that atoms N(1), H(1),
0(1), H(2) and 0(2) in IV are isosteric with the binding positions
in acetylcholine (XII) but that the rest of the structure of the
inhibitor is not related.

The chemical mechanisms for inhibition just given are not
likely to explain the type of specificity shown by the host-specific
toxins. A much greater selectivity must be shown by the inhibitor
including the ability to inhibit an enzyme isolated from one species
but not the same enzyme isolated from another species. Are there
any inhibitors that are this selective ? If so, could host-specific
toxins function in the same manner ?

Differences in allosteric sites of enzymes have been exploited
to develop highly selective enzyme inhibitors. Hitchings (10) developed a series of dihydrofolate reductase inhibitors related to
aminopterin(XIV)(Table 1). The benzyl pyrimidine derivative, TMP
(XV) specifically inhibits the bacterial reductases while the
triazine(XVI)is specific for the mammalian enzyme. Both enzymes are

Table 1[*]. *Inhibition of dihydrofolate reductases by folic acid antagonists*

Source of reductase	Inhibitory concentration x 10^{-8}M		
	Compound		
	XIV	XV	XVI
Escherichia coli	0.6	0.5	65 000
Staphylococcus aureus	0.1	1.5	50 000
Proteus vulgaris	0.5	0.5	10 000
Human liver	0.2	30 000	55
Rat liver	0.2	26 000	14

[*]   Data from FERONE, R., BURCHALL, J.J., and HITCHINGS, G.H. (1969). *Plasmodium berghei* dihydrofolate reductase : isolation, properties and inhibition by antifoliates. *Molec. Pharmacol., 5*, 49 - 59.

inhibited equally by aminopterin(XIV). The ratios between sensitive and resistant enzymes are of the order of magnitude for host-specific toxins and must arise from differences in the respective allosteric binding sites i.e. sites just outside of the binding site for folic acid. Note that since XV was designed to be used clinically as an antibiotic, no attempt was made to develop a compound that would inhibit only one bacterium. Nevertheless this probably could be achieved and the specificity thus generated would closely resemble that of the host-specific toxins. Another example of specificity produced by allosteric inhibition was reported by Baker (2). Compounds XVII and XVIII were found to be active-site directed, irreversible inhibitors of lactic acid dehydrogenase (LDH). Compound XVII inhibited only skeletal LDH whereas compound XVIII inhibited only heart muscle LDH. These dehydrogenases are isozymic. Possibly host-specific toxins inhibit an isozyme important to a plant defense mechanism.

The studies just cited support the idea that specificity of the type observed with host-specific toxins could be achieved by allosteric inhibition of enzymes or other important macromolecular receptors. Furthermore these studies suggest that chemical modification of a toxin may alter either the biological response or the specificity, or both. It is conceivable that a synthetic analog of a host-specific toxin might be found that binds to the receptor of a resistant variety but not the susceptible variety. This analog would specifically affect previously resistant varieties but not previously susceptible varieties . It might be possible to design selective herbicides using this strategy.

Studies designed to correlate structure with mechanism of action must be carefully planned and can succeed only if both the primary structure and the conformation of each analog in solution can be determined. Certainly it is necessary to establish first that the proposed structure of the toxin does cause the observed biological activity. Aside from the possibility that the tentative structure may be incorrect, a situation frequently encountered with extremely potent biologically active substances isolated in small quantity, the possibility also exists that the active compound may be only a contaminant of the isolated and characterized material. With regard to this point the controversy over whether all host-specific toxins must be of the same class of structure such as peptide, saccharide, or sesquiterpene is certainly premature. Only X-ray crystallographic analysis of a biologically active derivative or an unambiguous synthesis of the proposed structure with full biological activity will prove that the proposed structure produces the observed biological response.

The synthesis of the phytotoxin, tentoxin(XIX) illustrates the criteria needed to prove that the proposed structure causes the observed response (19). Tentoxin is a phytotoxin produced by *Alternaria tenuis*(27). Although not a host-specific toxin, tentoxin does exhibit specificity. It causes chlorosis in germinating seedlings of many flowering plants except members of the Cruciferae and some members of the Solanaceae and Gramineae. The basis for this specificity is unknown.

Tentoxin (XIX) is a cyclic tetrapeptide which contains the α, β-unsaturated amino acid, N-methyldehydrophenylalanine (MeΔPhe)(15). Tentoxin is base labile as are many peptides containing unsaturated residues. In this case the peptide ring is rapidly cleaved into a linear tetrapeptide by treatment with dilute base (11). Interestingly, many of the peptide-like host-specific toxins are also base labile; furthermore dehydroleucyl and dehydroalanyl residues are found in(*Helminthosporium carbonum*) HC-toxin and alternariolide respectively. Dehydro residues are known to alkylate nucleophilic groups and to be base labile. It is thought that the mechanism of action of nisin and subtilin, two antibiotics containing dehydroalanyl-residues, involves alkylation of sulfhydryl groups (7). Thus, an additional objective of the study of tentoxin was to determine the contribution of the double bond (in ΔPhe) to the mechanism of action of tentoxin and to test if this group functioned as an alkylating agent.

To synthesize tentoxin, linear tetrapeptides were prepared by standard methods and converted to dehydrophenylalanyl (ΔPhe) peptides XX and XXI. This route was chosen so that the configuration about the double bond in XX or XXI could be assigned on the basis of nuclear magnetic resonance (nmr) and ultraviolet (uv) spectroscopy. Furthermore, studies of model systems had established that

these residues could be N-methylated selectively without altering
the double bond geometry (22). This method was used to convert XX
to XXII. In this way the double bond configuration in XXII and,
therefore in the synthetic cyclic peptide XIX, was established unambiguously. Peptide XX was cyclized to give a 17% yield of ten-
toxin(19). Direct comparison of synthetic with natural tentoxin
by nmr, ir, uv, mass spectrometry and chromatography established
that both compounds were identical. In addition the biological
activities of synthetic and natural tentoxin were identical (Table
2). These data confirm that the structure proposed for tentoxin
is correct and that full biological activity is found for the
structure.

In order to correctly interpret structure-activity relation
(SAR) studies it is necessary to know the conformation of the molecule in addition to its primary structure. In general structure-
activity studies of complex natural products are difficult to interpret if each synthetic modification of the parent compound can
produce an analog differing from the parent in conformation as well
as primary structure. Tentoxin is an excellent molecule for the
study of structure-activity relationships because its conformation
can be determined readily in solution by nmr spectrometry. Only
biological data of analogs which retain the peptide ring conformation and substitution pattern found in tentoxin can be used to
interpret SAR studies. The following example illustrates how this
approach works.

Hydrogenation of tentoxin over palladium on carbon forms D-
MePhe dehydrotentoxin (DHT)(XXIII)as the sole product (14, 15).

Table 2.   *Effects of synthetic and natural tentoxin
on chlorosis of lettuce plants*

Toxin (μg)	mg chlorophyll/g wet weight of tissue	
	Synthetic cyclo (N-Me-L-Ala-L-Leu-N-MeΔPhe-Gly)	Tentoxin
0	0.052	0.052
1	0.030	0.033
3	0.017	0.015
5	0.0072	0.0075

No L-MePhe-DHT is formed. Bioassay showed that even at high concentrations D-MePhe-DHT(XXIII)did not cause chlorosis when applied
to germinating lettuce seeds. Thus reduction of the ΔPhe double
bond eliminated biological activity. However it is not possible
to conclude from this result that the double bond is essential for
activity; reduction of the double bond could also have changed the
conformation of the peptide-ring to an inactive conformer.

Nmr studies established that the conformation of XXIII differed from that of XIX. It was concluded that D-MePhe-DHT(XXIII)
existed in the conformation shown in which Phe- and Leu- were both
on the side of the 12-membered ring system and not at the end of
the system as they are in tentoxin. Thus either the incorrect
placement of substituents about the 12-membered ring caused by the
new conformation or the loss of the double bond could be responsible
for the loss of biological activity.

In order to design tentoxin analogs conformationally similar
to tentoxin, it was necessary to consider the forces causing D-
MePhe-DHT(XXIII)to adopt its conformation. The tentoxin-like conformation XXIV of D-MePhe-DHT is unstable because the benzyl group
is forced into a hindered position over the peptide ring, about
2.3 Å from the α-carbon of leucine. This severe steric interaction
is absent in the stable conformation XXIII and would also be relieved if the α-carbon of phenylalanine were epimerized from D- to L-.
Thus the L-MePhe-DHT(XXV)would be expected to resemble closely the
conformation of tentoxin and to provide a useful analog to test the
effect of the double bond on activity.

L-MePhe-DHT(XXV)was prepared by the method used to prepare
tentoxin except that L-MePhe replaced MeΔPhe (22). Cyclo (L-MeAla-
L-Leu-L-MePhe-Gly)(XXV)was obtained in 0.3% yield based on an
average of ten cyclization experiments.

The conformation of XXV was assigned (22) by comparing the α-
proton chemical shift and coupling constant data with the corresponding values reported for model compounds(4,5). The results
established that the peptide ring confirmation XXV is essentially
the same as in tentoxin(XIX). However, the shapes of the molecules
are not identical since the α-carbon of Phe is tetrahedral in XXV
but planar (trigonal) in tentoxin(XIX). Thus, the relative orientation of the phenyl-group to the peptide ring is changed; in XXV
the phenyl group is displaced about 1 Å out from the plane of the
phenylalanine carbonyl group whereas in XIX the phenyl group and the
carbonyl group are in the same plane.

In the lettuce seedling bioassay, L-MePhe-DHT(XXV)did not
cause chlorosis even at very high concentrations (200 μg/ml). It
is possible that this analog could lack "intrinsic activity" (1)
yet bind to the receptor. In this circumstance the analog would

be a competitive inhibitor. Compound XXV has not been tested yet
to determine if it antagonizes the effect of tentoxin; larger
quantities of it are needed for this experiment than are available
from our present method of preparation.

Cyclo (D-MeAla-L-Leu-L-MePhe-Gly)(XXVI)was chosen next for
synthesis because changing MeAla from an L- to D- configuration
was expected to facilitate the cyclization reaction, and because
this analog was expected to have the tentoxin-like peptide ring
conformation. Peptide XXVI was synthesized in the same manner as
XXV except that D-MeAla was used instead of L-MeAla. Analog XXVI
was obtained in 10% yield from the cyclization reaction. The nmr
spectrometry studies established that the conformation of XXVI is
the one shown (22). Analog XXVI did not cause chlorosis in germinating seeds.

D-MeAlaTentoxin(XXVII)was synthesized to check the effect on
biological activity of changing the configuration of MeAla from
L- to D-. Analog XXVII was synthesized by a route similar to that
used to prepare tentoxin(XIX)except that D-MeAla replaced L-MeAla.
Preliminary nmr studies indicated that the conformation of XXVI
is essentially unchanged from that of tentoxin. Furthermore, analog
XXVII is biologically active, causing chlorosis in germinating lettuce seedlings. The relative potency of the analog to tentoxin has
not been determined. Since analog XXVII is active the epimerization
of MeAla from L- to D- does not eliminate biological activity, and
therefore, the inactivity of both dihydrotentoxins XXV and XXVI
must be caused by the reduction of the double bond. The data do
not indicate whether this inactivity is due to the loss of a potential alkylating group or to the non-planar shape about the phenylalanine residue.

An alternative method for establishing the structure of an
isolated substance is by direct comparison with known compounds.
The availability of the L- and D-MePhe analogs XXIII and XXV allowed us to assign the structure of an unknown dihydrotentoxin which
was contaminating tentoxin isolated from some cultures of A. tenuis.
Because the chromatographic properties of the analog were very similar to those of tentoxin, it was not possible to isolate the unknown in pure form. However, both nmr and mass spectral data indicated that the contaminant was a dihydrotentoxin that contained
a methylphenylalanine residue instead of MeΔPhe. Amino acid analysis by gas-liquid chromatography proved the presence of MePhe.
The configuration of the MePhe was assigned by comparing the nmr
chemical shifts of the unknown DHT with the chemical shifts of authentic D-MePhe-DHT(XXIII), L-MePhe-DHT(XXV) and D-MeAla-L-MePhe-DHT
(XXVI). On this basis we have concluded that L-MePhe-DHT(XXV) is the
contaminant in the tentoxin preparations although we have not yet
been able to isolate pure XXV from the cultures. We do not know

whether XXV is a biosynthetic precursor of tentoxin or a metabolite formed by reduction.

Attempts to correlate structure with biological activity and specificity must also consider the relationships between structure and partition coefficient for each analog. Clearly any modification that greatly alters solubility could be expected to change the amount of analog reaching receptor sites. Quantitative correlations between structure, and lipophilicity (and other physical-chemical properties of the analogs) can be determined (8, 9). Very often biological activity is a linear or parabolic function of partition coefficient. The variation in biological activity of many of the phytoalexins may correlate with their partition properties.

The partition behaviour of the tentoxin analogs deserves comment here. All analogs have been found to migrate very closely to tentoxin in several chromatographic systems. Since the partition behaviour of the analogs and tentoxin is similar, it is likely that all analogs are able to partition through biological membranes to reach the site of action of tentoxin. The inactivity of analogs XXIII, XXV and XXVI is probably not a result of these compounds failing to reach the site of action.

Several models for selective toxicity have been described. Many others are possible such as those based on unique metabolism, partition phenomena, biochemical constituents or biosynthetic pathways. The diversity of possibilities makes it unlikely that any one mechanism or any one class of structure would explain the activity of all toxins of one type such as host-specific toxins. It is more likely that all these mechanisms would be observed either alone or in varying combinations. For example, it is reasonable to expect that some host-specific toxins will be found that do not reach the receptor protein in resistant plants but would bind to this receptor *in vitro* where partition effects are minimized. Others may be found that reach the receptor molecule but do not bind to it. Still others may be found the low activity of which is caused by a combination of both inefficient binding and partition properties.

Studies designed to relate chemical structures of plant toxins with specificity are in their early stages. They require isolation and characterization of both major components in the interaction, the toxin and the receptor molecule. The biochemical function of the receptor must be identified so that comparisons between substances isolated from susceptible and resistant plants can be made.

In the case of tentoxin our preliminary results suggest that tentoxin binds tightly to a specific protein in a non-covalent manner (J. Steele, personal communication). The inactive tentoxin analogs do not appear to bind. If this protein can be further purified and characterized, it will provide the other major component needed to study in depth the chemical nature of the specificity of this phytotoxin and thus contribute to our general understanding of the chemical basis of specificity.

ACKNOWLEDGEMENTS

I wish to thank Mr. Tom Uchytil, Dr. John Steele and Professor Richard D. Durbin for evaluating the biological activities of all the analogs.

REFERENCES

1.   ARIENS, E.J. (1964). *Molecular Pharmacology*, Vol. 1, 136 - 286. Academic Press, New York and London.

2.   BAKER, B.R. (1967). *Design of active-site-directed irreversible enzyme inhibitors.* John Wiley and Sons, New York, 325 pp.

3.   BUCHANAN, J.M. (1958). Mechanism of action of azaserine. *In : Amino acids and peptides with antimetabolic activity.* (WOSTENHOLME, C.E.W. and O'CONNOR, C.M., Eds.). CIBA Foundation Symposium. Academic Press, New York and London.

4.   DALE, J., and TITLESTAD, K. (1969). Cyclic oligopeptides of sarcosine. *J. chem. Soc., Chem. Communs,* 656.

5.   DALE, J., and TITLESTAD, K. (1970). A common conformation for five cyclic tetrapeptides. *J. chem. Soc., Chem. Communs,* 1403.

6.   DARDENNE, G., CASIMIR, J., MARLIER, M., and LARSEN, P.O. (1974). Acide 2(R)-Amino-3-Butenoique (Vinylglycine) dans les carpophores de *Rhodophyllus nidorosus. Phytochemistry,* 13, 1897 - 1900.

7.   GROSS, E. and MORELL, J.L. (1971). The structure of nisin. *J. Am. chem. Soc.,* 93, 4634 - 4635.

8.   HANSCH, C. and CLAYTON, J.M. (1973). Lipophilic character and biological activity of drugs II : The parabolic case. *J. pharm. Sci.,* 62, 1 - 21.

9.   HANSCH, C. and DUNN, W.J. (1972). Linear relationships between lipophilic character and biological activity of drugs. *J. pharm. Sci.,* 61, 1 - 19.

10. HITCHINGS, G.H. and BURCHALL, J.J. (1965). Inhibitions of folate biosynthesis and function as a basis for chemotherapy. *Adv. Enzymol.*, 27, 417 - 468.

11. KONCEWICZ, M., MATHIAPARANAM, P., UCHYTIL, T.F., SPARAPANO, L., TAM, J., RICH, D.H. and DURBIN, R.D. (1973). The sequence and optical configuration of the amino acids in tentoxin. *Biochem. biophys. Res. Commun.*, 53, 653 - 658.

12. LEE, B. (1971). Conformation of penicillin as a transition-state analog of the substrate of peptidoglycon transpeptidase. *J. molec. Biol.*, 61, 463 - 469.

13. LIENHARD, G.E. (1972). Transition-state analogs as enzyme inhibitors. *A. Rep. Med. Chem.*, 7, 249 - 258.

14. MEYER, W.L., KUYPER, L.F., LEWIS, R.B., TEMPLETON, G.E. and WOODHEAD, S.H. (1974). The amino acid sequence and configuration of tentoxin. *Biochem. biophys. Res. Commun.*, 56, 234 - 240.

15. MEYER, W.L., KUYPER, L.F., PHELPS, D.W. and CORDES, A.W. (1975). Use of [1]H Nmr spectroscopy for sequence and configurational analysis of cyclic peptides. The structure of tentoxin. *J. Am. chem. Soc.*, 97, 3802 - 3809.

16. OKUNO, T., ISHITA, T., SAWAI, K., and MATSUMOTO, T. (1974). Alternariolide, host-specific toxin produced by *Alternaria mali* Roberts. *Chem. Lett.*, 635 - 638.

17. RANDO, R.R. (1974). Chemistry and enzymology of kcat inhibitors. *Science, N.Y.*, 185, 320 - 324.

18. RANDO, R.R. (1975). Mechanisms of action of naturally occurring irreversible enzyme inhibitors. *Acc. chem. Res.*, 8, 281 - 288.

19. RICH, D.H. and MATHIAPARANAM, P. (1974). The synthesis of the cyclic tetrapeptide, tentoxin. Effect of an N-methyl dehydrophenylalanyl residue on conformation of linear peptides. *Tetrahedron Lett.*, 46, 4037 - 4040.

20. RICH, D.H., TAM, J., MATHIAPARANAM, P., GRANT, J.A., and MABUNI, C., (1974). General synthesis of didehydro amino acids and peptides. *J. chem. Soc., Chem. Communs*, 897 - 898.

21. RICH, D.H., TAM, J., MATHIAPARANAM, P., and GRANT, J.A. (1975). Selective N-methylation of dehydro amino acids and peptides. *Synthesis*, 402 - 403.

22. RICH, D.H., MATHIAPARANAM, P., GRANT, J.A., and BHATNAGAR, P. (1975). Synthesis of cyclic tetrapeptides related to tentoxin. *In : Peptides - Chemistry, Structure and Biology.* WALTER, R. and MEIENHOFER, J., Eds.). *Proc. 4th Am. Peptide Symp.,* Ann. Arbor Science Publishers, Inc. (In press).

23. SINGER, S.J. (1967). Covalent labeling of active sites. *Adv. Protein Chem.,* 22, 1 - 54.

24. SPEK, A.L., PEERDEMAN, A.F., VAN WIDNGAARDEN, I., and SOUDIGN, W., (1971). The absolute configuration and crystal structure of the anticholinergic drug dexbenzetimide. *Nature, Lond.,* 232, 575 - 576.

25. STEINER, G.W., and STROBEL, G.A. (1971). Helminthosporoside, a host-specific toxin from *Helminthosporium sacchari. J. biol. Chem.,* 246, 4348 - 4355.

26. STROMINGER, J.L. and BLUMBERG, P.M. (1974). Interaction of penicillin with the bacterial cell : Penicillin-binding proteins and penicillin-sensitive enzymes. *Bact. Rev.,* 38, 291 - 335.

27. TEMPLETON, G.E. (1972). *Alternaria* toxins related to pathogenesis in plants. *In : Microbial toxins.* (KADIS, S., CIEGLER, A. and AJL, S.J., Eds.). Vol. VIII. *Fungal Toxins.* 169 - 192. Academic Press, New York and London.

28. WOLFENDEN, R., (1972). Analog approaches to the structure of the transition state in enzymic reactions. *Accts. Chem. Res.,* 5, 10 - 18.

CONTRIBUTIONS

KEEN, N.T.   Differential responses of fungi to various isoflavanoid compounds.

SUMMARY OF POINTS FROM CONTRIBUTIONS AND DISCUSSIONS
BY
D. H. RICH

Chairman and Discussion Leader

The majority of comments in the discussion session pertained to criteria for establishing specific binding to an isolated protein alleged to be toxin receptor. Attention was drawn to the necessity for running suitable controls in binding studies where binding is detected by subjecting toxin-receptor complexes to gel filtration. This procedure will detect binding only if the rate of dissociation of the complex is slow. Usually this means the Ka is large but since K association is an equilibrium constant it is possible that some complexes will not be detected with this technique because equilibration is rapid. However one participant noted that one toxin was found in the void column when toxin detergent buffers were subjected to gel filtration in the absence of receptor. It was also pointed out that victorin binding could not be detected by this technique although its potency and specificity suggest it must interact specifically with some receptor molecule. The evidence for non-covalent binding of tentoxin to the apparent receptor protein was re-examined. Tritiated tentoxin could be isolated from the void fractions containing protein and tentoxin by chloroform extraction. No labelled tentoxin remained in the aqueous phase proving that the interaction was not covalent. This experiment suggests that L-MePhe-DHT is inactive because the residue is no longer planar.

Further questions requested additional data on relationships between structure, activity and specificity. It was pointed out that some analogs of natural peptides are more active than the parent compound. For example deamino-oxytocin is twice as active and deamino-(4-threonine)-oxytocin perhaps five times as active as oxytocin. Furthermore in the oxytocin-vasopressin peptide hormone series single amino acid substitutions can produce analogs with much more selective activity. For example (1-deamino,8-D-Arginine)-vasopressin is highly specific in possessing high antidiuretic activity and little vasopressor activity.

Keen reported on the differences in fungitoxicity of various isoflavonoid compounds produced by plants in the family *Leguminosae*. Simple isoflavones e.g. daidzein, genistein, etc. have only low fungicidal activity against *Phytophthora vignae, Cladosporium cucumerinum, Verticillium dahliae* or *Colletotrichum lindemuthianum*. Of interest was the recent finding that isopentenyl substituted isoflavones and isoflavanones (e.g. luteone and kievitone) have roughly two orders of magnitude greater fungitoxicity than the corresponding compounds lacking this substitution. Whereas synthetic 2'hydroxy isoflavones (e.g. 2'hydroxy daidzein) are only weakly antifungal, the corresponding 2'hydroxy isoflavanone is much more active. This may be due to spontaneous dehydration to pterocarp-6a-ene. It was noted that the effect on fungicidal activity produced by the isopentenyl group might be correlated with the lipophilicity of the derived molecule (Hansch, 1972, 1973).

EVIDENCE FOR DIRECT INVOLVEMENT OF NUCLEIC ACIDS IN HOST-PARASITE

SPECIFICITY

ROLAND ROHRINGER

Research Station, Agriculture Canada

Winnipeg, Manitoba, Canada

This paper deals with specificity as illustrated by the interactions between cultivars of a host plant species and strains of a particular pathogen.

Nucleic acids are the genetic basis for specificity in biological systems. However, in most diseases their involvement apparently is indirect, regulating the production of cell components, enzymes, toxins, or phytoalexins, and it is through these agents that many hosts and pathogens interact with each other. Nucleic acids are directly involved, in the sense used in this paper, only when they are translocated from pathogen to host or *vice versa* and when it can be shown that they produce symptoms similar to those observed in the natural interaction between host and pathogen.

When viral and viroid disease agents are excluded from a discussion of host-pathogen interactions, only a few cases remain in which nucleic acids were said to be directly involved. One of these is the crown gall disease caused by *Agrobacterium tumefaciens*. In this disease, bacterial RNA, or the DNA of a phage associated with some strains of the bacterium, were viewed at various times as tumor-inducing substances (1, 2, 5, 8, 9). However, this neoplastic disease occurs in many kinds of dicotyledonous plants and is therefore of little interest in a discussion on specificity.

Another case for which evidence has been presented (4, 7) for the direct involvement of nucleic acids in the host-parasite interaction is the wheat/stem rust system. The biochemistry of resistance of wheat(*Triticum aestivum*) to stem rust (*Puccinia graminis* f. sp. *tritici*) has been the subject of numerous investigations in

185

which many metabolic differences between compatible and incompatible
interactions have been described. It was often not clear whether
the observed changes were the cause or the result of the resistant
reaction or whether they reflected changes in constituents of the
host or in those of the parasite. Until recently, it was not possible to correlate any of the post-infectional biochemical changes
with the presence of any specific gene for resistance. The theoretical basis for progress in this field of research was provided by
Flor (3) and others who developed the gene-for-gene concept. Briefly,
this concept proposes that for each gene for resistance in the host
there is a corresponding gene for avirulence in the parasite. The
gene-for-gene concept has been applied to several other host-parasite
systems and has so far withstood the test of time. It is clear that
theories on the biochemical basis of resistance in host-parasite
systems to which this concept applies must be consistent with genetical data and with concepts on the co-evolution of host and parasite.
If certain specific genes for resistance in the host "correspond"
to certain specific genes for avirulence in the parasite, it can
be assumed that "information" specific to this gene interaction must
pass in either direction across the host-parasite interface. While
it is theoretically possible that low molecular weight compounds,
such as gene-regulators or "effectors", may function as carriers of
this information, a concept based on this notion would have to postulate the existence of a large number of such compounds since many
genes for resistance have been identified. Alternatively, the
"information carrier" may be of large molecular weight and may be
capable of a high information content. In this case, protein, complex carbohydrates, or polynucleotides may be involved. If macromolecules are involved, they need not necessarily be transported
from one cell into another, because the interaction between the product of the gene for avirulence and the product of the gene for resistance may well occur in membranes at the host-parasite interface.

Apart from these theoretical considerations, recent progress
in this area was stimulated because of the availability of lines of
wheat near-isogenic for genes for resistance (6), and because a bioassay (4) developed at Winnipeg made it possible to detect compounds
that determine the resistant reaction. This bioassay, which is not
described in detail here, employs resistant and susceptible near-
isogenic lines of wheat inoculated with a virulent race of stem rust.
After treatment with an extract that is to be tested for activity,
leaf segments are examined by fluorescence microscopy for occurrence
of necrotic host cells that are typical of an incompatible interaction. Use of this bioassay does not require the investigator to
make assumptions regarding the nature of the gene product except
that it must be extractable and, after contact with bioassay plants,
it must be able to reach the active site with which it normally
interacts in a specific manner. Both low molecular weight compounds
as well as biopolymers can be tested for activity. In practice,
however, a decision has to be made as to the type of extract to be

tested. The work described in the following employed extracts en-
riched with nucleic acids.

Initially, the source of the extract was an incompatible host-
parasite complex in which the host contained the *Sr6* gene for resistance and the rust the corresponding *P6* gene for avirulence. This
type of material was chosen because it is the only one of four possible combinations of these genes in which resistance is expressed
and because it was not known whether the compound that the bioassay
was designed to detect originates from the host as a product of the
gene for resistance, or from the parasite as a product of the gene
for avirulence. Table 1 shows the results of several bioassays.
Crude nucleic acid extracts, and extracts treated with cetyltrimeth-
ylammonium bromide (CTA), deoxyribonuclease (DNase), or protease
were all active when tested in this system, but no activity was detected using extracts treated with ribonuclease (RNase). Crude extracts contained 50% RNA, 45% DNA, and 5% protein, by weight, as
well as undetermined amounts of polysaccharide. Extracts treated
with CTA and DNase contained 95% RNA, 4% oligodeoxyribonucleotides,
and 1% protein, by weight, and no detectable polysaccharide. These

TABLE 1.   *Characterization of active component in crude extract*
           from resistant-reacting wheat leaves (Sr6/P6 *interaction*)

Treatment of crude extract	Necrotic sites/cm^2 leaf of bioassay plants[c]		
	R-Line (R)	S-Line (S)	R/S
None	97.4	62.6	1.6[b]
CTA[a]	20.2	7.6	2.7[b]
CTA, DNase	44.4	12.2	3.6[b]
CTA, DNase, protease	48.5	8.1	6.0[b]
CTA, RNase	8.5	20.2	0.4

[a] Cetyltrimethyl-ammonium bromide precipitation.

[b] R significantly greater than S at 0.05 level of probability.

[c] Necrotic sites visualized by staining with trypan blue.

TABLE 2.   *Bioassay systems used in tests for gene-specificity*

Source of extract (Gene interaction)	Bioassay plants (Gene interaction)	System
Srn/Pn	*Srn/pn; srn/pn*	homologous
Srn/Pn	*Srm/pm; srm/pm*	heterologous

results indicated that activity may be associated with RNA. Except when otherwise noted, all later preparations were assayed after purification with CTA and DNase.

Availability of several near-isogenic lines of wheat with different genes for resistance made it possible to test extracts for gene-specificity. To accomplish this, extracts were tested using both homologous and heterologus bioassay systems (Table 2). For example, using a homologous bioassay system, an extract obtained from an *Sr6/P6*-containing host-parasite complex is tested with bioassay plants (containing *Sr6* or *sr6*) previously inoculated with a race of rust containing the *p6* gene for virulence. Using a heterologous system, this extract is tested with bioassay plants of different near-isogenic lines that involve other genes for resistance *(Sr5* or *Sr11)* and are inoculated with races of rust containing the corresponding gene for virulence *(p5* or *p11)*. If an extract is "active" and gene-specific, positive results are expected with the homologous bioassay, negative results with the heterologous assay. Table 3 shows that extracts obtained from the *Sr6/P6-* or *Sr11/P11-* containing host-parasite complexes were active and gene-specific. Tests with an extract obtained from the *Sr5/P5*-containing host-parasite complex were inconclusive, because the homologous assay revealed no activity. Lack of activity in this case was attributed to the fact that the *Sr5* gene limits the host-parasite interaction to a few host cells per infection focus, and that an active compound, if produced, would be expected to be present in smaller amounts than in other interactions. Later evidence showed that this assumption was probably correct.

Apart from the question of gene specificity, it was important to know whether host cell necrosis was a prerequisite for the production of the active component. In an effort to answer this question, use was made of the fact that expression of the *Sr6* gene is temperature-dependent. Plants containing this gene respond with a necrotic reaction after inoculation with an avirulent race containing the *P6* gene, provided the ambient temperature is maintained at or

TABLE 3. *Specificity of active component towards host genes for resistance*

Host-parasite complex from which RNA extract was obtained		Bioassay system[a]		Necrotic sites/cm^2 leaf of bioassay plants[b]		
Gene interaction	Avirulent race used	Gene interaction		R-line (R)	S-line (S)	R/S
		R-line	S-line			
Sr6/P6	C18(15B-1L)	*Sr6/p6*	*sr6/p6*	9.5	3.2	3.0[c]
		Sr11/p11	*sr11/p11*	1.8	3.0	0.6
Sr6/P6	C17(56)	*Sr6/p6*	*sr6/p6*	15.9	9.1	1.8[c]
		Sr5/p5	*sr5/p5*	3.9	5.8	0.7
Sr11/P11	C21(32)	*Sr11/p11*	*sr11/p11*	4.9	2.1	2.3[c]
		sr6/p6	*sr6/p6*	2.9	3.7	0.8
Sr5/P5	C5(29-1)	*Sr5/p5*	*sr5/p5*	3.6	3.6	1.0
		Sr6/p6	*sr6/p6*	1.0	5.6	0.2

[a] Virulent races of stem rust used were C22(32), C45(56A), C22(32), and C11(15B-4), C45(56A), respectively.

[b] Necrotic sites visualized by staining with trypan blue.

[c] R significantly greater than S at 0.05 level of probability.

below 20°C. At a temperature of 25°C or above, few if any necrotic cells are produced in this interaction. Extracts obtained from two sets of plants infected and temperature-treated in this way, both showed activity when tested in the homologous bioassay system, indicating that necrosis was not necessary for activity and that production of the active component is not a consequence of host cell necrosis in the *Sr6/P6* gene interaction.

A further question concerned the origin of the active component. Since the source of each extract consisted of a host-parasite complex, the active component detected with the bioassay may have been

TABLE 4. *Production of active RNA in rust-infected wheat leaves in relation to various allelic combinations at* Sr6 *and* P6 *loci*

Host/parasite complex from which RNA extract was obtained			Necrotic sites/cm^2 leaf of bioassay plants		
Gene interaction	Race used	Experiment	R-line(R)	S-line(S)	R/S
Sr6/P6	C17(56)	1	62.5	30.8	2.0[b]
		2	41.1	29.4	1.4[b]
		3	20.8	13.5	1.5
		4	62.2	26.9	2.3[b]
		5[a]	13.1	5.5	2.4[b]
Sr6/p6	C45(56A)	2	26.2	27.4	1.0
		3	9.4	12.5	0.8
		5[a]	4.6	10.6	0.4
sr6/P6	C17(56)	1	38.5	20.8	1.9[b]
		5[a]	10.8	5.9	1.8[b]
sr6/p6	C45(56A)	4	51.9	58.3	0.9

[a] Leaves stained with trypan blue for detection of necrotic sites: average of 2 - 4 separate experiments.

[b] R significantly greater than S at 0.05 level of probability.

a product of the host or a product of the fungus. In an experiment designed to distinguish between these two possibilities, extracts were prepared from four sets of inoculated plants characterized by the following gene interactions: *Sr6/P6* (incompatible), *Sr6/p6* (compatible), *sr6/P6* (compatible), and *sr6/p6* (compatible). The two stem rust races used were race 56 (containing the *P6* gene for avirulence) and race 56A (containing the *p6* gene for virulence). The results (Table 4) showed that activity was detected only in

extracts from material in which the *P6* gene for avirulence was present, indicating that the active component may originate from the
pathogen and may be the product of the gene for avirulence.

These results prompted a re-investigation of the failure to
obtain an active extract from the *Sr5/P5* gene interaction. If the
correct interpretation is that the active component is a product of
the gene for avirulence, it should then be possible to obtain an
active extract from a host-parasite complex between a strain of
rust that is avirulent on the *Sr5* gene for resistance and a host
that does not contain this gene. This complex is characterized by
the *sr5/P5* gene interaction. An extract obtained from such material
was active when tested with the homologous bioassay system and inactive when tested with the heterologous assay involving *Sr6-* and
sr6-containing bioassay plants (Table 5). These results lend additional support to the theory that the active component is a product
of the gene for avirulence and that it is gene-specific.

TABLE 5. *Activity and specificity of RNA prepared from genotypically*
susceptible wheat leaves infected with a race of stem rust
containing the P5 *gene for avirulence*

Host-parasite complex from which RNA extract was obtained		Bioassay system[a] Gene interaction		Necrotic sites/cm^2 leaf of bioassay plants		
Gene interaction	Avirulent race used	R-line	S-line	R-line (R)	S-line (S)	R/S
sr5/P5	C24(17)	*Sr5/p5*	*sr5/p5*	58.3	28.3	2.1[b]
		Sr6/p6	*sr6/p6*	55.0	62.2	0.9
Sr6/P6	C17(56)	*Sr6/p6*	*sr6/p6*	38.7	23.9	1.6[b]

[a]   Virulent race of stem rust used was C45(56A).

[b]   R significantly greater than S at 0.05 level of probability.

Early experiments had shown that plants used for the bioassay must be infected with a virulent race of stem rust before host cells will respond to an active extract with the formation of necrotic cells. This suggested that the active component is not a phytotoxin, at least not in the usual sense. It also raised the question : what developmental stage of the fungus used in the bioassay is necessary before activity can be detected ? When bioassay plants were treated at various times after inoculation with an extract previously shown to be active, activity was detected only when at least half of the infection sites had developed one or more haustoria. Under the conditions used, this threshold was reached approximately 24 h after the beginning of the first photoperiod that first synchronized fungal development. The notion that haustorial penetration may facilitate passive uptake of the active component into host cells proved incorrect since bioassay plants infected with leaf rust (*P. recondita* f. sp. *tritici*) did not respond to an active extract prepared from a wheat/stem rust complex. Why leaf rust was unable to substitute for stem rust in the bioassay is not known. It is possible that the so called active component is inactive at the time of injection, and that stem rust, but not leaf rust, is required to process it into an active form. It is also possible that a gene regulator produced by stem rust, but not by leaf rust, de-represses the gene for resistance in bioassay plants containing this gene, and permits the active component to interact with it or with its gene product.

Initial attempts to fractionate preparations containing the active component have been partially successful (Table 6). Activity was detected in RNA fractions that are soluble in 3M potassium acetate, when these were tested at RNA concentrations between 40 and 70 O.D. units/ml. When tested at lower concentrations, no activity was detected. This fraction was shown, by acrylamide gel electrophoresis, to contain largely transfer ribonucleic acid (tRNA), lesser amounts of oligodeoxyribonucleotides, and 5S RNA. In typical experiments, fractions containing high molecular weight rRNA (3M potassium acetate insoluble) showed no activity. The active component also co-fractionated with isopropanol-soluble RNA (Table 7), but again the dose-response curve was not linear, showing a maximum when these fractions were tested at RNA concentrations between 20 to 40 O.D. units/ml.

Further evidence is needed to support the contention that the gene-specific component is RNA. The present claim rests on the fact that activity is destroyed by treatment with ribonuclease, and that the so called "specific activity" of preparations increases when components other than RNA are removed or altered during the purification. The bioassay, although theoretically simple, is subject to many variables, and some of these are as yet unknown or can only be controlled insufficiently. Frequently, it has been difficult to reproduce results, but it is not clear whether this was due to the

TABLE 6.   *Response of bioassay plants to an active preparation*
            after fractionation with 3M potassium acetate

Fraction and Experiment no.	Nucleic acid concentration (O.D. units/ml)	Necrotic sites/cm^2 of bioassay plants[b]		
		R-line (R)	S-line (S)	R/S
3M Potassium acetate soluble				
1	70	23.8	13.3	1.8[a]
2	70	50.0	23.3	2.1[a]
3	70	51.7	35.8	1.4[a]
4	70	85.0	64.2	1.3[a]
5	40	10.0	3.8	2.4[a]
6	30	6.9	5.6	1.3
7	20	7.5	11.9	0.6
3M Potassium acetate insoluble				
1	70	17.8	14.2	1.3
2	70	45.0	25.8	1.7[a]
3	70	48.3	47.5	1.0
4	70	33.3	49.2	0.8
5	40	5.0	3.8	1.4
6	30	11.3	11.3	1.0
7	20	11.3	12.5	0.9

[a]   R significantly greater than S at 0.05 level of probability.

[b]   Necrotic sites visualized by fluorescence.

TABLE 7. *Response of bioassay plants to iso-propanol soluble
fractions of active preparations*

Number of experiments at each concentration	Nucleic acid concentration (O.D. units/ml)	Necrotic sites/cm^2 of bioassay plants [a]		
		R-line (R)	S-line (S)	R/S
2	80	3.45	5.18	0.7
8	40	3.78	3.03	1.2
5	20	3.43	2.05	1.7[b]
3	10	1.10	1.55	0.7

[a] Means of replicate experiments.

[b] R significantly greater than S at 0.05 level of probability.

variability inherent in the bioassay, or whether it was due to an
inability to consistently produce comparable extracts.

     The presently available evidence regarding the active component,
presumably RNA, is consistent with genetic data on this host-parasite system and can be summarized with the following scheme:

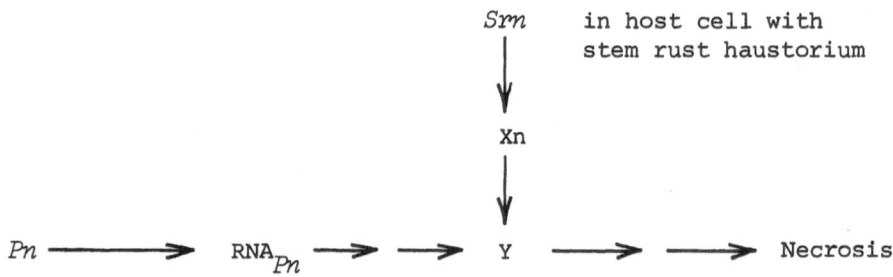

where *Pn* designates a gene for avirulence, RNA$_{Pn}$ the product of this
avirulence gene, *Srn* the corresponding gene for resistance, Xn the
hypothetical product of this gene for resistance, and Y the product
of the interaction between RNA$_{Pn}$ and Xn conditioning host cell
necrosis.

This scheme, if essentially correct, raises the following questions. How do gene-specific components detected with the bioassay differ from each other ? At what stage of fungal development is the gene-specific component produced ? What is the product of the gene for resistance with which it presumably interacts ? How do products of genes for resistance differ from each other ? What is the mechanism of the interaction between products of corresponding genes, and where does this interaction take place in the host-parasite complex ? The bioassay, or a modification of it, may be useful in research designed to answer some of these questions.

ACKNOWLEDGEMENT

The author thanks the Editor and publishers of *Nature* for permission to publish some results reported here that were previously published in that journal.

REFERENCES

1.   BELJANSKI, M., MANIGAULT, P., BELJANSKI, M. and AARON, Y. (1974). Genetic transformation of *Agrobacterium tumefaciens* by RNA and nature of the tumor inducing principle. *Abst. 1st Intersect. Congr. Int. Ass. Microbiol. Sci.*, Tokyo. Section A-2-1.

2.   EDEN, F.C., FARRAND, S.K., POWELL, J.S., BENDICH, A.J., CHILTON, M.-D., NESTER, E.W. and GORDON, M.P. (1974). Attempts to detect deoxyribonucleic acid from *Agrobacterium tumefaciens* and bacteriophage PS8 in crown gall tumors by complementary ribonucleic acid/deoxyribonucleic acid-filter hybridization. *J. Bact.*, 119, 547 - 553.

3.   FLOR, H.H. (1971). Current status of the gene-for-gene concept. *A. Rev. Phytopath.*, 9, 275 - 296.

4.   HOWES, N.K., SAMBORSKI, D.J. and ROHRINGER, R. (1974). Production and bioassay of gene-specific RNA determining resistance of wheat to stem rust. *Can. J. Bot.*, 52, 2489 - 2497.

5.   KADO, C.I. (1974). A concept on the role of *Agrobacterium tumefaciens* DNA in plant tumorigenesis. *Abstr. 1st Intersect. Congr. Int. Ass. Microbiol. Sci.*, Tokyo. Sections A-2-1.

6.   LOEGERING, W.Q. and HARMON, D.L. (1969). Wheat lines near-
      isogenic for reaction to *Puccinia graminis tritici*.
      Phytopathology, 59, 456 - 459.

7.   ROHRINGER, R., HOWES, N.K., KIM, W.K. and SAMBORSKI, D.J.
      (1974). Evidence for a gene-specific RNA determining
      resistance in wheat to stem rust. *Nature, Lond.*, 249,
      585 - 588.

8.   STROUN, M., ANKER, P., GAHAN, P., ROSSIER, A. and GREPPIN, H.
      (1971). *Agrobacterium tumefaciens* ribonucleic acid
      synthesis in tomato cells and crown gall induction. *J.
      Bact.*, 106, 634 - 639.

9.   SWAIN, L.W. and RIER, J.P. (1972). Cellular transformation in
      plant tissue by RNA from *Agrobacterium tumefaciens*. *Bot.
      Gaz.*, 133, 318 - 324.

CONTRIBUTIONS

VON BROEMBSEN, S., TEASDALE, J., ADAMS, M.J., and HADWIGER, L.H.
Disease resistance responses and their dependence on RNA
synthesis in peas.

STAPLES, R.C., APP, A.A., and RICCI, P.   DNA synthesis and nuclear
division during differentiation of bean rust uredospores.

SUMMARY OF POINTS FROM CONTRIBUTIONS AND DISCUSSIONS
BY
G. WOLF

Chairman and Discussion Leader

The discussion focussed mainly on the system used and the test
itself. Rohringer pointed out that the terms 'homology' and 'het-
erology' in their system do not refer to the rust used in the bioassay, but only to relate the genotype of the rust used for the
source of the extract with the genotype of the bioassay plants.
Mussell asked if the results could be interpreted as indicating two
differences between virulent and avirulent races i.e. the ability
to make the active factor and the ability to modify the factor in
the bioassay. One might be able to answer this question by assaying on an 'X' type leaf and observing the amount of change from 4
to 1 or of reaction types induced by the active principle, when
the 'X' leaf is inoculated by virulent and by avirulent races. If
both can modify the factor, one should see an equal amount of 4 to
1 shifts in both assays. Rohringer commented they never felt that
the effect of the injection would be detectable at a late stage

when reaction types can be determined. The bioassay system differs
from the normal system in several respects and is not designed to
show an effect on infection type. In bioassay, injection occurs
only once, and the active compound if labile, may have access to
an infection site only for a limited time. Conversely, in the natural system, the exchange of metabolites between host and parasite
occurs for a much longer period. It should also be realized that
infection densities in the bioassay are much higher than those used
in normal plant pathological work. In the bioassay, up to 20% of
all stomates are occupied by appressoria. The infection types
typical for specific interactions would not be seen at such high
infection densities.

Relating to Mussell's question, Daly asked whether a susceptible
bioassay was a requisite i.e. whether the same specificity can be
seen in a resistant response. Rohringer explained that the expected
effect of an active preparation on an incompatible interaction may be
difficult to detect because of the high "background" of necrotic
cells at each infection site. The discussion then turned to the
problem of specificity of the bioassay: differences in the nature
of fluorescence in incompatible and completely compatible tissues
(Daly); differences in number of fluorescent cells per infection
site between the R-line and the S-line (Heath). Rohringer commented
that cells which fluoresce because of mechanical injury can be distinguished from those that are associated with an infection after
staining with calcofluor. Fluorescing sites of collapsed host cells
are scored only if in contact with the fungus that has been made
visible with calcofluor. Sites are counted rather than individual
cells, since adjacent fluorescing cells may lie in the same line of
sight and cannot easily be distinguished from each other.

Asked by Bushnell for the percentage of experiments in which
the expected results were not obtained, Rohringer answered that
there were approximately 10% in early and about 30% in more recent
experiments in which several improvements had been introduced to
increase uniformity of spore deposition on bioassay plants and which
might have introduced variables that have not yet been identified.
Albersheim claimed that evidence for RNA as the active principle is
very limited. To rest the hypothesis some other simple experiments
could be done. The claim that the active component is RNA needs to
be substantiated much more fully was conceded. However, demonstration of gene-specificity of the active component appeared to be more
important than establishing its identity and has so far received
more attention.

Staples and co-workers had studied the requirements for DNA
synthesis and nuclear division during the formation of infection
structures by germinating bean rust uredospores. Actinomycin D and
hydroxyurea did not affect spore germination but inhibited DNA

synthesis, nuclear division and infection structure development.
Cordycepin did not affect germination, DNA synthesis, formation
of the appressorium, or initiation of the vesicle, but did inhibit
nuclear division and maturation of the vesicle. Although content
of polyadenylic acid in the spores increased 1.6-fold during formation of the appressoria and 6.1-fold during vesicle development,
cordycepin was not particularly effective in reducing either total
polyadenylic acid content or synthesis *in vitro*. It was concluded
that nuclear division need not accompany formation of appressoria
or vesicles, but synthesis of DNA is required.

Bramble asked for the significance of the differential synthesis of nuclear and mitochondrial DNA in germinating spores.
Staples replied that synthesis of nuclear and mitochondrial DNA
appears to be asynchronous in those fungi which form appressoria
since their germ tubes appear before nuclear division occurs. Fungi
which do not form appressoria usually produce their germ tubes at
the same time as nuclear division occurs.

Using actinomycin D in low concentrations as inducer, Hadwiger
presented a model for the induction of pisatin production in pea
pods. By application of different inhibitors he was able to show
that new synthesis of RNA and protein is a prerequisite for the
synthesis of pisatin. The observation of change in the texture of
chromatin after application of the pisatin inducer suggested that
the inducer acts as gene activator. Synthesis of pisatin was suppressed by the same inhibitors of RNA and protein synthesis after
infection with *Fusarium solani* f. sp. *phaseoli*. Hadwiger therefore
assumes that induction of pisatin production occurs by the same
mechanism in both cases.

Wood said that sucrose is one of the best "inducers" of pisatin
synthesis by discs of pea leaves and asked how this fits in with the
described scheme. Hadwiger replied that sucrose by itself does not
induce pisatin production in pea *pod* tissue, and assumed that if it
induces pisatin synthesis in leaves, the effect could be on membranes,
including the nuclear membrane.

Kuć pointed out that it is one thing to relate protein synthesis
to the accumulation of phytoalexins, but that it is quite another
thing to relate the accumulation of phytoalexins to the synthesis of
a single protein or a single enzyme. Hadwiger responded that inducers are gene specific and that it had been possible to examine
the synthesis of various species of proteins. The pattern of induced protein synthesis also varies with each of the inducers.

PROTEIN SPECIFICITY IN PLANT DISEASE DEVELOPMENT :

PROTEIN SHARING BETWEEN HOST AND PARASITE

JAMES E. DEVAY

Department of Plant Pathology, University of California

Davis, California, U.S.A.

INTRODUCTION

The phenomenon of protein sharing or common antigens between plant hosts and parasites has been interpreted in different ways in attempts to explain the basis of susceptibility of plants to disease or the reason for host specificity of plant pathogens. These views encompass immunogenetic responses, cytoplasmic reactions involving common (proteins) antigens, and lectin-protein reactions between host and parasite.

The idea that plants possess an immune system like the humoral system in vertebrate animals has been suggested many times. The literature on this subject has been reviewed by Chester (13), and more recently by Matta (34). There is no evidence to support the view that immunoglobulins are functional or even present in plants. However, plants do possess unique immune systems that are inducible and the immunogens of which may be translocated to distal plant organs. The induced immunity is somewhat persistent and the immune systems appear to involve multiple cell types in various tissues (4, 39, 43, 44). Considering the evidence for analogous immunological systems in plants and animals, it appears that the closest similarity would involve an agglutinin system in some plants which functions like the agglutinin system in crustaceans and other invertebrate animals (15).

Not discussed here but of major importance in protection phenomena in plants against disease, is the acquired resistance or immunity, either local or systemic, induced by viruses in hypersensitive hosts (39, 43).

Plants respond to incompatible associations with bacteria or fungi in various ways including the synthesis of highly specific protection factors (4) and the production of phytoalexins (3) which are usually nonspecific and which may function in disease resistance or in the hypersensitive reaction. On the basis of present knowledge it appears that in each host-pathogen relationship, host specificity depends on multiple factors. For a bacterial or a fungal disease of plants to occur, the pathogen must grow in, or on, or in proximity to a host plant so that its metabolites can interact with those of the host and give rise to disease symptoms. The growth of a pathogen is thus of major importance and any circumstance which inhibits growth is a limiting factor in disease development. Growth of the pathogen may be inhibited by many factors; however, the chemical nature of the host as a substrate in terms of nutritional or inhibitory substances, is probably the most important variable that affects the survival or death of a pathogen in host tissue.

It thus appears that for a micro-organism to invade a host, become established, and then grow to cause a pathological condition, it must adapt itself to a complex pattern of limiting circumstances in host tissue. The ability of a pathogen to cope with these various circumstances is reflected in the phenomenon of specificity in plant disease development.

When a parasite or pathogen invades a prospective host, its growth or lack of growth may depend on whether or not it possesses certain substances which mimic those of the host. The substances involved in this phenomenon, a sort of molecular mimicry, are antigenic when isolated and injected into experimental animals and have been given the name, common antigens which, then, are antigenic determinants in various biological materials that cause formation of antibodies (in experimental animals) with similar antigen specificity. This paper deals mainly with the possible role of protein-sharing or common antigens in host-parasite interactions.

AGGLUTININS (LECTINS)

The nature of the common antigens is largely unknown but it appears possible that certain agglutinins or lectins of plant hosts and of microbial parasites are antigenically similar and may account in some cases for the serological likeness of host and parasite. This concept is speculative but it warrants exploration because agglutinins are common to both plants (45) and fungi (5, 41) and because certain saccharides that react with host-specific lectins of plants are constituents of both bacterial and fungal cell walls. As such, lectins may function as recognition factors between host and parasite.

Lectins are a diverse group of proteins or glycoproteins which

agglutinate erythrocytes (9, 45). They are often referred to as
phytohemagglutinins when they are of plant origin. Besides plants
and micro-organisms lectins are present in invertebrate animals (15,
38). Extraction of lectins from source material and conditions
optimal for activity of lectins are variables that may account for
the apparent absence of lectins in some organisms. Seeds are the
best sources of lectins (8) but lectins also are present in smaller
amounts in roots, stems, and leaves (7, 45).

Lectins bind to mono- and oligo-saccharides specifically and
precipitate glycoproteins and polysaccharides; their reactions can
be visualized as precipitin bands in agar gel double-diffusion tests
(29). Binding of lectins to saccharides on surfaces of cells or
to cell membranes usually causes agglutination of cells (45). Moreover, the metalloprotein lectin from jack bean (*Canavalia ensiformis*)
named Concanavalin A (Con A) will precipitate several other lectins
from solution such as the glycoprotein lectins from wax beans (*Phaseolus vulgaris*), soybean (*Glycine max*), and lima bean (*Phaseolus
lunatus*), but not the lectin from wheat (*Triticum vulgare*). Most
lectins are proteins conjugated with a polysaccharide; however,
Con A and the wheat germ agglutinin (WGA) have no polysaccharide
components (26). The binding of lectins to receptor sites on cell
surfaces and to blood group substances is reversible and is competitively inhibited by the simple sugars involved in specific binding
sites in cell membranes or in purified blood group substances (6,
45).

In vertebrate animals a type of immunological tolerance may
occur because of the antigenic likeness of blood group substances
of the host with cell wall antigens or lectins of the parasite (14,
16). In vertebrates, immunoglobulins are necessary for specificity
in immunity. However, primitive animals such as invertebrates rely
on the activity of non-immunoglobulin molecules that function as
defense mechanisms against potential pathogens (15, 37). These non-
immunoglobulins include agglutinins.

An immune system similar to that in invertebrates that involves
agglutinin activity might be visualized in plants. However, evidence,
although limited in extent, suggests that agglutinins or lectins
in plants are more likely to have a selective influence on microbial
populations (7, 17), and thus facilitate rather than prevent the
establishment of parasites.

A possible role for lectins in host-parasite specificity was
found in a system involving soybeans and isolates of *Rhizobium* species (7). Soybean lectin isolated from seeds and roots was labelled
with fluorescein isothiocyanate, then tested for agglutinin activity with strains of *R. japonicum*, the soybean nodulating bacterium.
The agglutinin from soybean combined specifically with 22 out of 25

strains of *R. japonicum* tested. Twenty-three other strains of *Rhizobium* non-parasitic on soybeans, were not agglutinated by the soybean agglutinin. The authors hypothesized (7) that an inter-action between rhizobia and agglutinins of the soybean plants may account for the specificity among bacterial strains in the initiat-ion of nitrogen-fixing symbiosis.

COMMON ANTIGENS

The involvement of agglutinins or lectins as cross-reactive antigens in host-parasite interactions may be difficult to visualize but Dazzo and Hubbell (17) have detected common antigens on the surfaces of cells of *R. trifolii* and of clover roots. Moreover, antiserum to an extract of clover roots had a higher agglutinating titer with infective strains of *R. trifolii* than with non-infective strains. The antigen from infective strains of *R. trifolii* is con-tained in capsular material and is highly active in inducing root hair deformation or curling. They (17) have proposed a model in which preferential adsorption of infective versus non-infective cells of *R. trifolii* occurs on the surface of clover roots by a cross-bridging of their common surface antigens with a multivalent clover agglutinin. Whether or not the cross-reaction surface anti-gens as detected by antigen-antibody reactions are agglutinins and are directly cross-reactive with each other is unknown. However, a phytoprecipitin, specific for pathogenic cells of *Agrobacterium tumefaciens*, was isolated from bacteria-free sunflower (*Helianthus annuus*) crown gall tissue and in this case it was apparently diff-erent from the cross-reactive proteins shared by *A. tumefaciens* and sunflower (19).

Cross-reactive antigens also were found between eight legumes and three species of *Rhizobium* (12), whereas, in comparisons between *Rhizobium* species and eight non-leguminous plants in six genera, no cross-reactive antigens were detected. However, the presence of common antigens in the bacteria and legume plants was apparently not related to the host-specificity of the nodulating bacteria. Whether or not the cross-reactive antigens between *Rhizobium* species and legume hosts (12, 17) are similar to the agglutinins present in roots of host plants is still unclear.

Extracellular polysaccharides of five strains of *R. trifolii* contained the same proportions of glucose, galactose, glucuronic acid and slightly different amounts of acetyl and pyruvic acid (23); they were weakly cross-reactive but strongly reactive with homologous antisera in agar-gel double diffusion tests. Removal of pyruvate by mild acid treatment of the polysaccharide of *R. trifolii* strain TA1, abolished serological reactivity with homologous antiserum. Also, removal of acetyl from the polysaccharide by mild alkali treatment caused a loss of reactivity with homologous antiserum.

Pyruvate in polysaccharides of *R. trifolii*, *R. meliloti* and *R. radi-cicolum* functions as a major determinant in serological specificity; it also confers serological specificity to polysaccharides of other bacteria including *Xanthomonas* species (27, 36), *Corynebacterium insidiosum* (28), and a *Pseudomonas* species (28).

The various studies on *Rhizobium*-legume interactions (7, 12, 17) suggest the presence of at least two selective events: the first involves specific cellular receptor sites or recognition factors as indicated in both the soybean-*R. japonicum* system (7) and the clover-*R. trifolii* system (17); the second involves compatibility factors (common antigens) in legume hosts and parasitic bacteria which provide the basis for compatible and persistent parasitic relationships (12).

Common antigens between plants or animals and their parasites have been of special interest because of their coincidence in compatible host-parasite relationships (14, 16, 18, 20, 21, 40); their possible significance also has been indicated in cell-to-cell relationships in plant pollination (32). Proteins of pollen walls appear to have a major role as recognition factors in interspecific incompatibility interactions of pollen germ tubes and stigmatic tissue during pollination (32). Pollen grain germination and pollen tube growth in stigmatic tissue resembles certain plant-host parasite relationships where the ovary corresponds to the infection court for fungal pathogens. Pollen grains of the incompatible species of poplar *(Populus alba* and *P. deltoides)* have at least five saline-soluble cross-reactive antigenic proteins as detected by immunoelectrophoresis and agar-gel double-diffusion tests. The common proteins were located in the inner layer (intine) of the walls of pollen grains of both species. These proteins were released during pollen grain germination and, as shown by immunofluorescence, diffused on to the stigma surface where they were bound and affected the compatibility between the stigma and developing pollen tubes. Although the cross between *P. deltoides* and *P. alba* is usually incompatible, killed pollen of either species when mixed with viable pollen of the other, resulted in crosses that were always compatible and that resulted in successful hybridization in both directions. If viable pollen of *P. alba* was mixed with an equal volume of lyophilized proteinaceous material extracted from pollen of *P. deltoides* successful fertilization of *P. deltoides* also occurred (32). The exact physiological role of the pollen proteins is unknown.

The possible significance of common antigens between plant hosts and pathogenic micro-organisms in regard to disease susceptibility was recognized by Fedotova (24). Using antiserum of the pathogen and saline extracts of seeds she found that antibody titer and strength of the reaction in the equivalence zone in precipitin ring tests were directly related to the level of disease susceptibility of the host plant. Her conclusions were based on comparisons between selected varieties of crop plants such as beans, wheat, and

cotton differing in disease susceptibility, and bacterial and fungal
pathogens differing in virulence.

Sero-diagnosis based on seed protein also has been useful in
the selection of disease resistant materials in another plant breeding program. Seed proteins used in sero-diagnosis reflected the
relative susceptibility of watermelon (*Citrullus vulgaris*) to seedling blight caused by *Fusarium semitectum* (2). Serological eval-
uations were based on precipitin ring tests; antibody titer and
the rate of ring formation were related to disease susceptibility.
Resistant varieties also possessed a unique globulin fraction identified in immunoelectrophoresis.

The serological relatedness of cotton (*Gossypium hirsutum*) and
Xanthomonas malvacearum was reported by Fedotova (24) who used seed
extracts, and also by Schnathorst and DeVay (42) who used leaf extracts. In the latter study (42) compatible combinations of cotton
varieties and bacterial races had two to three additional minor
precipitin bands in agar gel diffusion tests and a greater intensity
of band development in contrast to serological reactions representing incompatible combinations of host and pathogen. However, single
gene differences in cotton which condition varietal susceptibility
to angular leaf spot (31) were not apparent in serological tests
involving homologous and heterologous reactions between preparations of plant antigens and antisera.

Cross-reactive antigenic substances are common among plants
(46) and also among fungi and bacteria (49). However, the close
serological relatedness of organisms of diverse phylogenetic origin
that can develop compatible host-parasite relationships is of special significance. This phenomenon supports the concept that tolerance of the parasite by the host increases with increasing antigenic
similarity, whereas incompatibility of host and parasite is characterized by an increasing disparity between antigenic determinants
(18, 20, 21).

If common antigens have an active role in host-parasite compatibility it would be especially evident during the infection phase
(11, 18). A basic cellular compatibility must exist between a host
and parasite in a persistent cell-to-cell relationship, otherwise,
during infection a range of events from rapid inhibition of growth
of the parasite to cellular disruption and death of the host and/
or parasite may occur. After infection and further growth of the
parasite, chemical changes of host cells and tissues often result
that are unfavourable to continued growth of potential parasites
or pathogens.

The possible role of a common antigen as a basic compatibility
or stabilizing factor was evident in a study involving cotton
(*G. hirsutum*) and the fungal pathogens *Verticillium albo-atrum*,

Fusarium oxysporum f. sp. *vasinfectum,* and *F. solani* f. sp. *phaseoli* which infect root and hypocotyl tissues of cotton; also included in the study were avirulent mutants of *F. oxysporum* f. sp. *vasinfectum* and two non-pathogens of cotton, *F. moniliforme* and *V. nigrescens* (11). Saline-soluble antigens and corresponding antisera for four varieties of cotton representing both disease-resistant and disease-susceptible varieties were compared for cross-reactivity with the *Fusarium* and *Verticillium* species in agar-gel double-diffusion and microprecipitin tests. In agar gel tests, a strong and distinct precipitin band was common in reactions between *F. oxysporum* f. sp. *vasinfectum,* *F. solani* f. sp. *phaseoli* and *V. alboatrum* but not with preparations of antigens from the non-pathogens, *F. moniliforme* and *V. nigrescens.* The antigen common to cotton, *V. albo-atrum,* *F. solani* f. sp. *phaseoli* and *F. oxysporum* f. sp. *vasinfectum* has been isolated and partially purified by disc-gel electrophoresis (11). These tests were the first to indicate that disease susceptibility or resistance were not always reflected in the sharing of major antigenic components by a plant host and parasite. In cotton both resistant and susceptible varieties shared the common antigen with both virulent and avirulent mutants of *F. oxysporum* f. sp. *vasinfectum.* These results were contrary to earlier views that common antigens have a direct relationship to disease susceptibility (20, 24). In certain cases this relationship appears to be true. However, in *Verticillium* and *Fusarium* wilt of cotton, the fungi colonize cortical tissues but resistance is not expressed by the plant until after its vascular system is invaded (10, 25); only those fungal isolates which became established in the vascular system caused disease. Thus, the concept relating antigen sharing to disease susceptibility which was proposed earlier (20) should be changed to focus primarily on the process of infection and establishment of the parasite or pathogen in the host without reference to the course of disease development. More clearly stated, the common antigen concept pertains to determinants which affect parasitism but not necessarily pathogenicity.* This conclusion also was reached by other workers (47) who found that a common antigen was shared by both avirulent and virulent isolates of *F. oxysporum* f. sp. *vasinfectum* with disease-resistant and susceptible lines of cotton but not with non-host plants.

Additional evidence which indicates that compatibility factors in plant host-parasite relationships are not necessarily reflected in disease development was found in a study of the antigenic affinity between the saline-soluble proteins of the soil-borne pathogen *Gauemannomyces (Ophiobolus) graminis* and wheat *(Triticum vulgare)* and oat *(Avena sativa)* roots (1). A single precipitin band was formed in agar gel diffusion tests when antisera of wheat and oat roots reacted with antigen preparations of both pathogenic and non-

---

* Or, as conceived by other authors ".... affect pathogenicity but not necessarily virulence." (R.K.S.W.)

pathogenic isolates of *G. graminis*. When compared with other fungi, the reactions observed between the wheat and oat antisera and antigens of isolates of *G. graminis* were specific for this host-pathogen relationship but apparently were not related to the pathogenicity of the fungi towards the two host species.

In contrast to the wilt diseases of cotton, and the associations of *G. graminis* with wheat and oats, a host-pathogen system where common antigens reflect disease development is the one involving sweet potato *(Ipomoea batatas)* and *Ceratocystis fimbriata*, the cause of black rot (20). The possible role of agglutinins in plant host specificity of *C. fimbriata* also has been reported (33).

Marked differences in common antigen relationships as related to disease development also have been found in other host-parasite systems where serological cross-reactions were compared between varieties of host plants and avirulent or virulent isolates of the pathogen. Examples are host-pathogen systems involving flax *(Linum usitatissimum)* and *Melampsora lini* (22), sunflower *(Helianthus annuus)* and *Agrobacterium tumefaciens* (19), and cotton and *Xanthomonas malvacearum* (20, 42).

Research on common antigens has been extended to serological relationships of fungi in the rhizosphere of wheat (30). Saline-soluble antigens from cell walls of rhizosphere and non-rhizosphere fungi after electrophoresis were cross-reacted with antisera of *G. graminis*. Out of 51 fungi isolated from the rhizosphere of wheat, antigen preparations from 27 isolates cross-reacted with whole cell antiserum of *G. graminis*, whereas only nine cross-reacted with cell wall antisera. Among the fungi with cross-reacting antigens with *G. graminis* were four species in the *Phycomycetes*, two species in the *Ascomycetes*, and 11 *Fungi Imperfecti*. Among 58 non-rhizosphere fungal isolates tested, 53 failed to cross-react with *G. graminis* antiserum. The results of this study (30) suggested that common antigenic relationships were possible determinants in the association of fungi in the rhizosphere of wheat roots. More recently, the antigenic affinity between the saline soluble proteins of *G. graminis* and wheat has been reported (1).

It has been hypothesized that the selective pressure of common antigen relationships for population changes in pathogens under natural conditions would favor those pathogens that were antigenically similar to the host (24). The selective effect of such an environment would be most influential on host-parasite systems in which the parasite is in intimate contact with host cells for significant periods to allow for growth or completion of its life cycle or reproductive phase. This hypothesis is supported by the studies on antigen-sharing by rhizosphere fungi of wheat (30), by the antigenic relationship which was found between *G. graminis* and wheat roots (1) and by the antigen shared by *F.solani* f. sp. *phaseoli*,

F. oxysporum f. sp. *vasinfectum*, *V. albo-atrum* and cotton (11).

Nematodes *(Meloidogyne incognita)* which cause the root knot disease in soybean and cotton roots also share a common antigen with these host plants (35). In agar-gel diffusion tests, antisera for eggs of *M. incognita* cross-reacted with preparations of antigens from root tips of three day old seedlings. Antisera for the plant roots also cross-reacted with antigens from larvae and eggs. In all cross-reactions at least one precipitin band was formed. In cross-reactions involving egg antisera and preparations of antigens from cotton, two precipitin bands were formed only one of which was shared with soybean. The common antigens were serologically similar since the precipitin bands joined completely in the gel diffusion tests.

The discussion so far has mainly concerned saline-soluble antigenic substances (post 105 000 x g supernatant fraction) isolated from plants and micro-organisms. In contrast, a protein shared by *Ustilago maydis* and maize *(Zea mays)* was associated with ribosomes (48). This relationship between a parasitic fungus and a host plant is unusual since in previous studies involving cytoplasmic ribosomes of spinach, bean, pea, tobacco and wheat, no cross-reactive antigens were found with those in ribosomes of microorganisms not parasitic on plants (49).

CONCLUSIONS

The occurrence of cross-reactive antigens in animals and plants with parasitic micro-organisms is widespread; this phenomenon has importance in the compatibility of animal hosts and parasites and perhaps in plant-parasite interactions. The role of common antigens in the parasitology of vertebrate animals appears to involve a type of immunological tolerance of the host that allows the establishment of the parasite. However, in invertebrates, agglutinins act as a defense against the establishment of parasites. Agglutinins or lectins in plants and micro-organisms may also have a role in host-parasite interaction but instead of acting as a defense mechanism in plants they may provide a mechanism for host-parasite specificity. An active physiological role for proteins shared by host and parasite in regard to host-parasite compatibility is still unknown.

REFERENCES

1. ABBOTT, L.K. (1973). Taxonomy and host specificity of *Ophiobolus graminis* Sacc.; an application of electrophoretic and serological techniques. Ph.D. Thesis, Monash Univ. Clayton, Victoria, Australia, 200 pp.

2.   ABD-EL-REHIM, M.A., IBRAHIM, I.A., MICHAIL, S.H. and FACEL,
       F.M. (1971). Serological and immunoelectrophoretical
       studies on resistant and susceptible watermelon varieties
       to *Fusarium semitectum* Berk.and Rav. *Phytopath. Z.*, 71,
       49 - 55.

3.   BAILEY, J.A. (1974). The relationship between symptom expres-
       sion and phytoalexin concentration in hypocotyls of
       Phaseolus vulgaris infected with *Colletotrichum lindemu-
       thianum*. *Physiol. Pl. Path.*, 4, 477 - 488.

4.   BERARD, D.F., KUĆ, J. and WILLIAMS, E.B. (1972). A cultivar-
       specific protection factor from incompatible interactions
       of green bean with *Colletotrichum lindemuthianum*. *Physiol.
       Pl. Path.*, 2, 123 - 127.

5.   BERNHEIMER, A.W. and FARKAS, M.E. (1953). Hemagglutinins
       among higher fungi. *J. Immun.*, 70, 197 - 198.

6.   BIRD, G.W.G. (1959). Hemagglutinins in seeds. *Br. med. Bull.*,
       15, 165 - 168.

7.   BOHLOOL, B.B. and SCHMIDT, E.L. (1974). Lectins : a possible
       basis for specificity in the Rhizobium-legume root nodule
       symbiosis. *Science, N.Y.*, 185, 269 - 271.

8.   BOYD, W.C. (1970). Lectins. *Ann. N.Y. Acad. Sci.*, 169, 168 -
       190.

9.   BOYD, W.C. and SHAPLEIGH, E. (1954). Specific precipitating
       activity of plant agglutinins (lectins). *Science, N.Y.*,
       119, 419.

10. BUGBEE, W.M. (1970). Vascular response of cotton to infection
       by *Fusarium oxysporum* f. sp. *vasinfectum*. *Phytopathology*,
       60, 121 - 123.

11. CHARUDATTAN, R. and DEVAY, J.E. (1972). Common antigens
       among varieties of *Gossypium hirsutum* and isolates of
       Fusarium and *Verticillium* species. *Phytopathology*, 62,
       230 - 234.

12. CHARUDATTAN, R. and HUBBELL, D.H. (1973). The presence and
       possible significance of cross-reactive antigens in
       Rhizobium-legume associations. *Antonie van Leeuwenhoek*,
       39, 619 - 627.

13. CHESTER, K.S. (1933). The problem of acquired physiological
       immunity in plants (continued). *Q. Rev. Biol.*, 8, 275 -
       324.

14. CLEGG, J.A., SMITHERS, S.R. and TERRY, R.J. (1971). Acquisition of human antigens by *Schistosoma mansoni* during cultivation in vitro. *Nature, Lond.*, 232, 653 - 654.

15. COOPER, E.L., LEMMI, C.A.E. and MOORE, T.C. (1974). Agglutinins and cellular immunity in earthworms. *Ann. N.Y. Acad. Sci.*, 234, 34 - 50.

16. DAMIAN, R.T. (1964). Molecular mimicry : antigen sharing by parasite and host and its consequences. *Am. Nat.*, 98, 129 - 149.

17. DAZZO, F. and HUBBELL, D. (1975). Presence and possible significance of cross-reactive antigens in the Rhizobium-clover symbiosis. *Proc. Am. Soc. Microbiol.* (in press).

18. DEVAY, J.E., CHARUDATTAN, R. and WIMALAJEEWA, D.L.S. (1972). Common antigenic determinants as a possible regulator of host-pathogen compatibility. *Am. Nat.* 106, 185 - 194.

19. DEVAY, J.E., ROMANI, R.J., MONADJEM, A.M. and ETZLER, M. (1970). Induction of phytoprecipitins in sunflower gall tissue in response to infection by *Agrobacterium tumefaciens*. *Phytopathology*, 60, 1289 (Abstr.).

20. DEVAY, J.E., SCHNATHORST, W.C. and FODA, M.S. (1967). *In : The dynamic role of molecular constituents in plant-parasite interaction*. (MIROCHA, C.J. and URITANI, I.,Eds.), 313 - 328. Bruce, Minneapolis, Minnesota, 372 pp.

21. DINEEN, J.K. (1963). Antigenic relationship between host and parasite. *Nature, Lond.* 197, 471 - 472.

22. DOUBLY, J.A., FLOR, H.H. and CLAGETT, C.O. (1960). Relation of antigens of *Melampsora lini* and *Linum usitatissimum* to resistance and susceptibility. *Science, N.Y.*, 131, 229.

23. DUDMAN, W.F. and HEIDELBERGER, M. (1969). Immunochemistry of newly found substituents of polysaccharides of *Rhizobium* species. *Nature, Lond.*, 164, 954 - 955.

24. FEDOTOVA, T.I. (1948). Significance of individual proteins of seed in the manifestation of the resistance of plants to diseases. *Trudÿ leningr. Inst. Zashch. Rast. Sborn.*, 1, 61 - 71.

25. GARBER, R.H. and HOUSTON, B.R. (1967). Nature of *Verticillium* wilt resistance in cotton. *Phytopathology*, 57, 885 - 888.

26.   GOLDSTEIN, I.J., REICHERT, C.M. and MISAKI, A. (1974). Inter-
      action of concanavalin A with model substrates. *Ann. N.Y.
      Acad. Sci.*, 234, 283 - 296.

27.   GORIN, P.A.J., ISHIKAWA, T., SPENCER, J.F.T. and SLONEKER,
      J.H. (1967). Configuration of the pyruvic acid ketals,
      4, 6-0-linked to D-glucose units, in *Xanthomonas campe-
      stris* polysaccharide. *Can. J. Chem.*, 45, 2005 - 2008.

28.   GORIN, P.A.J. and SPENCER, J.F.T. (1964). Isolation of 4,6-0-
      (1'-carboxyethylidene)-D-galactose from the exocellular
      polysaccharide of *Corynebacterium insidiosum*. *Can. J.
      Chem.*, 42, 1230 - 1232.

29.   GRUNDBACKER, F.J. (1973). Lectins in precipitin reactions
      with soluble H substance of human saliva and serum.
      Science, N.Y., 181, 461 - 463.

30.   HOLLAND, A.A. and CHOO, Y.S. (1970). Immuno-electrophoretic
      characteristics of *Ophiobolus graminis* Sacc. as an aid
      in classification and determination. *Antonie van
      Leeuwenhoek*, 36, 541 - 548.

31.   KNIGHT, R.L. (1956). Blackarm disease of cotton and its con-
      trol. *Proc. 2nd Int. Conf. Pl. Prot.*, Fernhurst Res.
      Stn, England. 53 - 59.

32.   KNOX, R.B., WILLING, R.R. and ASHFORD, A.E. (1972). Role of
      pollen-wall proteins as recognition substances in inter-
      specific incompatibility in poplars. *Nature, Lond.*, 237,
      381 - 383.

33.   KOJIMA, M. and URITANI, I. (1974). The possible involvement
      of a spore agglutinating factor(s) in various plants in
      establishing host specificity by various strains of black
      rot fungus,*Ceratocystis fimbriata*. *Pl. Cell Physiol.*,
      Tokyo, 15, 733 - 737.

34.   MATTA, A. (1971). Microbial penetration and immunization of
      uncongenial host plants. *A. Rev. Phytopath.*, 9, 387 -
      410.

35.   McCLURE, M.A., MISAGHI, I. and NIGH, E.L., JR. (1973). Shared
      antigens of parasitic nematodes and host plants. *Nature,
      Lond.*, 244, 306 - 307.

36.   ORENTAS, D.G., SLONEKER, J.H. and JEANES, A. (1963). Pyruvic
      acid content and constituent sugars of exocellular poly-
      saccharides from different species of the genus *Xanthomonas*.
      Can. J. Microbiol., 9, 427 - 430.

37. PAULEY, G.B. (1974). Physiochemical properties of the natural
     agglutinins of some mollusks and crustaceans. *Ann. N.Y.*
     Acad. Sci., 234, 145 - 160.

38. PEMBERTON, R.T. (1974). Anti-A and Anti-B of Gastropod origin.
     Ann. N.Y. Acad. Sci., 234, 95 - 121.

39. ROSS, A.F. (1961). Systemic acquired resistance induced by
     localized virus infections in plants. *Virology*, 14, 340 -
     358.

40. ROWLEY, D. and JENKIN, C.R. (1962). Antigenic cross-reaction
     between host and parasite as a possible cause of path-
     ogenicity. *Nature, Lond.*, 193, 151 - 154.

41. SAGE, H.J. and CONNETT, S.L. (1969). Studies on a hemagglut-
     inin from the meadow mushroom. II. Purification, compo-
     sition, and structure of *Agaricus campestris* hemagglutinin.
     J. biol. Chem., 244, 4713 - 4719.

42. SCHNATHORST, W.C. and DEVAY, J.E. (1963). Common antigens in
     Xanthomonas malvacearum and *Gossypium hirsutum* and their
     possible relationships to host specificity and disease
     resistance. *Phytopathology*, 53, 1142 (Abstr.).

43. SELA, I., HARPAZ, I. and BIRK, Y. (1966). Identification of
     the active component of an antiviral factor isolated from
     virus infected plants. *Virology*, 28, 71 - 78.

44. SEQUEIRA, L. and HILL, L.M. (1974). Induced resistance in
     tobacco leaves : the growth of *Pseudomonas solanacearum*
     in protected tissues. *Physiol. Pl. Path.*, 4, 447 - 455.

45. SHARON, N. and LIS, H. (1972). Lectins : cell-agglutinating
     and sugar-specific proteins. *Science, N.Y.*, 177, 949 -
     959.

46. VAN REGENMORTEL, M.H.V. (1963). Serologically related plant
     contaminants in preparations of partially purified plant
     viruses. *Virology*, 21, 657 - 658.

47. VENKATARAMAN, S., TAKSHMINARASIMHAM, C. and KALYANASUNDARAM,
     R. (1973). Antigenic determinants of host-pathogen
     specificity. *Abstracts of papers. 2nd Int. Congr. Pl.*
     Path., Abstr. No. 0958. Univ. of Minnesota, Minneapolis.

48. WIMALAJEEWA, D.L.S. and DEVAY, J.E. (1971). The occurrence
     and characterization of a common antigen relationship
     between *Ustilago maydis* and *Zea mays*. *Physiol. Pl. Path.*,
     1, 523 - 534.

49. WITTMANN, H.G., STOFFLER, G., KALTSCHMIDT, E., RUDLOFF, V.,
     JANDA, H.G., DZIONARA, M., DONNER, D., NIERHAUS, K.,
     CECH, M., HINDENNACH, I. and WITTMANN, B. (1970). Pro-
     teinchemical and serological studies on ribosomes of
     bacteria, yeast and plants. *FEBS Symposium,* 21, 33 - 46.

CONTRIBUTIONS

CALLOW, J.A., PALMERY, A. and PREECE, T.F.   The "common antigen"
     theory in relation to late blight of potatoes.

MANNERS, J.G.   The role of immunologically active proteins in re-
     actions to powdery mildews.

TANI, T., YAMAMOTO, H. and KADOTA, G.   Involvement of protein
     synthesis in resistant reaction of oat to *Puccinia coronata*
     var. *avenae.*

TOUZÉ, A., MAZAU, D., PLADYS, P., TOPPAN, A. and ESQUERRÉ-TUGAYÉ, T.
     Proteins in the interactions between *Colletotrichum lagenarium*
     and muskmelon.

SUMMARY OF POINTS FROM CONTRIBUTIONS AND DISCUSSIONS
BY
R. STAPLES

Chairman and Discussion Leader

     DeVay made the following points in response to questions from
or comments by Clarke, Yoder, Smith and Scheffer.

     Abbott (Monash University, Australia) has found that antigens
shared by wheat and oat roots and the isolates of *O. graminis* were
not related to the pathogenicity of the isolates to the two hosts.
So far as non-pathogens are concerned, in our earlier work we used
pathogenicity to compare antigenic relationships between microbial
pathogens and host plants. However, we soon found that there may be
common antigens between a pathogen and disease resistant or susceptible host plants. In the case of cotton, non-pathogenic strains
of *Fusarium oxysporum* f. sp. *vasinfectum* colonize the cortex of
roots and share the same common antigens with these tissues as those
shared by pathogenic isolates of this fungus with cotton roots. It
must be remembered that resistance to *Fusarium* wilt is not expressed
until the fungus invades the vascular system of disease resistant
varieties.

     Demonstrations of gene-for-gene relationships using serological
tests have been reported between antisera for certain races of

Melampsora lini and antigen preparations of nearly isogenic lines
of flax with single gene resistance to rust. Likewise, antigenic
relationships between *Xanthomonas malvacearum* and cotton *(Gossypium
hirsutum)* have been reported which reflect single gene differences
in cotton varieties for susceptibility to angular leaf blight. Although gene-for-gene relationships as a function of disease reaction
of host plants may be reflected in serological affinities between
host and pathogen, common antigen relationships have more to do with
basic compatibility of host and parasite. If a parasite finds a
favourable host environment, it may then grow enough to express a
pathogen potential. It now appears that a basic protein relationship exists between a plant and its various parasites, whether they
be bacteria, fungi or nematodes and, further, that this same protein relationship is common among the parasites but does not extend
to non-parasites. Thus, common antigens between host and parasites
can be considered as one of the basic conditions for disease susceptibility but resistance to disease can be due to a multitude of
other factors.

The common antigen relationship between *Ustilago maydis* and
maize involved a protein present in ribosomes of host and parasite.
It is not known whether or not this protein is involved in protein
synthesis in these organisms. However, a severe disruption in cellular metabolism in incompatible combinations of host and parasite
is probably reflected in an increased rate of movement of substances,
including proteins and other high molecular weight substances, in
and out of interacting cells.

After the discussion of DeVay's lecture, Callow described
briefly work by Callow, Palmerley and Preece on common antigens in
late blight of potatoes using race 4 of *Phytophthora infestans* and
tubers of potato demonstrating extremes of field resistance and
susceptibility (non-R gene cultivars). Antisera to total soluble
extracts of tubers and mycelium were cross-reacted by gel diffusion
in homologous and heterologous systems. In homologous systems several precipitin bands developed, but in heterologous systems involving fungal antigens and antisera against resistant or susceptible
tuber extracts, a weak common antigen relationship involving a
single precipitin band was detected. Reciprocal cross-reactions between fungal antisera and tuber antigens did not reveal common antigens. Fungal antiserum was cross-reacted against leaf antigens extracted from tomato, tobacco, maize and mung-beans. A single, strong
precipitin line was produced between fungal antiserum and tomato
antigens. A similar, but weaker precipitin line developed with tobacco antigens, but no lines developed with mung bean or maize antigens. Antigens of *Fusarium solani* f. sp. *coeruleum, Ustilago maydis,
Escherichia coli* and *Phytophthora cinnamomi* were also cross-reacted
with tuber antiserum. No precipitin lines developed except between
the resistant potatoes and *E. coli* and this showed partial identity

with one between *E. coli* antigens and *P. infestans* antiserum. Thus,
no convincing evidence was obtained for a role for common antigens
in resistance and susceptibility and the work emphasized the importance of investigating serological relationships between organisms
outside particular host-parasite combinations.

Manners summarized results of experiments on the role of immunologically active proteins in reactions to powdery mildews. Serological techniques, especially with fluorescent antibodies were
used for differentiating genotypes in *E. graminis tritici* race 30
and *E. graminis* race B 17 (Sultan) and other mildews. *E. graminis*
could be distinguished from other mildews, but in this species
formae speciales and races could not be separated. Antisera conjugated to fluoroscein isothiocyanate (FITC) and FITC conjugated
goat anti-rabbit sera were used to examine conidia of *E. graminis*
trapped over a season in a spore trap placed in a wheat field; these
conidia could be distinguished from those of other species. Techniques better than those now available will, however, be needed to
detect intraspecific differences.

Tani gave an account of work by Tani, Yamamoto and Kadota on
protein synthesis of the resistant reactions of oat plants to
Puccinia coronata avenae. In leaves of oat cultivar 'Shokan 1'
inoculated with incompatible race 226 incorporation of ^{14}C-leucine
into the acid insoluble fraction increased 1.3 x from 14 to 18 hours
after inoculation, compared to that in leaves inoculated with compatible race 203 and non-inoculated leaves. The increase was obscured
with leaves examined within 12 or more than 20 hours after inoculation. Treatment of leaves with 5 x 10^{-6}M blasticidin S (BcS), a
protein synthesis inhibitor, before inoculation with race 226 decreased incorporation of ^{14}C-leucine to one twentieth and mature
haustoria and uredosori were produced abundantly. Resistance was
decreased when BcS was applied within 12 hours after inoculation,
but there was less effect when treatment was delayed longer than
14 hours after inoculation and there was no effect when delayed for
24 hours. Cross-protection by race 226 against race 203 was nullified when leaves were treated with BcS before the first inoculation
with race 226; BcS treatment markedly decreased the increase of
antifungal activity in leaf exudate detectable from 20 hours after
inoculation of leaves with race 226. Direct treatment of the fungus
with BcS did not stimulate fungal development in leaves. These results suggest that induction of new synthesis of host proteins before
haustorial form is associated with the expression of the resistant
reaction.

Studies on the role of proteins in interactions between *Colletotrichum lagenarium* and muskmelon were described by Touzé on behalf of Mazau, Toppan, Pladys, Esquerré-Tugayé and Touzé. Proteins
in muskmelon seedlings infected by *C. lagenarium* differ from those

in uninfected seedlings. Some are of pathogen origin and are enzymes especially proteases, such that fungus mass and increase in
the proteolytic activities in infected tissue are closely related.
Other changes involve host proteins, mainly structural cell wall
glycoprotein (hydroxyproline-rich) and enzyme glycoproteins (cytoplasmic and cell wall bounded peroxidases). Host reactions following infection resemble physiological effects in uninfected plants
in which the hormonal balance has been modified.

A tentative scheme was proposed to explain how proteolytic
enzymes could cause the observed changes in glycoproteins of the
host and for significance of these host responses in specificity.

THE ROLE OF PHYTOTOXINS IN SPECIFICITY[*]

HARRY WHEELER

Department of Plant Pathology, University of Kentucky

Lexington, Kentucky, U.S.A.

SELECTIVE PATHOTOXINS AS SPECIFIC AGENTS

In his lecture, Brian indicated that phytotoxins which selectively damage plants susceptible to the pathogen involved provide the clearest examples of agents which determine specificity in plant-pathogen interactions. Five years ago, a preceding Institute was devoted entirely to the role of phytotoxins in plant disease. In the published proceedings of the Institute, Graniti (8) addressed problems of terminology and classification of toxins associated with plant diseases. Without retracing that ground, I should like to discuss briefly two terms, pathotoxin and host-specific toxin.

The term pathotoxin was introduced to identify toxins which play important causal roles in disease (29). It was intended to be a broad generic term for a toxic product of a pathogen, of a plant, or of a plant-pathogen interaction. The sole requirement for pathotoxicity was convincing evidence of an important causal role in pathogenesis. Although this was clearly spelled out, the term has been widely misused by restricting it to toxins which, like victorin, exhibit selective toxicity and reproduce all symptoms of a disease.

In Table 1, pathotoxins are separated into two main classes on the basis of origin with the organism producing the toxin indicated parenthetically. Sub-classes are based on whether or not the pathotoxin acts selectively when applied to plants susceptible and resistant to the disease involved. These examples were selected primarily on the strength of evidence for a causal role in pathogenesis

[*] Kentucky Agricultural Experiment Station Journal Series no. 75-11-115. Supported in part by Grant 216-15-21 from CSRS, USDA.

Table 1.   *Classification of pathotoxins*[a]

Class I. Pathogen produced

      Subclass A. Selective

          Examples : Victorin *(Helminthosporium victoriae)*,
                    T-toxin *(H. maydis* race T)*,* HC-toxin
                    (H. carbonum), HS-toxin*(H. sacchari)*, PC-
                    toxin *(Periconia circinata)*, PM-toxin
                    (Phyllosticta maydis), phytoalternarins
                    (Alternaria spp.)

      Subclass B. Non-selective

          Examples : Tabtoxin *(Pseudomonas tabaci)*, syringomycin
                    (P. syringae), phaseotoxin *(P. phaseolicola)*,
                    tentoxin*(Alternaria tenuis)*, fusicoccin
                    (Fusicoccum amygdali), fumaric acid
                    (Rhizopus spp.)*,* marticin *(Fusarium* spp.)

Class II. Plant or plant-pathogen produced

      Subclass A. Selective

          Example : Amylovorin *(Erwinia amylovora* on *Rosaceae)*

      Subclass B. Non-selective

          Example : Juglone*(Juglans nigra)*

---

[a] Substances for which there is evidence for a causal role in plant
disease. More information on these substances can be found in
recent reviews (11, 18, 23, 27).

but secondarily to illustrate the system of classification which is
similar to that proposed by Graniti (8). Table 1 is not intended to
be a definitive list. The pathotoxin status of some of these substances must be considered tentative, and many more examples could
be added, especially if as some prefer, growth regulators and enzymes are included in the toxin category. Non-selective pathotoxins
listed in Table 1 will not be discussed since there is little evidence that these play a role in disease specificity. I agree,

however, with Patil (13) who has argued persuasively that non-selective toxins play important causal roles in disease.

The term host-specific toxin was introduced by Pringle and Scheffer (14) for agents classified as pathogen-produced, selective pathotoxins in Table 2. Since toxins do not have hosts, the term is a literal misnomer. More important the word specific implies that the toxin has no effect on plants resistant to the pathogen or on non-hosts. Actually, so called host-specific toxins are those which at certain concentrations damage only susceptible plants but at higher concentrations also damage those that are resistant as the data in Table 2 show. These toxins are comparable to chemicals which, properly applied, kill certain plants without injuring others. Such chemicals are called selective herbicides. Toxins with similar properties can be properly termed selective toxins or selective pathotoxins if good evidence for a causal role in disease is available.

Victorin is, by far, the most potent and selective pathotoxin known. During the past 25 years, results from a number of laboratories have established victorin as the agent which enables *Helminthosporium victoriae* to attack specifically oat cultivars derived from the variety 'Victoria' (11). Victorin, substituted for *H. victoriae*, has made it possible to follow metabolic changes in diseased plants without complications introduced by metabolic activities of a living pathogen. The earliest effect detected in victorin-treated, susceptible tissues was a marked change in permeability. This led to the hypothesis that disruption of cell permeability may be an initial event which triggers subsequent pathological changes in many diseased plants (28). The implications of this hypothesis in relation to disease specificity will be discussed later.

Of the other pathotoxins listed in Table 2, three are produced by foliar pathogens of maize. These are T-toxin produced by *Helminthosporium maydis*, race T, PM-toxin, a less active substance produced by *Phyllosticta maydis*, and HC-toxin produced by *Helminthosporium carbonum*, with still lower activity. T-toxin and PM-toxin produce similar effects and the two toxins may be the same chemically (4, 29). If so, this will be an interesting case of two unrelated pathogens evolving the same chemical mechanism for attacking their host. The PC-toxin is produced by *Periconia circinata*, a pathogen of sorghum. The effects of this toxin are similar to those of victorin but the PC-toxin is much less active (18).

Amylovorin (Table 1) is of interest since it is the first selective toxin thought to be produced by a plant-pathogen interaction. Goodman *et al.* (7) have reported that the slime or "ooze" exuded from apple slices inoculated with the fire-blight bacterium,

Table 2. *Potency and differential toxicity of some pathogen-
produced selective pathotoxins*

| Pathotoxin | Plant | Dilution end-point on plants[a] | | Selectivity S/R |
		Susceptible(S)	Resistant(R)	
Victorin	Oat	1:10 000 000	1:25	400 000
T-toxin	Maize	1:5 000	1:200	25
PM-toxin	Maize	1:1 000	1:25	40
HC-toxin	Maize	1: 200	<1:10	> 20
PC-toxin	Sorghum	1:3 000	<1:10	>300

[a] Highest dilutions of culture filtrates or equivalent volumes
of partly refined materials toxic to plants susceptible (S) and
resistant (R) to the pathogens involved (see Table 1).

Erwinia amylovora, is selectively toxic to plants susceptible to
fire-blight. In their hands, the toxin is not produced by the
bacterium grown *in vitro*. It appears to be primarily a polymer of
galactose with a molecular weight of about 165 000. It should be
noted that the selective toxicity of amylovorin is based solely on
wilting which cannot be considered a distinctive symptom of fireblight. Pending further evidence, the pathotoxin status of amylovorin is tentative.

During the past five years, attention has centered on selective
pathotoxins produced by *H. maydis* and by *H. sacchari*. Only these
two will be discussed in some detail. Additional information on
others can be found in recent reviews cited in the footnote to
Table 1.

T-Toxin and Southern Corn Leaf Blight

Although a pathotoxin produced by *H. maydis* had been reported
earlier (19), the disastrous outbreak in the U.S.A. of southern
corn leaf blight in 1970 made this pathogen a subject of intense
interest. The disease was quickly brought under control by elim-
inating the use of the cytoplasmic Texas male sterility factor in
the production of hybrid seed corn, but the epidemic served to spur

legislative action which provided funds to about a dozen laboratories for research on this disease. A major part of this concen-
trated research effort has been devoted to the nature and mode of
action of T-toxin.

Some of the responses of susceptible maize tissues to T-toxin
and the dilution of toxin required to elicit the response are listed in Table 3. Loss of chlorophyll, loss of respiratory control
in mitochondria, and inhibition of dark CO_2 fixation are much more
sensitive responses than inhibition of root growth or photosynthesis and these in turn are more sensitive than loss of electrolytes
or transpiration inhibition. The formation of a well-defined lesion is the least sensitive response - it requires a 10-fold concentration rather than a dilution of the toxin.

Initial investigations of *H. maydis* T-toxin, led by a group
at the University of Illinois, have been reviewed by Hooker (10).
They reported that T-toxin caused rapid swelling and loss of respiratory control when added in low concentrations to mitochondria
isolated from susceptible plants. At the same concentrations, T-
toxin had no effect on mitochondria from resistant plants. These

Table 3. *Sensitivity of various responses to* H. maydis *T-toxin*

Response	Toxin dilution[a]
Loss of chlorophyll	1:50 000
Mitochondrial swelling and uncoupling	1:10 000
Dark CO_2 fixation inhibition	1:10 000
Root growth inhibition	1: 3 000
Photosynthesis inhibition	1: 1 200
Loss of electrolytes	1: 50
Transpiration inhibition	1: 50
Lesion formation	10: 1

[a] Highest dilution of culture filtrates or equivalent volumes
of partly refined materials which will induce the response.
Based on data from (2) and from the author's laboratory.

results have been confirmed by several other investigators, most
recently, by Scheffer and by Yoder at this Institute. Work in my
laboratory indicates that mitochondria from resistant plants are
approximately 10 000 times less sensitive to T-toxin than those from
susceptible plants. These selective effects led to the hypothesis
that mitochondria were the site of action of T-toxin (10).

Further work recently reviewed (23) failed to provide evidence
of early effects on mitochondria *in situ* or on respiratory activity when intact tissues were treated. Instead, T-toxin was found
to cause rapid and selective changes in K^+ fluxes, in transpiration,
and in cellular electrochemical potentials. These results were
interpreted as evidence for an initial effect of T-toxin on cell
permeability (1). A report of selective inhibition by T-toxin of
a membrane-bound ATPase in microsomal preparations provided support
for this idea (26). Bhullar *et al.* (2) have pointed out that these
rapid changes were obtained with very high toxin concentrations.
They pose the question of why an initial effect should require a
much higher toxin concentration than a dependent secondary disturbance. As the data in Table 3 show, changes in transpiration and
in losses of electrolytes which should reflect permeability changes
are much less sensitive responses than are effects on chlorophyll
or dark CO_2 fixation which are thought to be secondary events. The
issue has been further clouded by Scheffer's report at this Institute that he has been unable to confirm claims that T-toxin inhibits
ATPase activity in microsomes.

The hypothesis that mitochondria are the site of action of T-
toxin is attractive; first, because the response is highly sensitive (Table 3) and, second, because an effect on a cytoplasmic
organelle would account for the specificity and cytoplasmic inheritance of disease reactions (10). On the basis of this hypothesis,
an early effect on respiration would be expected, but such an effect has not been found. One laboratory reported no significant
effects on respiration in tissues exposed to toxin for three hours
(1). In my laboratory, Carroll Rawn obtained significant increases
in respiration with susceptible but not resistant tissues but only
after several hours of exposure to a T-toxin preparation that inhibited root growth by 90%. Although these results do not rule out
mitochondria as the site of action of T-toxin, data currently available are too inconsistent to justify conclusions about the mode of
action of this toxin.

Several compounds, thought to be terpenoid in nature, have
been isolated from filtrates from cultures of *H. maydis* and are reported to exhibit selective toxicity in leaf lesion tests which were
used as bioassays (23). These materials, which did not exhibit
selective toxicity in root growth tests, were applied in leaf bioassays at a very high concentration, approximately 770 µg/ml. This
is in sharp contrast to a report that a partially refined toxin

preparation inhibited dark CO_2 fixation by 50% at a concentration
of 0.02 µg/ml (2). These results suggest that a highly active
material was degraded into several less active compounds during
isolation, or that the pathogen produces several toxic materials
which act synergistically.

In summary, despite extensive investigations in many laboratories the chemistry of T-toxin and its role in disease specificity
remain undetermined.

HS-Toxin and Eye Spot of Sugarcane

This toxin, produced by *H. sacchari* which causes eye spot disease of sugarcane, was studied first by Steiner and Byther in Hawaii
where the disease is of some importance. They reported that fractions of filtrates from cultures of *H. sacchari* applied to susceptible leaves produced reddish-brown streaks, called runners, which
resembled the symptoms seen in advanced stages of disease caused by
the fungus. Using a bioassay for the toxin based on the length of
the red streaks or runners, these workers reported that in tests of
182 clones of sugarcane, susceptibility to *H. sacchari* was significantly correlated with sensitivity to the toxin (20). Although
evidence that HS-toxin plays a causal role in disease would be
greatly strengthened if toxin production had been shown to be correlated with pathogenicity, Steiner and Byther's results justify
the conclusion that a selective pathotoxin is involved in eye spot
disease of sugarcane.

Steiner and Byther's work was carried out with a partially refined toxin preparation. They used a mixture of active materials
eluted under two broad peaks from a Sephadex column. In further
work by Strobel and his associates (23), only material eluted under
the first peak was used. They have reported isolation of a pure
toxin which has been named helminthosporoside (proposed structure :
2-hydroxycyclopropyl-α-D-galactopyranoside). They have also reported that helminthosporoside binds specifically to a single protein present in susceptible but not in resistant clones of sugarcane. On the basis of a number of lines of indirect evidence,
Strobel (24) has postulated that this binding protein occupies a
specific site on the plasma membrane. When this protein binds to
helminthosporoside, a membrane bound ATPase is activated. This is
thought to disrupt the ionic balance of the cell and lead to water
soaking and the development of discoloured streaks or runners.

These results from Strobel's laboratory are very exciting. If
verified, they will provide, for the first time at the molecular
level, a specific mechanism for virulence in a pathogen and an equally specific mechanism for susceptibility in a plant. Because of
its great potential significance, Strobel's work merits careful and

critical review. Such a review has revealed points which require correction and inconsistencies which must be resolved before the conclusions which have been drawn can be accepted as valid. These have been discussed with Dr. Strobel in private and he is fully aware of the reservations about his work which will be expressed here.

In most cases, Strobel has reported results of single experiments without evidence of the range of variation or estimates of reliability. This is of special concern in regard to the binding of labeled helminthosporoside *in vivo* where recovery of a small number of disintegrations per minute served as evidence of binding activity (22). Another difficulty has been pointed out by Scheffer who reported at this Institute that apparent binding, similar to that observed by Strobel, occurred when control preparations containing only toxin and detergent were used.

When experiments with helminthosporoside, especially those involving leaf bioassays, were repeated, results were often highly variable. The variation, covering most of the reaction range from resistant to susceptible, which occurred in seven repeated bioassays has been discussed by others (2). In transferring bioassay data to a semilog scale it seems that two critical points were misplotted (21). Use of the slope of the line obtained resulted in over estimation of toxin activity in later experiments. A similar error seems to have been made in a Scatchard plot of toxin binding activity (22).

A further question arises from results obtained when 1 µl of ^{14}C-labeled helminthosporoside (1 600 disintegrations per minute) was applied to a small wound in the basal portion of a sugarcane leaf. Twelve hours later, approximately 95% of the label was recovered from a 5 x 15 mm area located 5 mm above the point of application (25). By this time a watersoaked lesion had developed which extended to the tip of the leaf, a distance of more than 10cm. These results indicate that helminthosporoside is highly immobile and suggest that some other agent may be responsible for lesion development. This possibility is enhanced by evidence of an unknown impurity in the nuclear magnetic resonance spectrum of the sample of helminthosporoside used for structural determinations (21).

Hopefully, further work will eliminate these reservations. If confirmed, results with helminthosporoside will provide the first direct support for the long standing hypothesis that specificity in diseases in which selective pathotoxins are involved depends on specific receptors present only in susceptible plants (14).

MYCOVIRUSES AND TOXIN PRODUCTION

The possibility that viruses or virus-like particles which have been found in a number of fungi might affect toxin production and hence play an indirect role in disease specificity has been explored with several pathogens that produce pathotoxins. In a preliminary study, virus-like particles were found in six virulent isolates of *H. maydis* race T, whereas none were found in three weekly pathogenic isolates of this pathogen (3). I have heard, however, that further work by the same investigators has failed to confirm the existence of a correlation between virulence and the presence of virus-like particles in other isolates of *H. maydis*. If so, the situation is similar to that in *P. circinata* where no correlation between toxin production and the presence or absence of virus-like particles could be demonstrated (5).

Virus-like particles (VLP) containing double-stranded RNA (ds-RNA) have been reported in four non-toxin-producing strains of *H. victoriae* (17). Two of these strains were normal in appearance and could not be distinguished on cultural characters from the high toxin-producing isolate from which they were derived. The other two were highly abnormal in cultural characters. One species of VLP was found in all four strains, and a second was present in the two abnormal strains. Neither virus-like particles nor ds-RNA could be found in the high toxin-producing strain of *H. victoriae*. Although these results are too preliminary for firm conclusions, they suggest that infection by virus-like agents may play a role in the loss of toxin producing ability which often occurs when pathogens which produce toxins are maintained on artificial media.

DISRUPTED PERMEABILITY AND DISEASE SPECIFICITY

The dominance of membrane theory in the Western Hemisphere led many to equate early changes in permeability with effects on the plasmalemma. Despite extensive searches, especially with the victorin-oat model, direct evidence of early effects on the plasmalemma has not been forthcoming. Time-course studies of responses to victorin revealed that changes in permeability were virtually instantaneous whereas increases in respiration and other changes clearly linked to metabolism occurred only after a lag of about 30 minutes. These results were thought to indicate an initial disruption of the plasmalemma followed after about 30 minutes by disruption of the tonoplast. Salts, phenolics or other materials leaking from the vacuole would account for the rise in respiration and subsequent pathological events (28). Although this hypothesis has not been ruled out, recent work suggests that the initial effect of victorin is not disruption of protoplast permeability.

In short term tests, victorin has been reported to cause a
rapid but transient increase in cell elongation of susceptible oat
coleoptiles (16). In contrast to auxin-induced elongation which
requires an induction period of 15 - 20 minutes, the response to
victorin began after five minutes and persisted for about 30 minutes.
These results indicate a rapid effect of victorin on cell walls.
Losses of electrolytes from victorin-treated coleoptiles were also
followed (16). An initial rapid loss occurred during the first
five minutes. This was followed for the next 20 - 30 minutes by
fluctuations which gave a slow average rate of loss, then a new
rapid rate began and continued until the test was terminated after
60 minutes. A similar pattern of changes was observed by Novacky
and Hanchey (12) who measured changes in electrochemical potentials
in victorin-treated cells. A sharp depolarization occurred during
the first five minutes of exposure. During the next 20 - 30 minutes there was an actual recovery of potential before a second pro-
longed depolarization took place.

The pattern of these short term effects, and especially the
recovery in electrochemical potential after the first five minutes
which has been confirmed, is not that expected if the initial effect
of victorin caused disruption of the plasmalemma. Instead, these
results suggest an initial effect on the cell wall or some other
readily accessible surface component and disruption of protoplast
permeability only after a lag of about 30 minutes. This 30 minute
lag would allow ample time for victorin to enter and to act within
the protoplast.

Further evidence of victorin-induced changes in cell walls
has come from a study of the incorporation of ^{14}C-labeled glucose
into various cell fractions of susceptible oat leaves treated with
victorin. The results, summarized in Table 4 show no effect of
victorin on materials soluble in ethanol or water. However, the
final three fractions which represent cell wall components all show
about twice as much label in victorin-treated as in control tissues.
Since victorin-treated tissues were respiring twice as fast as controls, these data indicate that synthesis of cell wall components,
but not other cellular materials, is geared to the victorin-induced
rise in respiratory metabolism. In other words, the victorin-treated
cell seems to have received a signal that its walls are under attack.
In response, it mobilizes all of its metabolic resources in an attempt to repair the damage.

The hypothesis that a signal originating in the cell wall triggers the physiological changes observed in diseased plants raises
several questions. What kind of a signal could be generated by such
diverse agents as pathotoxins, enzymes, viruses, and various pathogenic organisms ? How is the signal sent ? How is it received ?
Why does it require 30 minutes for a metabolic response ? One possible answer involves fixed negative charges at cell surfaces which

Table 4.   *Effect of victorin on incorporation of ^{14}C-labeled glucose into various fractions of susceptible cells*

Fraction soluble in	DPM/g fresh wt. [a]	
	Control	Victorin
Ethanol (hot)	22 000	20 000 NS
Water	1 200	1 300 NS
Dilute acid	700	1 400 **
Alkali	1 400	5 700 **
Concentrated acid	3 400	8 600 **

[a] Based on triplicate samples in three experiments by Rawn (15).
NS - not significantly different from control value.
** - difference from control highly significant (P < 0.01).

appear to be essential for the maintenance of normal cell permeability. There is considerable evidence that these fixed charges consist of anionic polysaccharides or polysaccharide-protein complexes secreted by the Golgi apparatus. Since victorin is highly cationic, it would be expected to neutralize surface anions and, as a result, discharge the cell surface. This discharge could provide the signal which triggers increased secretion of new anionic polymers by the Golgi apparatus. Such secretory activity would require large amounts of energy and, by relieving normal restraints, produce the observed increase in respiratory metabolism. The time lag of 30 minutes for the metabolic response might be expected since, under normal conditions, most plant cells do not exhibit secretory activity. The synthesis of secretory materials, formation of vesicles, and vesicular migration would require some time to be manifest.

Cell wall modifications and increased secretory activity were the earliest ultrastructural changes found in victorin-treated tissues (9). Similar changes have been observed in plants attacked by a wide variety of pathogenic agents (27). At this Institute, Hanchey has presented histochemical evidence of rapid and drastic increases in extracellular polysaccharides in victorin-treated tissues. Such materials may provide the fixed surface charges required for the maintenance of normal cell permeability.

In conclusion, I should like to comment briefly on the genetic implications of work with pathotoxins. In genetic terms, pathotoxins can be viewed as products of genes for virulence which are essential for pathogenicity. In the absence of such gene products, all plants are passively resistant. Such gene-for-gene interactions are exactly the opposite of those postulated by Flor (6) in which resistance must be active and can be expressed only through the interaction of a gene for avirulence in the pathogen with one for resistance in the plant. The inability of Flor's gene-for-gene hypothesis to account for genetic interactions involving pathogens which employ toxins, enzymes, or growth regulators as mechanisms for attacking plants has been discussed in detail elsewhere (27).

REFERENCES

1.   ARNTZEN, C.J., KOEPPE, D.E., MILLER, R.J. and PEVERLY, J.H. (1973). The effect of pathotoxin from *Helminthosporium maydis* (race T) on energy linked processes of corn seedlings. *Physiol. Pl. Path.*, 3, 79 - 90.

2.   BHULLAR, B.S., DALY, J.M. and REHFELD, D.W. (1975). Inhibition of dark CO_2 fixation and photosynthesis in leaf discs of corn susceptible to the host-specific toxin produced by *Helminthosporium maydis*, race T. *Pl. Physiol., Lancaster*, 56, 1 - 7.

3.   BOZARTH, R.F., WOOD, H.A. and NELSON, R.R. (1972). Virus-like particles in virulent strains of *Helminthosporium maydis*. *Phytopathology*, 62, 748 (Abstr.).

4.   COMSTOCK, J.C., MARTINSON, C.A. and GENGENBACH, B.G. (1973). Host specificity of a toxin from *Phyllosticta maydis* for Texas cytoplasmically male-sterile maize. *Phytopathology*, 63, 1357 - 1361.

5.   DUNKLE, L.D. (1974). Double-stranded RNA mycovirus in *Periconia circinata*. *Physiol. Pl. Path.*, 4, 107 - 116.

6.   FLOR, H.H. (1971). Current status of the gene-for-gene concept. *A. Rev. Phytopath.*, 9, 275 - 296.

7.   GOODMAN, R.N., HUANG, J.S. and HUANG, PI-YU. (1974). Host-specific phytotoxic polysaccharide from apple tissue infected by *Erwinia amylovora*. *Science, N.Y.*, 183, 1081 - 1082.

8.   GRANITI, A. (1972). The evolution of the toxin concept in plant pathology. *In : Phytotoxins in Plant Diseases* (WOOD, R.K.S., BALLIO, A. and GRANITI, A., Eds.), 1 - 18. Academic Press, London and New York.

9.   HANCHEY, P., WHEELER, H. and LUKE, H.H. (1968). Pathological
        changes in ultrastructure : effects of victorin on oat
        roots. *Am. J. Bot.*, 55, 53 - 61.

10. HOOKER, A.L. (1972). Southern leaf blight of corn - present
        status and future prospects. *J.envir. Quality 1*, 244 - 249.

11. LUKE, H.H. and GRACEN, V.E., Jr. (1972). *Helminthosporium*
        toxins. *In : Microbial Toxins, Vol. 8, Fungal Toxins*
        (KADIS, S., CIEGLER, A. and AJL, S.J., Eds.), 139 - 168.
        Academic Press, New York and London.

12. NOVACKY, A. and HANCHEY, P. (1973). Depolarization of mem-
        brane potentials in oat roots treated with victorin.
        Physiol. Pl. Path., 4, 161 - 165.

13. PATIL, S.S. (1974). Toxins produced by phytopathogenic bact-
        eria. *A. Rev. Phytopath.*, 12, 259 - 279.

14. PRINGLE, R.B. and SCHEFFER, R.P. (1964). Host-specific plant
        toxins. *A. Rev. Phytopath.*, 2, 133 - 156.

15. RAWN, C.D. (1974). Victorin-induced changes in carbohydrate
        metabolism in oat leaves. Ph.D. dissertation, University
        of Kentucky Library, Lexington.

16. SAFTNER, R.A. and EVANS, M.L. (1974). Selective effects of
        victorin on growth and the auxin response in *Avena*. *Pl.*
        Physiol., Lancaster, 55, 382 - 387.

17. SANDERLIN, R.S. and GHABRIAL, S.A. (1975). Virus-like part-
        icles containing double-stranded RNA in normal and dis-
        eased *Helminthosporium victoriae*. *Phytopathology*. (In
        press).

18. SCHEFFER, R.P. and YODER, O.C. (1972). Host-specific toxins
        and selective toxicity. *In : Phytotoxins in Plant*
        Diseases (WOOD, R.K.S., BALLIO, A. and GRANITI, A., Eds.),
        251 - 272. Academic Press, London and New York.

19. SMEDEGÅRD-PETERSON, V. and NELSON, R.R. (1969). The product-
        ion of a host-specific pathotoxin by *Cochliobolus hetero-*
        strophus. *Can. J. Bot.*, 47, 951 - 957.

20. STEINER, G.W. and BYTHER, R.S. (1971). Partial characterizat-
        ion and use of a host-specific toxin from *Helminthosporium*
        sacchari on sugarcane. *Phytopathology*, 61, 691 - 695.

21. STEINER, G.W. and STROBEL, G.A. (1971). Helminthosporoside,
        a host-specific toxin from *Helminthosporium sacchari*.
        J. biol. Chem., 246, 4350 - 4357.

22.   STROBEL, G.A. (1973).   The helminthosporoside-binding protein
      of sugarcane. *J. biol. Chem.*, 284, 1321 - 1328.

23.   STROBEL, G.A. (1974). Phytotoxins produced by plant parasites.
      A. Rev. Pl. Physiol., 25, 541 - 566.

24.   STROBEL, G.A. (1975).   A mechanism of disease resistance in
      plants. *Scient. Am.*, 232, 81 - 88.

25.   STROBEL, G.A., HESS, W.M. and STEINER, G.W. (1972).   Ultra-
      structure of cells in toxin-treated and *Helminthosporium
      sacchari*-infected sugarcane leaves. *Phytopathology*,
      62, 339 - 345.

26.   TIPTON, C.L., MONDAL, M.H. and UHLIG, J. (1973).   Inhibition
      of the K^+ stimulated ATPase of maize root microsomes
      by *Helminthosporium maydis* race T pathotoxin. *Biochem.
      biophys. Res. Commun.*, 51, 525 - 528.

27.   WHEELER, H. (1975). *Plant Pathogenesis*. Springer-Verlag,
      Heidelberg, New York, 106 pp.

28.   WHEELER, H. and LUKE, H.H. (1963).   Microbial toxins in plant
      disease. *A. Rev. Microbiol.*, 17, 223 - 242.

29.   YODER, O.C. (1973).   A selective toxin produced by *Phyllosticta
      maydis*. *Phytopathology*, 63, 1361 - 1366.

CONTRIBUTIONS

ALBERSHEIM, P.   A comment on the work of G. Strobel.

DALY, J.M.   Remarks on P. Albersheim's comment.

DANIELS, M.J.   Chemistry and mode of action of a phytotoxin pro-
    duced by the plant-pathogenic mycoplasma, *Spiroplasma citri.*

PATIL, S.S.   The mechanism of suppression of resistance in bean
    plants to *Pseudomonas phaseolicola* by phaseotoxin.

SUMMARY OF POINTS FROM CONTRIBUTIONS AND DISCUSSIONS
BY
S. S. PATIL

Chairman and Discussion Leader

    Albersheim opened the discussion by commenting that most re-
search workers make a mistake one time or another during their

careers and that Dr. Strobel is no exception. However, any mistakes
which may have been made by Strobel in reporting experimental protocol
or data and in interpreting data are minor. Albersheim believed
Strobel's basic claim that clones of sugar cane susceptible to *Helminthosporium sacchari*, but not those which are resistant, possess
a helminthosporoside binding protein and claimed that Strobel has
proved the existence of such a protein in his equilibrium dialysis
and affinity chromatography experiments which showed binding of the
toxin to the sugar cane protein.

Daly[a] then commented on certain aspects of work described and
published by Strobel as follows:

I (that is, Daly) appreciate Dr. Albersheim's comments and his
warmth toward a colleague and friend Dr. Strobel. Criticism never
is pleasant, either for a critic or for an author. It was not easy
for me to add to them. It should not be forgotten that critics have
just as much at stake, in terms of credibility and professional
reputation, as does any recipient of criticism. As Dr. Wheeler
stated, if Dr. Strobel's evidence is confirmed his findings are a
landmark in studies of host-parasite relations.

It is, however, important to underscore the fact that no con-
ceptual issues are at stake. The idea of host-specific toxins which
may bind to specific sites on membranes has been implicit in the
papers of Wheeler, Scheffer and others for at least ten years. The
issue that has been raised concerns the evidence for a *specific*
chemical structure and for its binding to a *specific* protein obtained
from susceptible tissue. Until there is additional confirmation,
most convincingly by other investigators, it is an intrinsic part
of the scientific endeavor to examine the available evidence in
order to ensure that the protocol was adequate and that the evidences
developed are consistent internally. If there are uncertainties,
there is a debt to science to voice them as Dr. Wheeler has done.
The onus for scientific proof never rests on the questioner.   It
is up to the investigator to fulfil the necessary experimental demands whenever possible. Undoubtedly, there are situations where
time and/or technology prevent this but only the lack of technology
is a valid excuse.

Independently of Dr. Wheeler, I have had the same reservations
about the bioassay, a crucial part of any isolation of biologically
active material but especially of host-specific toxins. A paper
by Bhullar, Daly and Rehfeld will appear in Plant Physiology (1975,
July issue). The same thoughts that Dr. Wheeler expressed are part
of the discussion. The criticism will then be part of the open
scientific literature and opinions expressed at this conference
cannot be thought improper or out of place because Dr. Strobel is
not present. It is surprising to me that, as far as I know, less
than a handful of readers up to now have detected the quantitative

[a] Account of Daly's comments prepared by Professor J.M. Daly.

inconsistency in the bioassay method, among others, employed in the work on *H. sacchari* toxin. I am not sure what the message should be at either the reviewer or the readership level.

Dr. Wheeler pointed also to another uncertainty that I found which bears on the binding protein. It has been stated (Strobel, G.A. *J. Biol. Chem.*, 248, 1321 - 1328, 1973)) that, for the binding of toxin to protein, dissociation constants of 6.8 and 1.19×10^{-5}M were obtained by a Lineweaver-Burk plot (not shown) and by a Scatchard plot (Fig. 9). Both plots were derived from data apparently presented first in Fig. 8 in a different form. Because these values are not particularly indicative of a strong binding affinity which seems necessary for a compound active in small amounts, I examined the Scatchard plot with interest and noted that the slope of the line was given as -193.90. Since the ordinate had micromolar units, by definition the dissociation constant must be 1.9390×10^{-4}M and this binding was approximately 19-fold less than indicated by the stated dissociation constants.

A first reaction was that the line had been visually approximated, although five significant figures for a slope obtained by such a method would be very unusual. Rather than replot the datum points of Fig. 8, I converted them into the units of Fig. 9, and calculated a regression coefficient by the method of least squares. The obtained value of +22 was unexpected and, after several attempts with the same results, I assumed I no longer knew how to determine a correlation coefficient properly. I turned to a colleague, R. Klucas, and asked him to run the regression line since he was in the process of publishing a paper for which he had been required to do this routinely. Klucas was shaken a bit when he obtained the same result because his paper was in review, and he now questioned either his original formulae or the functioning of the components of a statistical procedure which he had developed for a programable calculator. So we turned to a computer scientist whom I know reasonably well, my wife Sally, and she relieved us both tremendously by reporting the University's computer turned in the same answer.

The result was a bothersome one, because in theory at least a positive slope indicates no binding. After several days of puzzlement I decided to plot the data and found, as Dr. Wheeler indicated, that apparently either the data had been miscalculated or misdrawn in the original Fig. 9 if the data of Fig. 8 had been used. If a different set of data was used for Fig. 9, it clearly is contradictory with the data given in Fig. 8. Although not mentioned by Dr. Wheeler, I also calculated the dissociation constant by the Lineweaver-Burk method from the data of Fig. 8, but obtained a value of only 2.5×10^{-4}M, not 6.8×10^{-5}M.

The value of the binding constant may appear to be a minor

point, at least at first glance. But it does bear on the mechanism
of action *in vivo* and certainly it bears on the probability of successful isolation of protein detected with the aid of bound radioactive toxin. For toxin having a dissociation constant of 10^{-7} or
$10^{-10}M$, there technically is no question that it could be done with
the small amounts of toxin available.

I agree with Dr. Albersheim that an equilibrium dialysis experiment is simple to perform - once the parameters have been
clearly resolved. A difficulty which I have had in reading papers
concerned with *H. sacchari* toxin has been in following details of
certain experiments, especially those involving apparently small
levels of radioactivity. In the binding experiments, for example,
one half of the dialysis cell received 0.14 "µmole" of toxin "of
varying specific activities". In an earlier paragraph it is stated
that specific activities varied from 0.78 nanocuries (nci) to 7.4
nci per µmole. No significant procedural details are given for
the binding experiments. The major table on binding gives only the
final desired result of nmoles toxin bound to a mg of protein. For
the two specific activities given above, a nmole of toxin would
possess either 1.7 or 17 disintegrations per minute (dpm). With
such levels of activity, it would be reassuring if some indications
were given that experimental counting rates were sufficient to determine the degree of specific binding in biological materials.

It is not clear, for example, how much protein was used in
each assay although the figure of 4.3 mg is mentioned several times
in conjunction with these experiments. No mention is made of the
volume of dialysing solution which was assayed for radioactivity,
nor is the efficiency with which the samples were counted given.
To illustrate the uncertainty, the maximum binding reported for
membranes in the paper is 15.6 nmoles for 4.3 mg protein in an experiment in which membranes were treated with NaCl. This corresponds to 3.63 nmoles/mg of protein. Only four other values from
about 40 assays are over two nmoles/mg protein.

If the following optimizing conditions are assumed, then the
counting rates for the maximum binding of 3.63/mg can be calculated:
highest specific activity for 0.14 µmole of toxin (7.4 nci/nmole),
entire contents of the half cell removed and with 100% efficiency,
70% counting efficiency, and no less than 4.3 mg of protein. On
these assumptions, dialysis half cells in the absence of binding
would have 810 counts per minute (cpm) but binding of 15.6 nmoles
to 4.3 mg of protein should result in only 630 cpm. Again with
the same optimal assumptions, the mean value of 1.43 nmoles for
binding of susceptible lines would represent experimentally the
difference between 800 cpm and 733 cpm, or 67 cpm.

In studies with larger dialysis cells we found that pipetting

errors can account for differences as high as 50 dpm out of 4000
in replicate assays of 100 µl aliquots. Non-specific binding of
components of relatively pure corn blight toxin preparations by
bean homogenates or by synthetic membranes was observed. In order
to be convinced about binding, the purity of the toxin must be un-
questioned and sufficient detail presented of the original data to
reassure the skeptic that the low levels of radioactivity are in
fact adequate for accurately detecting binding. Further, it would
be nice to know that equilibrium was obtained.

Hadwiger suggested that Strobel and others should work together
to resolve the controversy between them; this arrangement has been
used successfully to resolve similar controversies.

Staples spoke in support of Strobel's work and pointed out
that in a recent paper Strobel had reported that tobacco protoplasts
normally resistant to helminthosporoside became susceptible when
they were treated with helminthosporoside binding protein isolated
from susceptible sugar cane cultivars. He thought that this supported Strobel's contention that the toxin acts by binding to the
protein in membranes of susceptible sugar cane clones.

In further contributions, Daniels described isolation of a
toxin from *Spiroplasma citri*, the causal agent of citrus stubborn
disease. The toxin is a carbohydrate (mol. wt. 200 - 300) and inhibits growth of the algae *Chlorogonium euchlorum* and *Chlamydomonas
reinhardii*, and of cells and protoplasts of tobacco. It has weak
activity toward yeasts and none against bacteria and cultured am-
phibian or insect cells. A role of the toxin in the expression of
disease symptoms can be inferred from these findings but nothing
is known about its role in the specificity of the mycoplasma for
its host.

Patil described studies on the possible role of phaseotoxin
in specificity of *Pseudomonas phaseolicola* for its host and reported
that in resistant plants, in spite of substantial multiplication of
P. phaseolicola, no phaseotoxin is detected indicating that toxin
production in such plants may be under regulatory control. In resistant plants treated with phaseotoxin prior to inoculation, the
pathogen grew more, by several orders of magnitude, than in non-
treated plants, and the hypersensitive reaction which normally occurs in untreated resistant plants was suppressed. Concomitantly,
accumulation of isoflavonoid phytoalexins was also suppressed in
toxin treated plants. Patil said that four isoflavonoids were
identified from hypersensitive host tissues and that these, including phaseollin (reported to be non-toxic to the pathogen),
suppressed growth of colonies of *P. phaseolicola*. From this work
phaseotoxin appears to be necessary for establishment of the pathogen in host plants and the basis of resistance may involve

the ability of bean plants to suppress toxin production or to degrade toxin if it is produced.

Keen said that the work on phaseotoxin strongly supports the
induced susceptibility concept and Rahe suggested that the argument that phaseotoxin suppresses host resistance will be strengthened if suppression of phytoalexin synthesis in beans can be demonstrated in plants infected with a pathogen other than *P. phaseolicola*. In response to questions from Deverall and Lyon regarding
inhibition of bacterial growth by phaseollin, Patil replied that
the different results reported may depend on differences between
strains of test organisms.

ROLE OF PREFORMED FACTORS IN SPECIFICITY

FRITZ SCHÖNBECK

Institut für Pflanzenkrankheiten und Pflanzenschutz der Technischen Universität Hannover
W. Germany

INTRODUCTION

Specificity in plant diseases based on preformed factors means that one pathogen is prevented from infecting a potential host by one or more factors occurring in the plant before, and independently of infection. Another pathogen, however, may have the ability to overcome this established barrier. Also included are diseases in which infection is prevented by the deficiency of essential factors in the plant or in the parasite.

SUPERFICIAL STRUCTURES

Apart from periderm or cork, the intact healthy plant is covered and protected by a layer of epidermal cells. Only pathogens that can penetrate this barrier can gain access to nutrients in the plant. Bacteria and many potentially parasitic fungi which cannot penetrate this surface either directly or through natural openings depend on wounds of various sorts for entry.

The epicuticular wax layer may act as a barrier by repelling the water film required by a pathogen for germination and growth on the surface. Wetting of a plant surface is influenced by the physical structure of the wax layer and by its chemical composition. These features and consequently the wettability differ considerably in varieties of one plant species, and may often affect the incidence of infection.

There are also differences in the wettability of fungal spore

surfaces (9). Readily wettable asexual spores are found in species
of *Fusarium*, *Verticillium*, *Cladosporium*, *Aspergillus*, whereas,
spores of species of *Penicillium* are much less readily wetted.
Wettable spores readily enter water droplets whereas non-wettable
spores remain on the surface of droplets. When a droplet, carrying
spores, falls on and moves over a hydrophobic plant surface, non-
wettable spore brought into contact with the surface will tend to
be deposited to the rear of the moving droplet, whereas wettable
spores will either be washed away or will be deposited where the
droplet comes to rest. Plants or parts of plants with water rep-
ellent surfaces enhance deposition of non-wettable spores. The
surface of the plant and of the spore can, therefore, greatly affect the deposition of inoculum.

A well developed cuticle does not necessarily prevent infection.
Its role as an effective mechanical barrier for invading fungi is
still in doubt. The thickness of the cuticle, however, may have
another role. Substances secreted by plants and present on or in
the surface layers stimulate the activities of some pathogens. Certain of the substances may be triggering agents that induce germination of propagules, and therefore initiate fungal growth. Other
substances may increase rate of germination, break the spore dormancy,
shorten the ripening time, or stimulate the formation of appressoria.

Differences occur in the amount of these substances between
plant species and between cultivars of one species. The differences
are apparently related to conditions within the plant and to the
physical properties of the surface layers. Leaves with a thick,
waxy surface which are difficult to wet, are usually less subject
to loss of such substances than are leaves with a thin cuticle.

Many fungal spores will germinate only when external supplies
of nutrients or particular activators are available (7). The availability of these substances to spores may result in specific host-
parasite-relationships. Arens (1) reported that zoospores of the
downy mildew *Plasmopara viticola* collect and germinate around stomata of grape plants, the normal host, and also around stomata of
many other species from a number of plant families. The chemotactic
response of the zoospores did not occur in species with heavy wax
deposits unless these were removed.

Kerr and Flentje (13) claimed that the attachment to, organi-
sation on, and penetration of a host, by the radish strain of
Pellicularia filamentosa is controlled by the nature of the cuticular surface and by diffusible material secreted to the surface from
underlying epidermal cells. This implies that the specificity of
the reaction of the different fungal strains to different hosts depends on differences in the cuticular surface and the diffusible
material in various hosts. Micro-organisms in the rhizosphere and

phyllosphere including plant pathogens also may be affected by this
secreted material.

   The restricted movement of substances through the outermost
layers of the epidermis can lead to nutrient deficiencies which
may affect certain parasites before penetration. Only if the plant
tissue supplies all nutrients necessary for growth or if the pathogen can synthesize those which are lacking, will it be able to
penetrate the tissue.

   There can be no doubt that the nutrition of a pathogen is important for successful infection. Is specificity, particularly of
obligate or nearly obligate parasites, related to their specialized
requirements of nutrients that are present or available in adequate
quantities only in certain host plants ? Different races of a pathogen may differ in their nutrient requirements. However, in review-
ing the literature, I found little convincing evidence to support
the idea that such factors control resistance or specificity.

   Amino acids may affect the metabolism of certain fungi. Thus,
Stover (18) studied production of chlamydospores on media containing different sources of nitrogen by a grey strain of *Thielaviopsis
basicola* from cotton, and a brown strain from tobacco. Chlamydospore production by both strains varied with the nitrogen sources.
Production was stimulated by certain amino acids but not by others.
The grey strain formed chlamydospores when arginine or arginine
plus leucine were sources of nitrogen, whereas the grey strain did
not. In other words the strains differed in their responses to
excess arginine.

   The content of amino acids in higher plants can vary considerably. This is well established for plants infected by mycorrhizal
fungi. From an ecological point of view, substances produced in
mycorrhizal plants can be considered preformed. For vesicular-
arbuscular mycorrhiza, the content of free amino acids differed
markedly between mycorrhizal and non-mycorrhizal plants. Also,
total amino acid was about 50% higher in the mycorrhizal plants,
with arginine and citrulline showing the greatest increases (4).
This seems to be a common feature for this form of mycorrhiza
(Fig. 1).

   The strong inhibition of chlamydospore production by *T. basicola*
in mycorrhizal roots probably depends on the high content of arginine and it may be inferred that mycorrhizal roots favour strains
of *T. basicola* that are not sensitive to excess arginine.

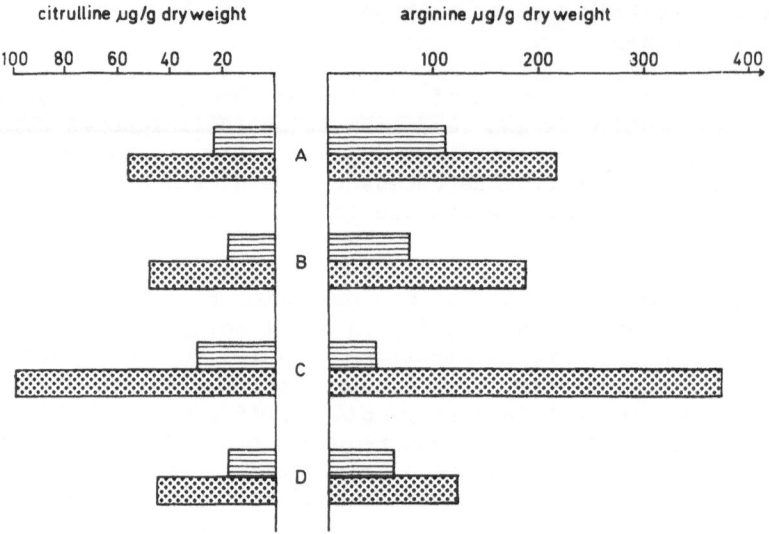

Figure 1. *Arginine and citrulline levels in the roots of several mycorrhizal* ▦ *and non-mycorrhizal* ▤ *plants.*

A. *Phaseolus vulgaris* 'Pinto'
B. *P. vulgaris* 'Saxa'
C. *Nicotiana tabacum*
D. *Daucus carota*

PREFORMED INHIBITORY SUBSTANCES

Many plants contain substances which inhibit growth of bacteria and/or fungi. There is an extensive literature attributing disease resistance to such substances. Much of it is based on inadequate evidence. In connexion with specificity I shall deal with only two groups of substances, saponins and unsaturated lactones.

Saponins

The only plant known to contain two types of saponins is *Avena sativa*. Roots contain triterpenes, whereas aerial parts contain furostanol derivatives. The root saponin is identical with the fluorescent "root tip glucoside", described from meristems of oat roots (10, 11). It was isolated, characterized and named avenacin (8). Structures were established by Tschesche *et al.* (19) who distinguished avenacin A and B. The sugar chain of avenacin A contains two molecules of glucose and one of arabinose, whereas the sugar chain of avenacin B contains two molecules of glucose.

Avenacin is biologically the most active saponin tested to date;
only 0.2 - 0.3 µg/ml is required for complete haemolysis. The haem-
olytic activity of the aglycons avenamin A and B, however, is much
lower (14).

It is claimed that avenacin is responsible for resistance in
oats to the wheat variety of the take-all fungus *Gaeumannomyces
(Ophiobolus) graminis* var. *graminis* (22). According to Turner (22),
resistance can be overcome by *G. graminis* var. *avenae*, first isolated from diseased oats in Wales (23). In the early stages of seedling growth the difference in infection of the seminal roots of
oats by *G. graminis* var. *graminis* and *G. graminis* var. *avenae* is
only one of degree, with considerable penetration in both cases.
Three weeks later, however, cells of oat roots overcome infection
by the wheat variety and only scattered cells show persistent hyphae
which are often enclosed in lignitubers. Infection by *G. graminis*
var. *avenae* persists and the hyphae grow through to the stele, resulting in brown lesions on the roots and often in the death of the
seedling.

Growth of *G. graminis* var. *graminis in vitro* is inhibited by
an extract from oat roots which supports good growth of *G. graminis*
var. *avenae*. This fact cannot be explained by lower toxicity of the
active substance to the oat variety of the fungus because partially
purified preparations of the inhibitor from the root extract are
almost equally toxic to both varieties (23). The clue to this problem, is that *G. graminis* var. *avenae* produces an enzyme avenacinase
when grown on extracts from oat plants. This enzyme hydrolyses
avenacin into products of much lower toxicity. The wheat variety of
G. graminis does not produce this enzyme.

Inactivation of avenacin by another pathogen of oat has been
examined *in vivo* by Lüning (14). He found that the amount of avenacin in oat roots decreased greatly six days after inoculation with
Fusarium avenaceum. Thus, healthy roots contained 171 ± 64 µg avena-
cin/g fresh wt. compared with 2 µg/g fresh wt. in infected roots.
A crude enzyme preparation from oat roots infected by *F. avenaceum*
rapidly inactivated avenacin. The enzyme produced by the fungus has
been characterized as an extracellular specific β-glucosidase. The
degradation products are less water soluble and less biologically
active than the original compound.

Turner (23) also found inhibitory substances in oat leaves which
she believed were not identical with the avenacin in oat roots. It
has now been reported that oat leaves contain the bisdesmosidic saponins avenacosid A and B (20, 21) which are biologically inactive.
Enzymatically they are quickly activated by wounding to their corresponding monodesmosidic 26-desglucoderivatives. Lüning (14) found up
to 4500 µg 26-desgluco-avenacosid / g fresh weight. The β-glucosidase,
26-desgluco-avenacosidase, catalysing this reaction, appears highly

specific, since it splits off only the glucose molecule attached at
C-26 but does not affect glucose molecules of the side chain attached
at C-3. This enzyme is produced by the oat plant.

After inoculation of oats with *Helminthosporium avenae* infected
leaves did not contain active saponins (Table 1). The fungus produced *in vitro* and *in vivo* a specific β-glucosidase which splits off
glucose molecules of the side chain at C-3. The fungal β-glucosidase does not affect the glucose at C-26.

Tomatine, a steroidal alkaloid, the role of which in resistance
has been investigated for many years, occurs in a number of species
of *Solanum* and is toxic to a broad range of fungi. Pathogens of
tomato, however, appear to be less sensitive than are most other
fungi. The minimum concentration of tomatine that completely inhibits growth of the tomato parasite *Septoria lycopersici* is about
2000 times higher than that which inhibits *S. lactucae* which is not
a parasite of tomato (2, 3). *S. lycopersici* produces *in vitro* and
in vivo a β-glucosidase that detoxifies tomatine by removing one
glucose unit.

A hypha of a potential pathogen near a plant cell containing
an active saponin in its vacuole may secrete membrane-lytic materials which impair the cell membrane, resulting in a release of the
saponin into the intercellular spaces. If an inactive saponin is
present in the vacuole it may be enzymatically converted to an
active derivative during release from the cell. In either case
exposure of cell membranes of hyphae to the active saponin may result in their death.

Pathogens such as *G. graminis* var. *avenae, S. lycopersici,
H. avenae* and *F. avenaceum* can produce enzymes which transform

TABLE 1.   *Content of 26-desgluco-avenacosid in water extracts
from oat leaves infected with* Helminthosporium avenae
(after Lüning (14))

	μg 26-desgluco-avenacosid/g fresh wt.
Healthy leaves	2.625 ± 1.264
Infected leaves	180

active saponins into inactive derivatives so that the pathogens are
not affected by the fungitoxic substances. The crucial point for
specificity is that the potential pathogens do produce the inactivating enzymes. The two varieties of *G. graminis* differ in this
respect *in vitro* but we can only infer that this also happens *in
vivo*. Production of enzymes *in vitro* and *in vivo* by *F. avenaceum*
has, however, been demonstrated. Detoxification, therefore, probably explains resistance of oat roots to this pathogen. The same
can be said for the toxic derivatives of avenacosid from *H. avenae*
and of tomatine from *S. lycopersici*. But it is not yet clear whether the inability of other closely related potential pathogens to
infect these plants depends on the presence of the saponin, and
whether this inability is correlated with the absence of saponin-
inactivating enzymes.

Unsaturated Lactones

Glycosides with lactone forming aglycones are another group
of widely distributed substances in plants particularly in species
of *Ranunculaceae, Liliaceae* and *Rosaceae*. In most cases unsaturated hydroxy carboxylic acids are part of the glycosides. After
hydrolysis many of these acids readily form unsaturated lactones,
some of which are volatile and almost always physiologically active.

Tuliposides which occur in tulips and related plants are
1-acylglucosides of the α-methylene-γ-hydroxybutyric acid (tuliposide
A) or of the α-methylene-β-γ-dihydroxybutyric acid (tuliposide B)
(Fig. 2). They are relatively stable below pH 5.2. Toxic lactones

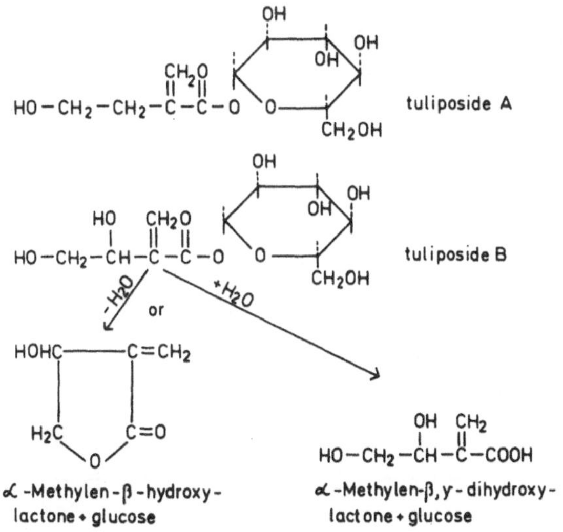

Figure 2. *Tuliposides A and B and the cleavage of tuliposide B.*

are formed under mild acid hydrolysis, whereas alkaline hydrolysis
liberates the corresponding non-toxic acids. (Fig. 2).

Field infection of tulips by *F. oxysporum* f. sp. *tulipae* usually occurs during the last weeks before harvesting (5), when the
outermost white scale, previously containing high amounts of tuliposides, turns into a brown leathery scale which is nearly devoid
of antimicrobial substances. In the underlying white, fleshy scales,
the tuliposide content at this time is very low but after a few days
storage it increases to about 2000 μg α-methylene-γ-butyrolactone/
g fresh wt. (6). Although fungal inoculum is present in the soil
during the entire growth period, the phase of susceptibility of
tulip bulbs to *F. oxysporum* f. sp. *tulipae* coincides with the drop
in tuliposide content. These facts lead to the conclusion that the
fungitoxic substances protect the white skin and, therefore, the
whole of the growing young bulbs against infection by *F. oxysporum*
f. sp. *tulipae* so long as they occur at sufficiently high concentrations. The fluctuating content of the toxic substances probably
explains why the pathogen can attack the tulip bulbs only at certain
times. Lack of toxic preformed substances causes a temporary susceptibility which may be regarded as a kind of specificity.

Botrytis tulipae is a specialized pathogen infecting only tulips; it attacks all parts of the plants. In contrast *B. cinerea*
is an unspecialized pathogen with a wide host range which does not,
however, include tulips. Because tuliposides are conspicuous compounds of tulips accounting for as much as 30% of the dry weight of
the pistil (12), it was of interest to investigate whether they
were implicated in the resistance of tulips to *B. cinerea*. Growth
of *B. cinerea* in liquid cultures containing water extracts of tulip
pistils is inhibited more than is the growth of *B. tulipae*. Concentration of these extracts that kill *B. cinerea* greatly stimulate
growth of *B. tulipae* (Fig. 3). The toxic effect on *B. tulipae* at

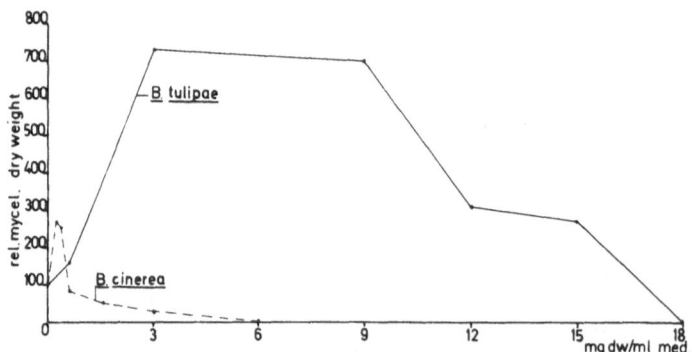

Figure 3. *Growth of* B. tulipae *and* B. cinerea *in nutrient solutions
containing various amounts of comminuted tulip pistils.*

the highest concentration was only fungistatic in contrast to the fungicidal effect on *B. cinerea*.

The correlation between increase in permeability, release of toxic substances and ability of the fungi to invade the tissue was studied in an experiment summarized in Fig. 4. Tulip pistils were subjected to treatment: A no special treatment, B storage at -25°C to destroy the semi-permeability of the membranes, and C, leaching of pistils with destroyed semi-permeability. The pistils were then inoculated with *B. tulipae* or *B. cinerea*, whereas uninoculated pistils were covered with paper disks which were later transferred to an agar medium seeded with *Bacillus subtilis* to detect anti-microbial substances which had diffused from uninoculated pistils.

After treatment A, only *B. tulipae* was able to invade the tis-sue, and no antimicrobial substances diffused into the paper disks. After loss of semi-permeability in treatment B, the paper disks be-came antimicrobial, and the fungi did not invade the pistils. Lea-ching of the pistils which had lost their semi-permeability in tr-eatment C resulted in the infection of pistils by *B. tulipae* and *B. cinerea* (16). These results showed that the amounts of fungi-toxic compounds in the pistils released by the destruction of semi-permeability are adequate to prevent infection by the specialized pathogen *B. tulipae*. Another important finding is that *B. cinerea* causes a much higher increase of permeability than does *B. tulipae* and this results in a higher release of fungitoxic substances.

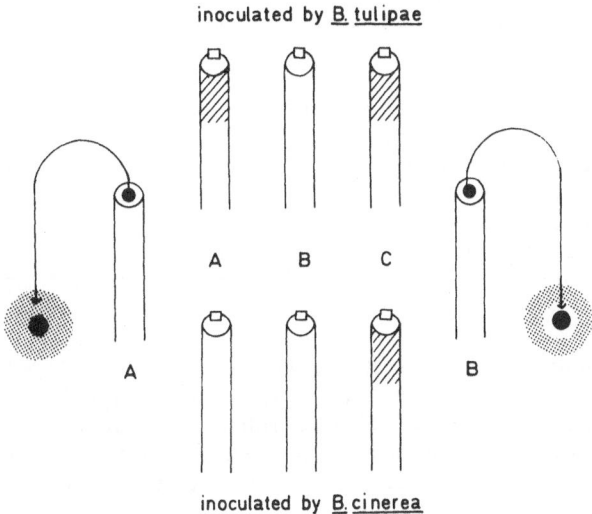

Figure 4. *Effects of permeability change in tulip pistils on release of inhibitory compounds, and on susceptibility of* Botrytis.

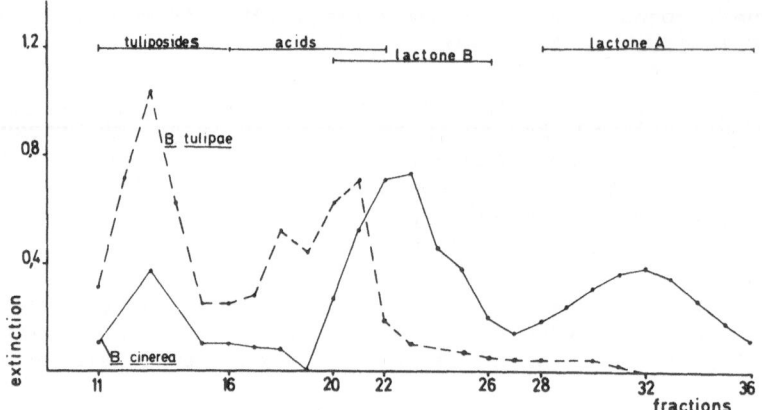

Figure 5. *Cleavage of tuliposides by* B. tulipae *and* B. cinerea
           in liquid cultures.

     What happens to tuliposides in infected tissue ? Tissue dis-
integrated by *B. tulipae* does not contain tuliposides and shows no
activity. This was demonstrated in an experiment in which mycelium
of *B. cinerea* and *B. tulipae* was grown for one week in a liquid medium containing tuliposides (2 000 µg/ml). Forty eight hours later
the tuliposides and their derivatives were determined. Tuliposides
in cultures of *B. cinerea* were converted almost completely into
lactones whereas cleavage of tuliposides in *B. tulipae* cultures
produced large quantities of acids and only traces of lactones
(Fig. 5).

     These results show that inhibition of *B. cinerea* and *B. tulipae*
by tuliposides may depend on differences in the ability of these
fungi to degrade the tuliposides rather than to a difference in
their toxicity. Further tests showed that there were marked differences in sensitivity of the fungi to the lactone. Growth of *B.
cinerea* was greatly decreased at most concentrations used, whereas the growth of *B. tulipae* was stimulated or only slightly decreased (Table 2).

     It was then found that *B. tulipae* converts tuliposides into
γ-hydroxylic acids and Schroeder (17) showed that these acids stimulate the growth of *B. tulipae* but inhibit growth of *B. cinerea*.
Moreover, addition of the acids to culture media induced sporulation of *B. tulipae* whereas *B. cinerea* was unaffected.

     In summarizing the work on *Botrytis* spp. and tuliposides it
can be said that there is close relationship between concentrations
of inhibitory substances in tulips and the inability of *B. cinerea*

Table 2. *Influence of lactone B on growth of* B. tulipae *and* B. cinerea *in Petri dishes (16)*

Concentration µg/cm^3	Relative growth (cm^2)	
	B. *cinerea*	B. *tulipae*
0	100	100
3	92	105
6	63	106
9	27	112
12	19	96

to invade plants. Only when tuliposides are removed can the fungus invade the tissue. Nevertheless, B. *cinerea* remains a potential parasite of tulips. Amounts of tuliposides in undamaged cells could prevent invasion by B. *tulipae;* nevertheless, the fungus can grow intercellularly. B. *tulipae* and B. *cinerea* affect tissues in quantitatively different ways when they secrete substances which increase the permeability of host cell membranes which results in release of tuliposides from the vacuoles. Because B. *cinerea* causes greater increases in permeability, more tuliposides are released than by B. *tulipae*. The tuliposides are converted to aglycones after contact with the fungi. In this respect, too, the two fungi differ. B. *cinerea* converts tuliposides mainly into lactones but B. *tulipae* causes the corresponding γ-hydroxylic acid to be formed. These acids stimulate the growth of B. *tulipae* but inhibit the growth of B. *cinerea*. Also, B. *cinerea* is more sensitive to the tuliposides than is B. *tulipae*. So far as we know, these factors explain why the specialized fungus B. *tulipae* but not B. *cinerea* infects tulips.

ATTRACTION

There are a number of ways in which preformed substances may act in specificity of parasites. If a species of a parasite contains a receptor for a particular compound in plant exudates and another species does not, the former would be attracted to the host whereas the latter would not. This mechanism sometimes occurs in parasites having motile phases, for example zoospores which can be attracted to hosts by exudates from various plant organs for example leaves through stomata, roots, or by sap released in wounds. There is some specificity for *Phytophthora cinnamomi* and exudates from roots of avocado trees (24). Thus, zoospores of this species but not of *P. citrophthora* are attracted to avocado roots. Dead roots and roots of related species have no effect. In most other cases however, attraction is non-specific.

Specific attraction of plants towards animal parasites might be expected more frequently. There is good evidence that receptors for specific stimulants occur in some insects for example the beetle *Chrysolina brunsvicensis*, a parasite of *Hypericum hirsutum* (15). Many of the so called secondary substances of plants appear to function as chemical signals for animal parasites. Insect species, adapted to a particular plant, may use such substances to recognize suitable plants; other species of insects may be repelled by the same compounds. This phenomenon rarely occurs among microbial parasites, possibly because animal parasites normally move actively towards their hosts under the influence of attractants, whereas microbial parasites are usually carried passively to their host plants.

REFERENCES

1.   ARENS, K. (1929). Physiologische Untersuchungen an *Plasmopara viticola*, unter besonderer Berücksichtigung der Infektionsbedingungen. *Jb. wiss. Bot.*, 70, 93 - 157.

2.   ARNESON, P.A. and DURBIN, R.D. (1967). Hydrolysis of tomatine by *Septoria lycopersici* : a detoxification mechanism. *Phytopathology*, 57, 1358 - 1360.

3.   ARNESON, P.A. and DURBIN, R.D. (1968). The sensitivity of fungi to α-tomatine. *Phytopathology*, 58, 836 - 837.

4.   BALTRUSCHAT, H. and SCHÖNBECK, F. Untersuchungen über den Einfluss der endotrophen Mycorrhiza auf den Befall von Tabak mit *Thielaviopsis basicola*. *Phytopath. Z.* (In press).

5.   BERGMAN, B.H.H. (1966). Presence of a substance in the white skin of young tulip bulbs which inhibits growth of *Fusarium oxysporum*. *Neth. J. Pl. Path.*, 72, 222 - 230.

6.   BERGMAN, B.H.H. and BEIJERSBERGEN, J.C.M. (1971). A possible explanation of variation in susceptibility of tulip bulbs to infection by *Fusarium oxysporum*. *In : First int. Symp. Flower Bulbs*, Noordwijk/Lisse, 1970. Ed. Int. Soc. Hort. Sc., Den Haag, Vol. 2, 225 - 229.

7. BROWN, R. (1946). Biological stimulation in germination. *Nature, Lond.*, 157, 64 - 69.

8. BURKHARDT, H.J., MAIZEL, J.V. and MITCHELL, H.K. (1964). Avenacin, an antimicrobial substance isolated from *Avena sativa*. II. Structure. *Biochemistry*, 3, 426 - 431.

9. DAVIES, R.R. (1961). Wettability and the capture, carriage and deposition of particles by raindrops. *Nature, Lond.*, 191, 616 - 617.

10. GOODWIN, R.M. and KAVANAGH, F. (1948). Fluorescing substances in roots. *Bull. Torrey bot. Club*, 75, 1 - 17.

11. GOODWIN, R.M. and POLLOCK, B.M. (1954). Studies on roots. I. Properties and distribution of fluorescent constituents in *Avena* roots. *Am. J. Bot.*, 41, 516 - 520.

12. KÄMMERER, F.J. (1967). *Über die antibiotischen Substanzen aus der Tulpe* (Tulipa hybrida). Dissertation, Bonn.

13. KERR, A. and FLENTJE, N.T. (1957). Host infection of *Pellicularia filamentosa* controlled by chemical stimuli. *Nature, Lond.*, 179, 204 - 205.

14. LÜNING, H.U. (1975). *Saponine in* Avena sativa, *ihre Bedeutung im Resistenzmechanismus gegenüber phytopathogenen Pilzen*. Dissertation, Bonn.

15. REES, C.J.C. (1969). Chemoreceptor specificity associated with choice of feeding site by the beetle *Chrysolina brunsvicensis* in its foodplant, *Hypericum hirsutum*. *Entomologia exp. appl.*, 12, 565 - 583.

16. SCHÖNBECK, F. and SCHROEDER, C. (1972). Role of antimicrobial substances (tuliposides) in tulips attacked by *Botrytis* spp. *Physiol. Pl. Path.*, 2, 91 - 99.

17. SCHROEDER, C. (1972). Die Bedeutung der γ-Hydroxysäuren für das Wirt-Parasit-Verhältnis von Tulpe und *Botrytis* spec. *Phytopath. Z.*, 74, 175 - 181.

18. STOVER, R.H. (1956). Effect on nutrition on growth and chlamydospore formation in brown and grey cultures of *Thielaviopsis basicola*. *Can. J. Bot.*, 34, 459 - 472.

19. TSCHESCHE, R., CHANDRA JHA, H. and WULFF, G. (1972). Über Triterpene. *Tetrahedron*, 29, 629 - 633.

20. TSCHESCHE, R. and LAUVEN, P. (1971). Avenacosid B, ein zwei-
     tes bisdesmosidisches Steroidsaponin aus *Avena sativa*.
     Chem. Ber., 104, 3549 - 3555.

21. TSCHESCHE, R., TAUSCHER, M., FEHLHABER, H.W. and WULFF, G.
     (1969). Avenacosid A, ein bisdesmosidisches Steroidsap-
     onin aus *Avena sativa*. *Chem. Ber.*, 102, 2072 - 2082.

22. TURNER, E.M.C. (1953). The nature of resistance of oats to
     the take-all fungus. *J. exp. Bot.*, 4, 264 - 271.

23. TURNER, E.M.C. (1956). The nature of resistance of oats to
     the take-all fungus. II. Inhibition of growth and respi-
     ration of *O. graminis* and other fungi by a constituent
     of oat sap. *J. exp. Bot.*, 7, 80 - 97.

24. ZENTMYER, G.A. (1961). Chemotaxis of zoospores for root
     exudates. *Science, N.Y.*, 133, 1595 - 1596.

CONTRIBUTIONS

DÉFAGO, G., GEESLER, C., KERN, H., MEMMEN, K.F. and RUFFNER, F.
Significance of saponins and sterols in disease resistance of
plants.

SCHLÖSSER, E.   Specificity and role of β-glycosidases in saponin
dependent host-parasite interactions.

SMITH, I.M.   Antifungal compounds in red-clover.

ZENTMYER, G.A., ZAKI, A., SIMS, J.J. and PETTUS, J.   Preformed
toxicant in *Persea* spp. involved in resistance to *Phytophthora
cinnamoni*.

SUMMARY OF POINTS FROM CONTRIBUTIONS AND DISCUSSIONS
BY
R. D. DURBIN

Chairman and Discussion Leader

The following points were made in the discussion following the
lecture. Schönbeck confirmed that 60 - 70% of the species he has
examined contain preformed antimicrobial substance(s). During evolution, pathogens of these plants seem to have evolved mechanisms
to circumvent such preformed toxicants. Relatively little is known
about their biosynthesis. Many may be present in plants in relatively non-toxic forms, e.g. as glycosides, but when the plant is

infected they are readily converted into inhibitory forms. We do
not know whether these secondary plant substances appeared during
evolution in response to parasites, or how effective they and phytoalexins are in limiting infection. But most probably, preformed
substances are more important as a barrier against a potential pathogen becoming established in particular species rather than as factors for varietal resistance. There is no indication, for example,
that differences in susceptibility of tulip cultivars to *Botrytis
tulipae* are related to tuliposide content or that tuliposide synthesis increases at or near the site of infection.

Many plants, e.g. tomato, apple, tobacco and bean, containing
preformed antifungal substances can also synthesize phytoalexins.
As an example, Smith said that red clover, *Trifolium pratense*, which
contains antifungal isoflavones (biochanin-A, genistein and form-
ononetin) as well as trifolirhizin (maackiain glycoside), also
synthesizes the phytoalexins medicarpin and maackiain. The behaviour of these compounds in leaves of four clover varieties susceptible to *Sclerotinia trifoliorum* and resistant to *Botrytis cinerea*
has been examined by Debnam. In leaves inoculated with either fungus,
the glycosylated forms of the isoflavones were almost completely hydrolysed. Generally, the content of the isoflavones was lower than
in healthy tissues, but the content of genistein increased. The
three isoflavones did not inhibit the germination of spores of
Botrytis cinerea but they did inhibit germ tube growth; none affected *S. trifoliorum*. Maackiain and medicarpin accumulated to inhibitory concentrations in tissues challenged by *B. cinerea* but much
lower concentrations were found in tissues challenged by *S. trifoliorum*. Probably this is because *B. cinerea* cannot metabolize these
compounds. *S. trifoliorum* was not inhibited by maackiain (up to
100 µg/ml) which it degraded to an unknown compound also found in
infected tissues. The fungus was sensitive to medicarpin but at
concentrations greater than those present in infected tissues. Inhibition of *B. cinerea* is more probably explained by phytoalexin
accumulation than by isoflavones already present. The specificity
of *S. trifoliorum* may be due to its ability to degrade phytoalexins
and its insensitivity to the isoflavones. Differences in varietal
susceptibility are not readily explicable on the basis of these
chemicals.

Steroidal saponins and alkaloids are other common examples of
secondary plant substances that may determine specificity. They
act similarly to polyene antibiotics by complexing with free sterols
present in fungal membranes. Défago and co-workers reported that
saponins can protect plants against *Pythium paroecandrum*, a cause
of seedling damping-off. This fungus like other *Pythium* spp., does
not contain sterols but it can rapidly incorporate exogenous sterols
into its plasma membranes. By placing mats of the fungus in solutions having different cholesterol concentrations, mycelia with

different sensitivity to the saponin digitonin were obtained, the
most sensitive combination being three molecules of digitonin per
cholesterol molecule. Hyphae containing more cholesterol are much
less sensitive to digitonin. When different plant species were inoculated with either cholesterol-containing or normal mycelia, the
roots of saponin-containing species (tomato and sugar beet) were
found to be more resistant to cholesterol-containing mycelia than
to normal mycelia. Conversely, roots of species not containing
saponins (peas and soybeans) were more susceptible to cholesterol-
containing inocula. However, in peas, quantitative and qualitative
changes in root sterols during the first week after germination may
be responsible for the increase in resistance which occurs during
this period. It was concluded that saponins can protect plants
against fungal invasion and that they can partially explain host
specificity.

Schlösser discussed the saponin-inactivating enzyme systems
of pathogenic fungi that are based upon β-glycosidases which hydrolyse saponins to their relatively non-toxic aglycones. These
enzymes can be either intra- or extra-mural, or both. Although
they catalyse the same reactions, they are different in several ways
from the enzymes found in fungal cell membranes which are required
for the lytic action of saponins on membranes. First, they are
highly specific toward a given saponin. Second, they hydrolyse,
and thereby inactivate, the saponins before they can reach their
site of action in the fungal membrane. Pathogens of tomato, oats,
and cyclamen were shown to possess β-glycosidases capable of hydrolysing the fungitoxic saponins of their respective hosts.

Zentmyer reported that a preformed substance, burbonol, is
present in *Persea* spp. resistant to *Phytophthora cinnamomi* the
cause of root rot. In these species it occurs in leaves, stems and
roots in concentrations up to 1 mg/g dry wt., but is either absent
or is in very low concentrations in susceptible varieties of *P.
americana* (avocado) and it is absent from *P. indica,* another susceptible species. The purified material significantly decreases
mycelial growth of the pathogen and a test organism, *Cladosporium
cucumerinum,* at 2 - 10 µg/ml. The compound may occur in several
closely related chemical forms and as a mixture of keto and enol
isomers. It is a long chain hydrocarbon with a furan ring and has
an empirical formula of $C_{19}H_{31}O_3$ (mol. wt. 308). A simplified quantitative test for burbonol is being developed for use in identifying
promising resistant plant material. It may also be possible to use
this compound as soil or systemic fungicide.

PHYTOALEXINS AND THE SPECIFICITY OF PLANT-PARASITE INTERACTION[*]

JOSEPH KUĆ

Department of Plant Pathology, University of Kentucky

Lexington, Kentucky, U.S.A.

INTRODUCTION

Phytoalexins are compounds which accumulate in plants following infection and various forms of stress. The term stress metabolites more accurately describes this diverse group of compounds since their accumulation after infection appears to be primarily a function of stress rather than of infection. Their relationship to stress, however, does not preclude their participation in mechanisms for disease resistance. Phytoalexins are produced by non-hosts as well as by resistant and susceptible cultivars of species. They accumulate in plants to concentrations that are inhibitory to the growth of infectious agents when tested *in vitro*. Phytoalexins have been characterized in the *Leguminosae, Solanaceae, Malvaceae, Rosaceae, Convolvulaceae, Umbelliferae* and *Compositae*. They have not been characterized in other families of major agronomic importance, e.g. , *Gramineae* and *Cucurbitacaceae*. Since phytoalexins accumulate at and around sites of infection to levels inhibitory to the growth of infectious agents, their relationship to the containment of microbial development in plants cannot be ignored. Opinions differ as to the extent of their contribution in determining susceptibility or resistance.

[*] Journal paper no. 75-11-140 of the Kentucky Agricultural Experiment Station, Lexington, Kentucky, U.S.A. The author's research reported in this paper has been supported in part by a grant from the Herman Frasch Foundation and grant no. 316-15-51, P. L. 89-106 of the Cooperative States Research Service of the United States Department of Agriculture.

The basis for disease resistance and susceptibility in plants has received attention for more than a century (7). It is only within the past 20 years or so that these studies have taken a bio-chemical and molecular orientation (8, 28). Studies of microbial toxins and phytoalexins have generated considerable interest though they have been reported in relatively few interactions. Many phytoalexins and recently a few host specific toxins have been chemically characterized and this has permitted intensive research into their role in determining host-parasite interactions. This paper will not present an exhaustive review of the literature pertinent to phytoalexins (9, 24, 27, 29, 30, 31, 44). It will be concerned principally with a discussion of the role of phytoalexins in determining specificity of interaction, that is, whether plants are hosts or non-hosts and whether cultivars are resistant, tolerant or susceptible.

ARE PHYTOALEXINS PRODUCED ONLY BY NON-HOST PLANTS OR RESISTANT CULTIVARS ?

In the major plant families in which phytoalexins have been chemically characterized, it is clear that their accumulation is not limited to non-host plants or to resistant cultivars. Thus pisatin accumulates in pea (11) and phaseollin in green bean (12) in response to inoculation with pathogens and non-pathogens, and phaseollin accumulates in hypocotyls of resistant and susceptible cultivars inoculated with *Colletotrichum lindemuthianum* (4, 38). The total quantity of phaseollin which accumulates in susceptible cultivars may equal or surpass that in resistant cultivars (4, 38). This relationship is also evident for accumulation of kievitone, phaseollidin and phaseollinisoflavan (1, 43). Cultivars of green bean apparently susceptible to all races of *C. lindemuthianum* accumulate phaseollin, phaseollidin, kievitone and phaseollinisoflavan when inoculated with the non-pathogen of bean *Colletotrichum lagenarium* (14). *C. lagenarium* protects green bean hypocotyls locally and systemically from disease caused by *C. lindemuthianum,* but the commitment for protection can clearly be differentiated from the accumulation of phytoalexin (this will be discussed later).

Potato tubers lacking major or *R* genes for resistance to *Phytophthora infestans* accumulate rishitin, phytuberin, lubimin and 16 - 18 additional terpenoids when treated with cell-free sonicates or cell wall preparations of the fungus (35, 40, 47, 49, 50, 51). The magnitude of accumulation in response to sonicates equals or surpasses that in incompatible interactions of fungus and host. Cell-free sonicates of all races of the fungus appear equally capable of eliciting accumulation of the terpenoids and all cultivars of potato appear capable of accumulating them. The nature of the interaction may be determined by the ability of the living compatible race of the fungus to actively suppress rapid cell collapse, necrosis and terpenoid accumulation (49, 50). Thus, specificity may depend upon preventing a general non-specific response from occurring soon enough or in sufficient magnitude for resistance.

Viruses, chemical toxicants, pesticides, UV radiation, fungal metabolites and antibiotics all elicit accumulation of phytoalexins (18, 19, 39). Since susceptibility and resistance are terms based on economic considerations or appearance rather than on metabolic function, it is not surprising to observe resistance mechanisms in susceptible plants. The limitation of lesion size may be deter-mined by phytoalexin accumulation, part of a mechanism for resist-ance, in susceptible plants, as defined by economic considerations (37, 43). Clearly resistance and susceptibility, host or non-host are not determined by the biosynthetic capability or incapability to produce phytoalexin.

DO DIFFERENT MICRO-ORGANISMS OR RACES OF THE SAME MICRO-ORGANISM ELICIT ACCUMULATION OF DIFFERENT PHYTOALEXINS ?

As stated earlier, many fungi elicit the accumulation of pisa-tin in pea (11), phaseollin in green bean (12) rishitin in potato (47) and this observation is valid for the other chemically char-acterized phytoalexins. In plant-parasite interactions, however, more than one phytoalexin has been reported to accumulate and quan-titative differences in different phytoalexins have recently been reported in response to inoculation with different fungi. Green bean hypocotyl inoculated with *Helminthosporium carbonum*, a pathogen of corn, accumulates little phaseollin, but other phytoalexins, a major one probably being kievitone, readily accumulate (38). Kiev-itone is the major phytoalexin accumulating in young and intermed-iate age lesions on green bean inoculated with *Rhizoctonia solani*, whereas phaseollin and kievitone levels are approximately equal in mature lesions (43). *C. lagenarium* elicits the accumulation of high levels of kievitone but little phaseollin in green bean hypocotyls (14). Phaseollin is the major phytoalexin accumulating in diffu-sates on green bean pods infected with many phytopathogens and sap-rophytes, but with many cultivars of bean inoculated with the non-pathogen of bean, *Monilinia fructicola*, phaseollidin is the major phytoalexin (10). The β, γ and δ races of *C. lindemuthianum* elicit phaseollin accumulation in green bean cultivars and it appears to be the major, but not the only phytoalexin (2, 4, 14). Since tobac-co necrosis virus also elicits accumulation of phaseollin, phaseol-lidin, phaseollinisoflavan and kievitone, degradation by fungal enzymes is not a sole consideration in explaining quantitative differences in amounts of these phytoalexins (2). Clearly, differ-ent phytoalexins are not produced by the host in response to inoc-ulation with each different infectious agent. The infectious agent can, however, influence the magnitude of accumulation by methods

other than degradation. The presence of a number of phytoalexins
associated with infection is consistent with their possible involvement in disease resistance. It is extremely unlikely that a single
phytoalexin could confer resistance to all non-pathogens. The control of the quantities of different phytoalexins which accumulate
in different interactions does introduce a degree of specificity
which may be pertinent to determining the nature of an interaction.

ARE NON-PATHOGENS OF A HOST SELECTIVELY SENSITIVE TO ITS PHYTOALEXINS ?

Extensive work by Cruickshank and his colleagues suggested
that pathogens of pea and bean were less sensitive to inhibition
of growth by pisatin and phaseollin, respectively, than were non-
pathogens (11, 12). Exceptions to this generalization, however, are
noted by Cruickshank and recently have become increasingly evident
in the literature (2, 43, 48). Cruickshank indicated (8) that pathogens of pea and green bean were generally associated with the accumulation, in pod diffusates, of concentrations of phytoalexins below
their ED_{50}, whereas non-pathogens were associated with concentrations in excess of their ED_{50}. Since the ED_{50} is an arbitrary and
highly variable value, the significance of which cannot be extra-
polated to development of an infectious agent *in vivo*, the importance of this observation remains to be established. Clearly, neither
sensitivity of an infectious agent to a phytoalexin nor the magnitude of its accumulation can individually determine whether an infectious agent will parasitize a host. A degree of specificity may
be evident in the sensitivity of infectious agents to phytoalexins
relative to the time and magnitude of phytoalexin accumulation.

ARE PHYTOALEXINS HOST SPECIFIC ?

Phytoalexins are at best specific for different plant families
though some phenolics, such as chlorogenic acid and its oxidation
products, appear ubiquitous in plants. Rishitin accumulates in potato and tomato (42). Lubimin has been isolated from infected potato, eggplant and *Datura stramonium* (45). Capsidiol has been isolated from pepper (5) and tobacco (3), and medicarpin is found in
red clover and alfalfa (23).

DOES PHYTOALEXIN DEGRADATION DETERMINE SPECIFICITY IN HOST-PARASITE INTERACTION ?

The accumulation of phytoalexin at the site of interaction
can be influenced by at least three phenomena : the rate of phytoalexin synthesis, the rate of its degradation by host enzymes, and

by the rate of its degradation by the parasite. Liberation of phytoalexin from preformed compounds by host or microbial enzymes is an
additional consideration. Rishitin and phytuberin in potato (13,
41) and glyceollin in soybean (25, 39) increase in infected or stressed tissues for 72 - 120 hours and then decrease to levels which
often approach those of uninfected tissues. This degradation occurs
as the result of host rather than microbial enzymes since chemicals
and fungal sonicates elicit the same pattern of accumulation. Specificity in plant-parasite interaction may be influenced, therefore,
by the time and efficacy with which host enzymes capable of degrading phytoalexins are activated or synthesized.

Infectious agents also degrade phytoalexins. Thus, the δ race
of *C. lindemuthianum* metabolizes phaseollin to yield 6a-hydroxy-
phaseollin and 6a,7-dihydroxyphaseollin (6). Both metabolites are
fungitoxic, but they disappear from a liquid medium within a twelve
hour period and may, therefore, be intermediates in a detoxication
process. Evidence has been presented from numerous other *in vitro*
studies that a number of pathogens are capable of degrading phytoalexins produced by their particular host. Many non-pathogens,
however, also degrade phytoalexins although, in a few reported
cases, to a more limited extent than do pathogens (20, 22). Phytoalexin degradation *in vitro* may have little relation to the rate of
phytoalexin degradation *in vivo*. Time studies of the accumulation
of phytoalexins and their degradation products in inoculated non-
hosts and susceptible and resistant cultivars of host plants are
necessary, and even these studies are difficult to interpret. The
alfalfa pathogen, *Stemphylium botryosum*, has been demonstrated to
degrade medicarpin *in vitro* and in leaf diffusates (21). The fungus also degrades pisatin and phaseollin in the non-hosts pea and
green bean, respectively (20). Early conversion products of the
phytoalexins appeared as inhibitory as the parent compounds, but
toxicity decreased markedly with further conversion. Both phytoalexins and their early conversion products were detected in pod
diffusates and in infected detached leaves. 1a-Hydroxyphaseollone,
a metabolic product of phaseollin produced *in vitro* by the bean
pathogen *Fusarium solani* f. sp. *phaseoli*, was isolated from bean
pod diffusates and hypocotyls infected with the fungus (48). Hydro-
xyphaseollone was less inhibitory to the growth of eight fungi tested than was phaseollin. Capsidiol is oxidized by many fungi *in
vitro* and *in vivo* to the less fungitoxic ketone capsenone (46).
The oxidation does not appear associated with pathogenicity on pepper to a number of *Fusarium* species. *Alternaria alternata* and *Fusarium oxysporum*, both pathogens of pepper, were inhibited *in vitro*
by concentrations of capsidiol and capsenone which apparently accumulate in the host (52).

The production of microbial enzymes to detoxicate phytoalexins
may be a factor in determining specificity of plant-parasite interaction. However, conclusive data that relate detoxication of
phytoalexin to susceptibility or resistance is lacking.

ARE THE TIME AND MAGNITUDE OF PHYTOALEXIN ACCUMULATION SPECIFIC
DETERMINANTS OF INTERACTION ?

In the potato-*Phytophthora infestans* (40, 41, 49 , 50), green
bean-*Colletotrichum lindemuthianum* (4, 38), soybean-*Phytophthora
megasperma* var. *sojae* (25) and other interactions of fungi with
resistant or susceptible hosts, rapidity of phytoalexin accumulation
may be as critical in determining the nature of an interaction as
is the maximum amount of phytoalexin which accumulates. Necrosis
and restriction of microbial growth in 'hypersensitive-type' resistance occurs quickly after penetration of the infectious agent
into the host and these are accompanied by a concomitant accumulation of phytoalexin. It is also clear that necrosis, only one of
many manifestations of stress, is not a requirement for phytoalexin
accumulation. Nevertheless, the rapidity and magnitude of phytoalexin accumulation may reflect the advent and extent of stress in
resistance associated with hypersensitive reactions. Clearly the
initial event in determining the interaction would not be the phytoalexin but rather the factor(s) controlling the compatibility of
host and infectious agent. The phytoalexin, though possibly important in restricting development of the infectious agent, would be
one of the end products of the interaction. Since viruses can cause
lesions and phytoalexin accumulation, it is possible that compatibility and incompatibility are based on an active response of the
host (host factor) and are not merely the effect of a microbial metabolite which causes necrosis. Does the unreplicating virus particle elicit phytoalexin accumulation or does the rapid replication
of virus particles place a stress on the host and is it this stress
that results in phytoalexin accumulation ? It is entirely possible
that cell wall fractions of infectious agents, as suggested by
Currier (13) and referred to by Albersheim during this meeting, are
active elicitors of phytoalexin accumulation. Susceptibility and
hypersensitive-type resistance appear dependent on the rate with
which processes occur and one of these processes controls phytoalexin accumulation. The presence of specific elicitors produced
by an infectious agent which determine the nature of an interaction
has been suggested (26). The role of specific elicitors, however,
is confused by the presence of non-specific elicitors. A case can
be made for specificity in some host-parasite interaction based on
the rate with which an event occurs, such as phytoalexin accumulation, rather than the presence or absence of a genetic, and hence
metabolic potential, for the event.

IS ALL RESISTANCE DEPENDENT ON PHYTOALEXINS ?

In the course of our studies we have encountered three host-
parasite interactions in which high resistance has been elicited
in plants without evidence for the participation of phytoalexins :
green bean-*Colletotrichum lindemuthianum* , pear-*Erwinia amylovora*,

and cucumber-*Colletotrichum lagenarium*.

Resistant interactions with *C. lindemuthianum* (race-specific resistance) and resistant interactions with *C. lagenarium*, a non-pathogen of bean, induce both local and systemic resistance against anthracnose (14, 15). These inducers clearly interact with host tissue. All of them penetrate host cells and stimulate hypersens-itive or resistant interactions. Penetrated host cells granulate, brown, and collapse, and the fungus is contained. The accumulation of phytoalexins is associated with these changes. Clearly the re-action of bean to these organisms involves metabolic changes asso-ciated with cell death and the synthesis and accumulation of com-plex organic molecules. This does not imply that cell death as such is responsible for containment of the fungi.

In experiments in which local resistance is induced[*], the organism used later as a challenge is applied to the same sites in which the inducing interactions occur. Thus, both inducer and the challenge interact with the same group of cells. Spores of the inducer germinate and the inducer penetrates the host. Spores of the challenge also germinate, and the challenge penetrates and de-velops within the tissue, giving rise in many sites to minute les-ions. However, the normally extensive development of the challenge within host tissue does not occur. Thus, protection does not ap-pear to be due to interference with pre-penetration development of the challenge. In locally induced resistance, the accumulation of phytoalexins in response to the inducers is probably a factor con-tributing to the containment of the challenge and prevention of disease.

In experiments in which systemic induced resistance is demon-strated, sites of interactions with inducer and challenge are sep-arated on the hypocotyl and the inducer is limited to the induction site. The inducer interacts with one group of cells; the challenge interacts with a different group of cells and both groups of cells respond resistantly to the fungus. Since phytoalexins are often associated with the expression of resistance to and containment of fungi, we investigated the possibility that systemic induced re-sistance is due to accumulation to toxic levels of phytoalexins in systemically protected tissue prior to interaction with the chal-lenge. Thin-layer chromatographic analysis (TLC) of extracts of tissues containing inoculated induction and uninoculated challenge sites indicated that phytoalexins accumulated in induction sites over the time span of the experiments, but they were not detected in tissue that was systemically protected but had not been inoculated

[*] Editors' note. Resistance is induced by inoculating plants with an avirulent pathogen, the inducer; it is expressed against the challenge which is a potentially virulent pathogen inoculated later.

with the challenge. Phytoalexins did accumulate in systemically
protected tissue inoculated with the challenge. This accumulation
may account for the containment of the fungus in the tissue.

The absence of phytoalexins in systemically protected, unchal-
lenged tissue, and their presence in protected, challenged tissue,
suggest that the process of induction of resistance may involve a
conditioning of cells, an activation of a potential to respond resistantly rather than the response itself. The phytoalexins associated with resistance may be one of the end products arising
from the realization of this potential, but accumulation is a phenomenon that is distinct from the conditioning of cells to accumulate
phytoalexin.

Although phytoalexin accumulation appears to be associated with
both local and systemic induced resistance, the mechanisms of the
two forms of induced resistance appear to be distinct. This conclusion is based primarily on results of experiments in which protection of heated and unheated plants were compared. Heat treatment
consisted of exposure of bean seedlings to a temperature of 37°C
for 12 hours prior to inoculation, a treatment insufficient to break
race-specific resistance to *C. lindemuthianum* or resistance to *C.
lagenarium*. Local protection was equally effective with heated and
unheated plants. The accumulation of phytoalexins in response to
the inducers was not diminished by heating. In fact, the accumulation of these compounds appeared to be enhanced in induction sites
of heated plants. In contrast, heat applied either before inoculation with inducers or after inducing interactions were well under-
way, greatly reduced the effectiveness of systemic protection.
Large lesions developed in many challenge sites of heated plants.
Resistant reactions occurred in corresponding sites of unheated
plants. Heat treatment had a similar effect on resistance of mature
hypocotyl tissue. Instead of containment of the fungus in a small
volume of tissue, as occurs in unheated plants, large lesions developed. This similarity suggests that resistance of mature tissue
and systemic induced resistance should be grouped together and distinguished from race-specific resistance to *C. lindemuthianum*, resistance to *C. lagenarium*, and local protection. This conclusion
is supported by observations that the challenge develops to a similar extent in mature tissue and in systemically protected tissue,
that both forms of resistance are race non-specific, and that both
appear to be effective in all cultivars of bean.

Protection of 'Bartlett' pear and 'Jonathan' apple against fireblight, incited by *E. amylovora*, has been reported with avirulent
E. amylovora, *E. herbicola*, and *Pseudomonas tabaci* (17, 33, 53).
Recently it was demonstrated that cell-free sonicates of avirulent
and virulent *E. amylovora* protect against fireblight (33). The
sonicates were not inhibitory to the pathogen and did not alter its
virulence. Further investigations (34) demonstrated that DNA from

virulent or avirulent *E. amylovora* is the active protectant.

Sonicates of *E. amylovora* protected the shoots of young trees, etiolated and green germinated seedlings of pear, but protection was lost when nucleic acids were precipitated with protamine sulfate. The growth of virulent *E. amylovora* in protected and unprotected etiolated seedlings was approximately equal. Nucleic acids purified from sonicates protected, and treatment with DNase but not with RNase destroyed activity. The nucleic acids isolated from sonicates were separated into two 258 nm absorbing peaks by linear log sucrose gradients. The two peaks were found to be RNA and DNA by treatment of nucleic acid preparations with DNase or RNase prior to centrifugation. DNA reisolated from the sucrose gradients protected as did DNA isolated by the Marmur technique. Protection of etiolated seedlings against fireblight was found to depend on DNA concentration and DNA did not affect the growth of *E. amylovora in vitro*. *E. amylovora* remained virulent when grown in culture with DNA from the avirulent bacterium. Melting curves indicated the DNA was essentially native DNA with a relatively high molecular weight. Isolated DNA from cesium chloride centrifugation also protected against fireblight.

DNA was infused into seedlings from which the radicle had been excised and the seedlings were subsequently often observed to form roots. Rooted seedlings were transplanted into vermiculite and soil and kept between baking dishes containing wet filter paper. Seedlings maintained in this manner for one week were challenged 0.5 cm below the cotyledons. Control seedlings showed fireblight symptoms within two days of challenge, whereas seedlings protected with DNA showed no symptoms for at least one week. This demonstrates the persistence of protection with DNA and that protection occurs even at sites removed from the point of initial DNA application.

Lack of protection by bovine serum albumin, RNA from *E. amylovora*, apoferritin, polypeptides, polyanions, polycations, salmon sperm DNA, nitrogen bases, ribose, deoxyribose, nucleosides and nucleotides indicates that protection is not directly related to size or charge. Increased protection by relatively unsheared DNA (Marmur technique) suggests some relationship between native composition of the DNA and protection. Protection of pear with DNA from virulent *E. amylovora* suggests that the virulent bacterium initiates interactions with the host which are responsible for symptom expression before its DNA can protect. Protection requires a time period between injection of DNA and challenge. Aside from the possibilities for the control of fireblight, the work with DNA and protection also raises some questions concerning specificity. In protected tissue the virulent pathogen multiplies but does not cause disease. *E. herbicola* and avirulent *E. amylovora* also multiply in apple and pear, do not cause disease, and protect against virulent *E. amylovora*.

Specificity in these instances is not dependent solely upon whether
the bacterium multiplies, so that phytoalexin involvement in resistance appears unlikely.

Eight cultivars of cucumber, susceptible to *C. lagenarium*
race 1, were systemically protected against disease caused by the
pathogen by prior inoculation with race 1 of the pathogen (32).
Inoculation of a single leaf of young cucumber plants with a suspension of spores of *C. lagenarium* protected the leaf above against
disease caused by a subsequent inoculation with *C. lagenarium*. Repeated inoculations of protected plants enhanced protection and
plants remained protected throughout the ten week duration of experiments. New growth of protected and repeatedly inoculated plants
was symptomless. Protection was also elicited by low levels of
inoculum which caused two to ten lesions/plant. Damage to a leaf
by dry ice did not protect the leaf above against *C. lagenarium*.
After the unprotected leaf developed symptoms, it elicited protection
for the leaf above and this leaf was often symptomless. Though
Cladosporium cucumerinum protected cucumber cultivars resistant to
C. cucumerinum against *Colletotrichum lagenarium*, *C. lagenarium* protected cucumber varieties against *C. lagenarium* regardless of their
resistance or susceptibility to *C. cucumerinum*. The phenomenon of
protection reported suggests a remarkable resiliency and versatility
inherent in a plant's mechanism for disease resistance.

In summary, phytoalexins appear part of a plant's mechanism for
disease resistance. Specificity of interaction may be influenced by
quantitative differences in the phytoalexins which accumulate. These
differences may in turn be influenced by rate of synthesis, rate of
degradation by host and rate of degradation by infectious agents.
Sensitivity of the infectious agent to the phytoalexin, due to factors other than detoxication, may also influence specificity. Rapid
collapse of host cells and phytoalexin accumulation are characteristic of resistance in tissues which exhibit hypersensitive reactions.
In these cases, the phytoalexin is unlikely to be the elicitor or
controlling factor in determining the rapidity of response. Suppression of the resistance response by a successful pathogen may determine specificity in some interactions. Certainly mechanisms
for active susceptibility are as feasible as those for resistance.
The cellular commitment for interaction occurs soon after penetration of the host by the infectious agent; cell wall or membrane
components of the infectious agent and host may influence the character of the interaction and quantitative and qualitative differences
in the accumulation of phytoalexins.

Non-host resistance and the resistance or susceptibility of host
cultivars have not yet been explained on the basis of specific elicitors in infectious agents which control phytoalexin accumulation.
Many problems relating to the role of phytoalexins in resistance
remain unanswered. Are they host products produced after death of

the fungus, and hence not of major importance in determining the resistance in the *P. infestans*-potato interaction (16) ? The presence of substantial quantities of pisatin in young expanding lesions containing the phytoalexin-sensitive root-rot pathogen of pea, *Aphanomyces euteiches* is not consistent with the suggested role of pisatin as a disease resistance factor (36). Economic definitions of disease have been confused with biochemical and physiological definitions. Clearly, some cases of non-host and cultivar resistance, as well as induced resistance, have not been explained within the framework of the 'phytoalexin theory'. Nevertheless, the research literature strongly suggests that phytoalexins have a role in containing infectious agents and hence a role in disease resistance. It is not evident to the author that they are the initial determinants of specificity in interactions.

REFERENCES

1.   BAILEY, J. (1974). The relationship between symptom expression and phytoalexin concentration in hypocotyls of *Phaseolus vulgaris* infected with *Colletotrichum lindemuthianum*. *Physiol. Pl. Path.*, 4, 477 - 488.

2.   BAILEY, J. and BURDEN, R. (1973). Biochemical changes and phytoalexin accumulation in *Phaseolus vulgaris* following cellular browning caused by tobacco necrosis virus. *Physiol. Pl. Path.*, 3, 171 - 177.

3.   BAILEY, J., BURDEN, R. and VINCENT, G. (1975). Capsidiol : An antifungal compound produced in *Nicotiana tabacum* and *Nicotiana clevelandii* following infection with tobacco necrosis virus. *Phytochemistry*, 14, 597.

4.   BAILEY, J. and DEVERALL, B. (1971). Formation and activity of phaseollin in the interaction between bean hypocotyls (*Phaseolus vulgaris*) and physiological races of *Colletotrichum lindemuthianum*. *Physiol. Pl. Path.*, 1, 435 - 449.

5.   BIRNBAUM, G., STOESSL, A., GROVER, S. and STOTHERS, J. (1974). The complete stereostructure of capsidiol. X-ray analysis and ^{13}C nuclear magnetic resonance of eremophilane derivatives having trans-vicinal methyl groups. *Can. J. Chem.*, 52, 993 - 1005.

6.   BURDEN, R., BAILEY, J. and VINCENT, G. (1974). Metabolism of phaseollin by *Colletotrichum lindemuthianum*. *Phytochemistry*, 13, 1789 - 1791.

7.   CHESTER, K. (1933). The problem of acquired physiological
      immunity in plants. *Q. Rev. Biol.*, 8, 129 - 154, 275 -
      324.

8.   CRUICKSHANK, I. (1963). Phytoalexins. *A. Rev. Phytopath.*, 1,
      351 - 374.

9.   CRUICKSHANK, I., BIGGS, D. and PERRIN, D. (1971). Phytoalexins
      as determinants of disease reaction in plants. *J. Indian
      bot. Soc.*, 50A, 1 - 11.

10.   CRUICKSHANK, I., BIGGS, D., PERRIN, D. and WHITTLE, C. (1974).
      Phaseollin and phaseollidin relationships in infection
      droplets on endocarp of *Phaseolus vulgaris*. *Physiol. Pl.
      Path.*, 4, 261 - 276.

11.   CRUICKSHANK, I. and PERRIN, D. (1963). Studies on phytoalexins.
      VI. Pisatin : the effect of some factors on its formation
      in *Pisum sativum* L., and the significance of pisatin in
      disease resistance. *Aust. J. biol. Sci.*, 16, 111 - 128.

12.   CRUICKSHANK, I. and PERRIN, D. (1971). Studies on phytoalexins.
      XI. The induction, antimicrobial spectrum and chemical
      assay of phaseollin. *Phytopath. Z.*, 70, 209 - 229.

13.   CURRIER, W. (1974). *Characterization of the induction and
      suppression of terpenoid accumulation in the potato -
      Phytophthora infestans interaction.* Ph.D. Thesis, Purdue
      University, 114 pp.

14.   ELLISTON, J. (1975). *A histological and biochemical study of
      local and systemic protection of* Phaseolus vulgaris
      against Colletotrichum lindemuthianum *as elicited by
      fungi.* Ph.D. Thesis, Purdue University, 271 pp.

15.   ELLISTON, J. and KUĆ, J. (1975). Metabolic control of the re-
      sistance of bean to *Colletotrichum lindemuthianum*. *Kagaku
      to Seibutsu*, 13, 522 - 525.

16.   ÉRSEK, T., BARNA, B. and KIRÁLY, Z. (1973). Hypersensitivity
      and the resistance of potato tuber tissues to *Phytophthora
      infestans*. *Acta phytopath. Acad. Sci. hung.*, 8, 3 - 12.

17.   GOODMAN, R. (1967). The protection of apple stem tissue a-
      gainst *Erwinia amylovora* infection by avirulent strains
      and three other bacterial species. *Phytopathology*, 57,
      22 - 24.

18. HADWIGER, L. (1972). Increased levels of pisatin and pheny-
    lalanine ammonia lyase activity in *Pisum sativum* treated
    with antihistaminic, antiviral, antimalarial, transquil-
    izing or other drugs. *Biochem. biophys. Res. Commun.*,
    46, 71 - 79.

19. HADWIGER, L. and SCHWOCHAW, M. (1971). Ultraviolet light-
    induced formation of pisatin and phenylalanine ammonia
    lyase. *Pl. Physiol., Lancaster*, 47, 588 - 590.

20. HEATH, M. and HIGGINS, V. (1973). *In vitro* and *in vivo* con-
    version of phaseollin and pisatin by an alfalfa pathogen
    Stemphylium botryosum. *Physiol. Pl. Path.*, 3, 107 - 120.

21. HIGGINS, V. and MILLAR, R. (1969). Degradation of alfalfa
    phytoalexin by *Stemphylium botryosum*. *Phytopathology*, 59,
    1500 - 1506.

22. HIGGINS, V. and MILLAR, R. (1970). Degradation of alfalfa
    phytoalexin by *Stemphylium loti* and *Colletotrichum phomo-
    ides*. *Phytopathology*, 60, 269 - 271.

23. HIGGINS, V. and SMITH, D. (1972). Separation and identification
    of two pterocarpenoid phytoalexins produced by red clover
    leaves. *Phytopathology*, 62, 235 - 238.

24. INGHAM, J. (1972). Phytoalexins and other natural products as
    factors in plant disease resistance. *Bot. Rev.*, 38, 343 -
    424.

25. KEEN, N. (1971). Hydroxyphaseollin production by soybeans
    resistant and susceptible to *Phytophthora megasperma* var.
    sojae. *Physiol. Pl. Path.*, 1, 265 - 275.

26. KEEN, N. (1975). Specific elicitors of plant phytoalexin pro-
    duction : determinants of race specificity in pathogens ?
    Science, N.Y., 187, 74 - 75.

27. KOSUGE, T. (1969). The role of phenolics in host response to
    infection. *A. Rev. Phytopath.*, 7, 195 - 222.

28. KUĆ, J. (1966). Resistance of plants to infectious agents.
    A. Rev. Microbiol., 20, 337 - 370.

29. KUĆ, J. (1972). Phytoalexins. *A. Rev. Phytopath.*, 10, 207 -
    232.

30. KUĆ, J. (1973). Metabolites accumulating in potato tubers
    following infection and stress. *Teratology*, 8, 333 - 338.

31. KUĆ, J. (1975). Phytoalexins, plants and human health. *Adv. Chem.* (In press).

32. KUĆ, J., SHOCKLEY, G. and KEARNEY, K. (1975). Protection of cucumber against *Colletotrichum lagenarium* by *Colletotrichum lagenarium*. *Physiol. Pl. Path.* (In press).

33. MCINTYRE, J., KUĆ, J. and WILLIAMS, E. (1973). Protection of pear against fireblight by bacteria and bacterial sonicates. *Phytopathology*, 63, 872 - 877.

34. MCINTYRE, J., KUĆ, J. and WILLIAMS, E. (1975). Protection of bartlett pear against fireblight with deoxyribonucleic acid from virulent and avirulent *Erwinia amylovora*. *Physiol. Pl. Path.* (In press).

35. METLITSKY, L., OZERETSKOVSKAYA, O., VULFSON, N. and CHALOVA, L. (1971). Lubimin in potato resistance. *Mikol. i Fitopatol.*, 5, 439 - 443.

36. PUEPPKE, S. and VAN ETTEN, H. (1974). Pisatin accumulation and lesion development in peas infected with *Aphanomyces euteiches*, *Fusarium solani* f. sp. *pisi* or *Rhizoctonia solani*. *Phytopathology*, 64, 1433 - 1440.

37. RAHE, J. (1973). Occurrence and levels of the phytoalexin phaseollin in relation to delimitation at sites of infection of *Phaseolus vulgaris* by *Colletotrichum lindemuthianum*. *Can. J. Bot.*, 51, 2423 - 2430.

38. RAHE, J., KUĆ, J., CHUANG, C. and WILLIAMS, E. (1969). Correlation of phenolic metabolism with histological changes in *Phaseolus vulgaris* inoculated with fungi. *Neth. J. Pl. Path.*, 75, 58 - 71.

39. REILLY, J. and KLARMAN, W. (1972). The soybean phytoalexin, hydroxyphaseollin, induced by fungicides. *Phytopathology*, 62, 1113 - 1115.

40. SATO, N., KITAZAWA, K. and TOMIYAMA, K. (1971). The role of rishitin in localizing the invading hyphae of *Phytophthora infestans* in infection sites at the cut surfaces of potato tubers. *Physiol. Pl. Path.*, 1, 289 - 295.

41. SATO, N. and TOMIYAMA, K. (1969). Localized accumulation of rishitin in the potato tuber tissue infected by an incompatible race of *Phytophthora infestans*. *Ann. phytopath. Soc. Japan*, 35, 202 - 207.

42. SATO, N., TOMIYAMA, K., KATSUI, N. and MASAMUNE, T. (1968).
    Isolation of rishitin from tomato plants. *Ann. phytopath.
    Soc. Japan*, 34, 344 - 345.

43. SMITH, D., VAN ETTEN, H. and BATEMAN, D.F. (1975). Accumulation
    of phytoalexins in *Phaseolus vulgaris* hypocotyls following
    infection by *Rhizoctonia solani*. *Physiol. Pl. Path.*, 5,
    51 - 64.

44. STOESSL, A. (1970). Antifungal compounds produced by higher
    plants. *Recent Adv. Phytochem.*, 3, 143 - 180.

45. STOESSL, A., STOTHERS, J. and WARD, E.W.B. (1974). Lubimin :
    A phytoalexin of several solanaceae. Structure revision
    and biogenetic relationships. *J. chem. Soc. Chem. Commun.*,
    709 - 710.

46. STOESSL, A., UNWIN, C. and WARD, E.W.B. (1973). Postinfectional
    fungus inhibitors from plants : fungal oxidation of cap-
    sidiol in pepper fruit. *Phytopathology*, 63, 1225 - 1231.

47. TOMIYAMA, K., SAKUMA, T., ISHIZAKA, N., SATO, N., KATSUI, N.,
    TAKASUGI, M. and MASAMUNE, T. (1968). A new antifungal
    substance isolated from resistant potato tuber tissue
    infected by pathogens. *Phytopathology*, 58, 115 - 116.

48. VAN ETTEN, H. and SMITH, D. (1975). Accumulation of antifun-
    gal isoflavonoids and 1a-hydroxyphaseollone, a phaseollin
    metabolite, in bean tissue infected with *Fusarium solani*
    f. sp. *phaseoli*. *Physiol. Pl. Path.*, 5, 225 - 237.

49. VARNS, J. and KUČ, J. (1971). Suppression of rishitin and
    phytuberin accumulation and hypersensitive response in
    potato by compatible races of *Phytophthora infestans*.
    Phytopathology, 61, 178 - 181.

50. VARNS, J. and KUČ, J. (1972). Suppression of the resistance
    response as an active mechanism for susceptibility in
    the potato - *Phytophthora infestans* interaction. *In :
    Phytotoxins in Plant Diseases*. (WOOD, R.K.S., BALLIO, A.,
    and GRANITI, A. Eds.), 465 - 468. Academic Press, London
    and New York.

51. VARNS, J., KUČ, J. and WILLIAMS, E. (1971). Terpenoid accumu-
    lation as a biochemical response of the potato tuber to
    Phytophthora infestans. *Phytopathology*, 61, 174 - 177.

52. WARD, E.W.B., UNWIN, C.H. and STOESSL, A. (1973). Postinfec-
        tional inhibitors from plants. VII. Tolerance of capsidiol
        by fungal pathogens of pepper fruit. *Can. J. Bot.*, 51,
        2327 - 2332.
53. WRATHER, J., KUĆ, J. and WILLIAMS, E. (1973). Protection of
        apple and pear fruit tissue against fireblight with non-
        pathogenic bacteria. *Phytopathology*, 63, 1075 - 1076.

CONTRIBUTIONS

FUCHS, A.   A possible role of aschochitine and pisatin in *Ascochyta
    pisi*-infected pea plants.

HIGGINS, V.J.   Detoxification of phytoalexins by fungal pathogens
    in relation to host specificity.

KEEN, N.T.   Specific phytoalexin elicitors of pathogen origin.

KEEN, N.T.   Specificity of production of phytoalexins by plants to
    various pathogens.

KEEN, N.T. and RICH, J.   Phytoalexins against nematode pathogens.

MANSFIELD, J.W. and HARGREAVES, J.A.   Phytoalexins and the speci-
    ficity of *Botrytis fabae* toward *Vicia faba*.

RAHE, J.   Studies on phytoalexins and the specificity of *Colleto-
    trichum lindemuthianum* on beans.

TJAMOS, E.C.   Mechanisms of resistance in near-isogenic varieties
    to *Verticillium* wilt.

VAN DEN HEUVEL, J.   Sensitivity to and metabolism of phaseollin by
    different strains of *Botrytis cinerea*.

WARD, E.W.B.   The range and interrelationships of sesquiterpenoid
    phytoalexins in the *Solanaceae*.

SUMMARY OF POINTS FROM CONTRIBUTIONS AND DISCUSSIONS
BY
N. T. KEEN

Chairman and Discussion Leader

Kuć's report on the "induced protection" of cucumber leaves by
prior inoculation with *Colletotrichum lagenarium* attracted much

comment. It was noted that the results may have little practical
value since the protective response has never been seen in the field
with fungus diseases. Although, as pointed out in the discussion,
the protection phenomenon has been observed for many years by other
workers in response to injury, virus infection, or heat treatment,
its practicality in disease control appears minimal. Kuč stated
that the cucumber protective response thus far has only been elicited with *C. lagenarium,* lasts for several days, is systemic throughout the plant, protects against other pathogens, and does not at
present appear to be related to phytoalexin synthesis.

Kuč's summary of data linking phytoalexin production to the
expression of varietal resistance in plants was amplified by a number of papers. Rahe reported that freeze-injured *Phaseolus vulgaris*
hypocotyls produced phaseollin at an appreciable rate. He postulated that perhaps accumulation of phaseollin in response to incompatible races of *Colletotrichum lindemuthianum* represents a similar
response of the plant to wounding but that compatible fungus races
in some way suppress elicitation of phaseollin caused by wounding.

Several short papers presented evidence on degradation of phytoalexins by pathogens of a particular plant. Van den Heuvel observed that *Botrytis cinerea* strains from hosts other than *Phaseolus
vulgaris* degraded phaseollin very little, but that *P. vulgaris* strains were efficient degraders. The initial degradation product was
identified as 6a-hydroxyphaseollin, which is less antifungal to *B.
cinerea* than is phaseollin. Fuchs presented evidence that virulent
races of *Ascochyta pisi* were able to degrade pisatin, but avirulent
races were not. Pisatin was degraded initially to 6a-hydroxyinermin.
Of interest was the finding that pisatin inhibited production of the
phytotoxin ascochitine by the fungus so that pisatin may function
in defense by inhibiting production of toxins as well as growth of
the pathogen. Higgins showed that *Stemphylium botryosum* grew well
on media containing medicarpin and maackiain, the phytoalexins produced by its host plant, alfalfa, but not when supplied with phaseollin and pisatin, present in the non-host plants *Phaseolus vulgaris* and pea. This supports other evidence indicating that the
ability to degrade the alfalfa pterocarpans could be a prime factor
allowing pathogenesis by *S. botryosum.* Mansfield and Hargreaves
reported that *Botrytis fabae* degraded the broad bean phytoalexin
wyerone acid to the reduced form which is not markedly antifungal;
this ability may again contribute to pathogenic potential. The
authors also utilized the fluorescent properties of the broad bean
polyacetylene phytoalexins by obtaining fluorescence spectra from
live cells in a defensively responding area of the leaf that were
indistinguishable from fluorescence spectra of authentic wyerone
and wyerone acid. The fluorescing cells were restricted to the de-
fensively-responding portion of the leaf, but cells that were already dead had no detectable fluorescence. The results are useful

since they offer convincing proof that phytoalexins can occur in
live plant cells and confirm that localization of phytoalexin production to defensively reacting tissue.

Keen and Rich stated that the hypersensitive response of lima
bean roots to the nematode *Pratylenchus scribneri* involved the accumulation of four coumestanes, two of which were identified as
coumestrol and psoralidin. *Phaseolus vulgaris*, which is susceptible
to the nematodes, did not undergo a similar chemical response. Coumestrol was found to inhibit the motility of *P. scribneri in vitro*
at 10 - 25 ppm. It is, therefore, possible that the inducibly
formed lima bean coumestanes represent phytoalexins important in
restricting the multiplication of *P. scribneri*.

Keen reviewed his research on the detection of a substance from
culture fluids of race 1 of *Phytophthora megasperma* var. *sojae*
(*Science*, 187, 74, 1975) that elicits greater production of the soybean phytoalexin glyceollin in hypocotyls of monogenically resistant Harosoy 63 soybeans than in the near isogenic susceptible line
Harosoy. The substance from race 1 has been called a *specific phytoalexin elicitor*, since it appears to be differentially recognized
by the two soybean genotypes. The race 1 specific elicitor was not
detected in culture fluids of race 3 of the pathogen which gives
susceptible reactions on both Harosoy and Harosoy 63. In addition
to the race 1 specific elicitor, gel filtration on Bio-gel P-2 columns yielded several fractions from both race 1 and race 3 that
gave identical phytoalexin responses in both soybean genotypes.
These substances are, therefore, referred to as *non-specific elicitors*. The available data support the hypothesis that specificity
in the *Phytophthora*-soybean host-parasite system may be determined
by the production or non-production of specific elicitors by various
races of the pathogen that specifically interact with the plant resistance genes to dictate high or low rates of phytoalexin production. Attempts to isolate the race 1 specific elicitor have thus
far been hampered by variability in the bioassay and by the instability of the specific elicitor during storage.

Keen also presented results with soybean plants which suggested
that the phytoalexin responses to various pathogenic and abiotic
agents are chemically different and indicate a possible specificity
with respect to the phytoalexin response of a single plant species
to various stimuli. For example, the defense reaction of soybeans
to incompatible races of *Phytophthora megasperma* var. *sojae* and to
tobacco necrosis virus involve the production of high amounts of
glyceollin and lesser amounts of coumestrol and daidzein. On the
other hand the reaction to incompatible races of *Pseudomonas glycinea* and insect damage cause less accumulation of glyceollin and
relatively greater amounts of daidzein and coumestrol. Finally,
the response of soybean leaves to injurious levels of ozone causes
large amounts of daidzein and coumestrol to accumulate but no

glyceollin was detectable. With respect to the latter data it was
noted that ozone can oxidize glyceollin.

Ward discussed recently work on the chemistry of phytoalexins
from the Solanaceae. He and Stoessl have recently isolated several
open ring compounds from pathogen-challenged Solanaceous plants that,
although not particularly antibiotic themselves, may be intermediates in the biosynthesis of the more potent phytoalexins such as
capsidiol, lubimin, and rishitin. One of these was 2, 3 dihydroxy-
germacrene, isolated from *Datura stramonium*.

Tjamos described recent work with Smith with *Verticillium albo-
atrum* and cultivars of tomato isogenic except for the *Ve*-gene for
resistance. In comparisons between stem segments from resistant
(*Ve*) and susceptible *(ve)* plants infiltrated with conidia it was
found that responses of xylem parenchyma interpreted as hypersensitive occurred more rapidly and were more extensive in resistant
than in susceptible segments and concentrations of rishitin in resistant segments (*c*. 4.4 µg/g fresh weight) were about five times
higher than in susceptible segments four days after inoculation.
The significance of these figures in relation to the striking differences in growth in and reactions of resistant and susceptible
plants to infection remains to be assessed.

INDUCTION OF HYPERSENSITIVE RESPONSES TO FUNGAL PATHOGENS

ANTJE KAARS SIJPESTEIJN

Institute for Organic Chemistry of the Organization for Applied Scientific Research in the Netherlands TNO Utrecht, The Netherlands

INTRODUCTION

Hypersensitivity as a resistance reaction covers a field so large that I will not review it generally but will deal with one disease only, tomato and *Cladosporium fulvum*, which was studied in our Institute for a doctoral dissertation by Van Dijkman (13). This disease was a fortunate choice and could well serve as a model for explaining hypersensitivity in other diseases.

I shall first make some general remarks about natural resistance before I deal with our work on biochemical mechanisms of resistance of tomato to *C. fulvum*. Finally I shall speculate on the basis of hypersensitivity in this disease and its relation to the phenomenon in other diseases.

NATURAL RESISTANCE TO DISEASE

I often feel that the ability of fungi to attack plants is strongly over-estimated. Only a few have the appropriate means to attack particular plants. Therefore, to most fungi, plants to not need a mechanism of resistance.

A fruitful approach to the understanding of natural resistance may be the study of the crucial differences between varieties of one species which are either susceptible or resistant to a particular pathogen. This approach has the advantage that genetic data can be a helpful tool in the search for the crucial biochemical differences. In some diseases the genetic data has been well established by geneticists and plant breeders. Ultimately, they reflect

all information on the biochemical mechanism of resistance so that
any mechanisms which are proposed must accord with them.

Resistance and susceptibility in certain diseases depend on
single genes. Resistance is often dominant. A difference in one
gene means primarily a difference in one protein which may or may
not be an enzyme. Examples of such diseases are cucumber and
Cladosporium cucumerinum, tomato and *C. fulvum*, apple and *Venturia
inaequalis* and oats and *Helminthosporium victoriae*.

We were interested in the biochemical factors controlled by
single genes in various diseases bearing in mind that resistance of
a host plant to its various pathogens is so far always controlled
by different genetic factors so that resistance to different pathogens has different biochemical bases.

One type of biochemical mechanism of resistance/susceptibility
dealt with elsewhere in this meeting is that of the host-specific
toxins which has been intensively studied by Wheeler, Scheffer and
their co-workers. A toxin of a pathogen is supposed to combine
with a receptor of the susceptible host and so disorganize host cell
membranes. In the resistant variety this receptor is not present
so that disease does not develop. I mention this mechanism because
in hypersensitivity with which I shall be concerned the mechanism
of resistance seems comparable but at the same time the reverse of
that postulated for diseases involving host-specific toxins. The
rest of my lecture will be concerned mainly with hypersensitivity,
characterized by a local killing of host cells following invasion
of resistant plants by the pathogen and containment of the pathogen
to the zone containing the dead cells.

RESISTANCE OF TOMATO TO *CLADOSPORIUM FULVUM*

When this work was started we had a hypothesis on the mechanism
of resistance and we were encouraged by the fact that at various
Institutes in Holland there was a wealth of suitable plant and fungal material and expert knowledge on the genetics of the disease (4,
5).

Most tomato varieties are susceptible to *C. fulvum* which attacks the leaves, especially the lower sides. A tomato plant resistant to *C. fulvum* reacts to inoculation with a hypersensitive
reaction and in such resistant plants there is little development
of the pathogen. The pathogen enters leaves of susceptible and resistant plants through stomata.

There are many physiological races of *C. fulvum*. Table 1 gives
the reaction of various tomato cultivars to some of these races.

Table 1. *Relation between tomato varieties and races of* Cladosporium fulvum [a]

Tomato variety	Dominant resistance genes	Races of *C. fulvum*						
		0	1	2	4	1.2	1.2.3	1.2.4
'Money-maker'	None	S	S	S	S	S	S	S
'LMR'[b]	*Cf1*	R	S	R	R	S	S	S
'Vetomold'	*Cf2*	R	R	S	R	S	S	S
'V473'	*Cf1 Cf2*	R	R	R	R	S	S	S
'V121'	*Cf3*	R	R	R	R	R	S	R
'Purdue 135'	*Cf4*	R	R	R	S	R	R	S
'Vagabond'	*Cf2 Cf4*	R	R	R	R	R	R	S

[a]   R - Resistant, incompatible; S - Susceptible, compatible

[b]   'Leaf mould Resister No. 1'

It shows a complicated pattern of susceptibility (compatibility) and
resistance (incompatibility), and suggests a gene-for-gene relation
(2, 5).

Genetically, resistance of tomato to *Cladosporium* is assumed
to be determined by four genes, *Cf1*, *Cf2*, *Cf3* and *Cf4*, located on
three chromosomes. Lenhardt and Kerr (6) have evidence, however,
that *Cf4* may be an allele of *Cf1*. I shall return to this later.
For the moment we will accept Table 1 which is largely according to
Hubbeling (4). Each *Cf* gene confers resistance only to certain
races of the pathogen and is ineffective for other races. This
scheme can be extended to tomatoes with other combinations of resistance genes, and to other races of the pathogen. These are indicated by the indices of all the resistance genes that are not effective against them. Thus gene *Cf1* is not effective against race
1, 1.2, 1.2.3 and 1.2.4.

Now to our hypothesis on the biochemical basis for pattern of
resistance. In earlier work at our Institute on natural resistance
of apple scab, Raa obtained strong indications that *V. inaequalis*

produced a toxin active on resistant but not on susceptible apples.
This toxin was assumed to trigger the hypersensitive reaction in resistant apples and so cause resistance (9, 10). We reasoned that
the hypersensitive reaction of tomato to *C. fulvum* might be caused
in a similar way and this is what we have tried to prove (13, 14,
15). Reduced to its simplest form (Table 2) the problem is : Why is
'LMR' susceptible to race 1, but not to race 2 and why is 'Vetomold'
resistant to race 1 and susceptible to race 2 ? Our hypothesis suggested that race 1 of *C. fulvum* produces a compound toxic to, and
causes hypersensitive reactions in tomato varieties with the gene
Cf2. This compound should not be toxic to susceptible varieties
such as 'Moneymaker' or 'LMR' which lack the gene *Cf2*. The pattern
of resistance towards race 2 is the opposite. Therefore, we must
suppose that this race produces a different toxin which is selectively toxic to tomato 'LMR' or 'V473' with gene *Cf1* but is not toxic to the susceptible varieties 'Moneymaker' or 'Vetomold' which
lack gene *Cf1*.

In Table 2 A_1 indicates the presumed ability to form a compound
toxic to tomato with gene *Cf1* and A_2 represents the ability to form
a compound toxic to tomato with gene *Cf2*. These abilities are expressed tentatively as genotypes for toxin production. They may also be
called presumed gentoypes for avirulence or, perhaps more correctly,
for incompatibility. Race 0 which cannot attack tomatoes with gene
Cf1 or *Cf2* may produce both toxins A_1 and A_2 and race 1, 2 should

Table 2.   *Presumed toxin production by* Cladosporium fulvum

Tomato variety	Dominant resistance gene	Races of *C. fulvum*			
		0	1	2	1.2
		Presumed genotype for toxin production			
		A_1A_2	a_1A_2	A_1a_2	a_1a_2
'Moneymaker'	None	S	S	S	S
'LMR'	*Cf1*	R	S	R	S
'Vetomold'	*Cf2*	R	R	S	S
'V 473'	*Cf1 Cf2*	R	R	R	S

not produce toxins active on tomatoes with gene *Cf1* or *Cf2*. In the
same way we can visualize the genes A_3 and A_4 for production of toxins selectively active on tomatoes with the gene *Cf3* or *Cf4* respectively. Thus, our hypothesis was that the various physiological
races would produce one or more toxins which selectively cause hypersensitivity in tomatoes with the corresponding resistance gene. In
the following the various races are codified by their presumed genotype for toxin production instead of by their traditional indices.

To obtain evidence for our hypothesis, Van Dijkman studied the
toxicity of filtrates from cultures of the various races towards
different tomato varieties. The first experiment investigated whether race $a_1A_2A_3A_4$ produces a toxin for 'Vetomold' and race $A_1a_2A_3A_4$
for 'LMR'. It was carried out as follows. Shake cultures of the two
races were grown in a protein-free medium for about three weeks.
Culture filtrates were freed from low molecular material by filtration over Sephadex G25 and the high molecular fraction (mol. wt. >
1 000) was freeze-dried. Mature leaves of 'LMR' and 'Vetomold' were
placed for several hours in a solution of $^{32}PO_4$, to label leaf cells.
Disks (8 mm diam.) from these leaves were transferred to a solution
of the fungal material in Tris-HCl buffer (10^{-2}M, pH 6.0). The protein concentration was adjusted to that of the unpurified culture
filtrate (2-4% w/v). Buffer alone served as control. For proper
contact the solutions were slowly infiltrated at low temperature by
evacuation of the flask. Thereafter air was slowly readmitted. Samples of the ambient fluid were taken at intervals and their radioactivity was measured.

It was expected that the infiltrated fungal material would contain a compound selectively toxic to cells of resistant host plants
but not to those of susceptible host plants. Thus resistant host
cells would be damaged and labelled compounds would escape whereas
there would be no such leakage from susceptible cells. Fig. 1 shows
leakage of labelled compounds after infiltration of products of the
two races. Leakage is expressed as a percentage of maximum leakable
from disks infiltrated with 2% (v/v) chloroform in buffer (14, 15).

In 'LMR' as well as in 'Vetomold' products of compatible races
(S) caused leakage no greater than that caused by the buffer control.
But products of incompatible races (R) increased leakage from both
plants. Thus, the fungal products apparently contain a toxin specific or selective for incompatible tomatoes. Now the race compatible to 'LMR' is incompatible to 'Vetomold' and conversely, the race
compatible to 'Vetomold' is incompatible to 'LMR'. The two races
thus produce a different toxin; one acts on 'Vetomold' and the other
acts on 'LMR'. These toxins presumably damage cell membranes, causing leakage of cell contents.

Similar experiments gave corresponding results. Thus, resistance *in vivo* and leakage were correlated for the cultivar 'V 473'

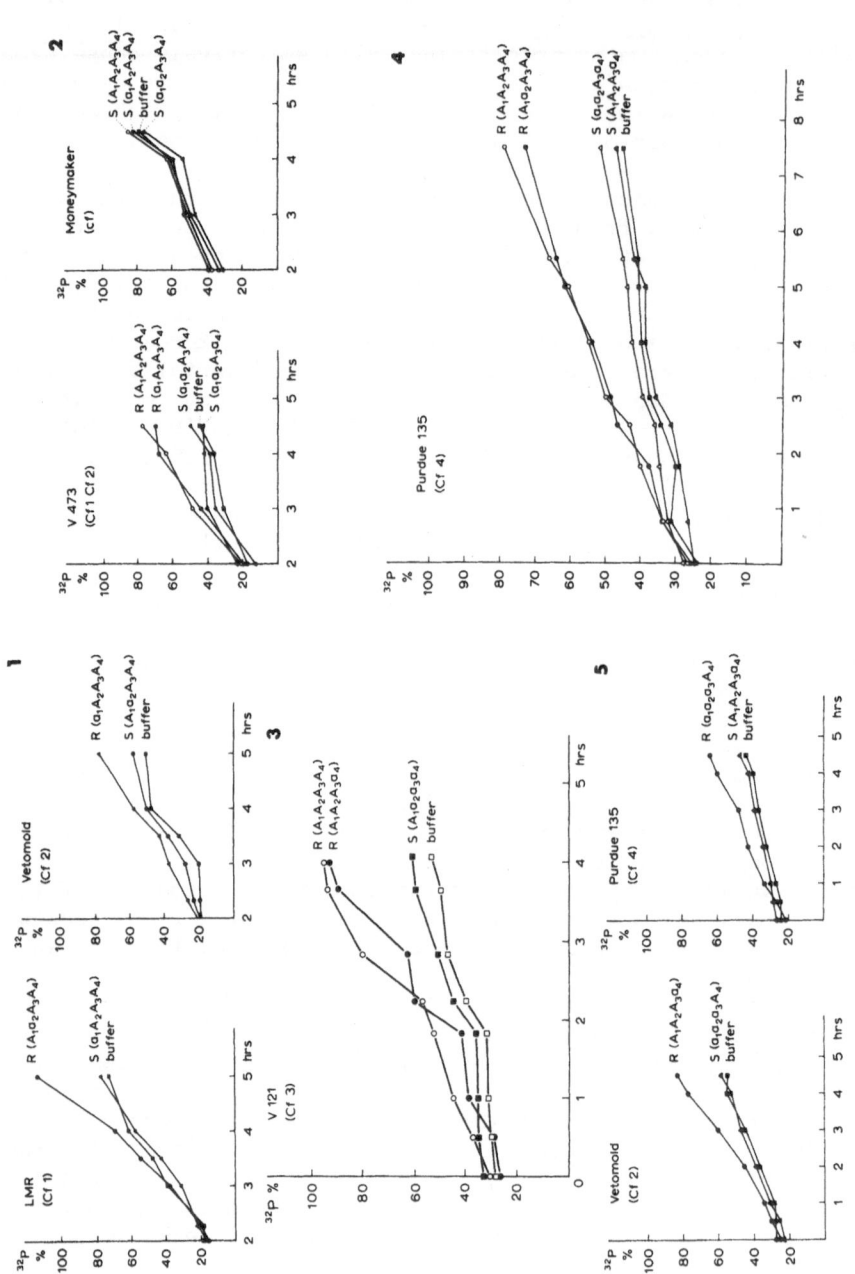

Figures 1 – 5. Leakage of ^{32}P from leaf disks of tomato cultivars following infiltration with a solution of the void volume fraction of filtrates from cultures of races of Cladosporium fulvum. Abscissa : Period of infiltration. Ordinate : ^{32}P-leakage as percentage of maximum leakable ^{32}P. R, incompatible combination; S, compatible combination; buffer, buffer control.

with resistance genes *Cf1* and *Cf2* (Fig. 2). Material from the two races causes leakage from resistant plants in excess of that from control plants. In contrast, high molecular material from the two other races which attack this tomato variety did not cause excess leakage.

In the same experiment products from three of these races were infiltrated into 'Moneymaker' (cf_1 cf_2 cf_3 cf_4) susceptible to all races of the pathogen. As expected, leakage did not exceed that of controls (Fig. 2). Fig. 3, 4 illustrate similar results with the cultivars 'V 121' (*Cf3*) and 'Purdue 135' (*Cf4*).

Leaf disks of cultivar 'Vetomold' (*Cf2*) and 'Purdue 135' (*Cf4*) were exposed to purified filtrates from cultures of races with opposite genotypes. The results (Fig. 5) again suggest that different toxins are produced by these races.

So far these and similar experiments have confirmed our hypothesis that the fungal races each produce compounds which are selectively toxic to tomatoes with resistance genes *Cf1*, *Cf2*, *Cf3* and *Cf4*. Table 3 summarizes the results of the leakage experiments.

Table 3. *Observed correlation between resistance* in vivo *and leakage by selective toxins*

Tomato variety	Dominant resistance genes	Races of *Cladosporium fulvum*							
		0	1	2	4	1.2	1.2.3	2.3.4	1.2.4
		Presumed genotype for toxin production (avirulence)							
		A_1 A_2 A_3 A_4	a_1 A_2 A_3 A_4	A_1 a_2 A_3 A_4	A_1 A_2 A_3 a_4	a_1 a_2 A_3 A_4	a_1 a_2 a_3 A_4	A_1 a_2 a_3 a_4	a_1 a_2 A_3 a_4
'Money-maker'	None	S/n	S/n	–	S/n	S/n	–	–	S/n
'LMR'	*Cf1*	–	S/n	R/l	–	–	–	–	–
'Vetomold'	*Cf2*	R/l	R/l	S/n	R/l	–	S/n	–	–
'V 121'	*Cf3*	R/l	–	–	R/l	–	–	S/n	–
'Purdue 135'	*Cf4*	R/l	–	R/l	S/n	–	R/l	–	S/n
'V 473'	*Cf1 Cf2*	R/l	R/l	–	–	S/n	–	–	S/n
'Vagabond'	*Cf2 Cf4*	R/l	–	R/l	R/l	–	R/l	–	S/n

R - Resistant, incompatible; S - Susceptible, compatible;
l - leakage in excess of control;n - no leakage in excess of control

Without exception a strict correlation was found between resistance *in vivo* (R) and the occurrence of excess leakage (l) and between susceptibility (S) and the absence of excess leakage (n) in the model experiments. The differences are significant.

These experiments suggest four genes for toxin production or avirulence (A_1 to A_4). Each avirulence gene corresponds to one resistance gene or allele in the tomato host such that each toxin increases the permeability of the cell membranes only of host plants with the corresponding resistance gene. The resistance genes or alleles in the host then may control selective receptor sites for the toxins which presumably are in host cell membranes or cell walls. There is the possibility that the fungal product is not itself toxic but becomes so only after combining with a receptor in the plant.

As mentioned above Lenhardt and Kerr (6) assume that *Cf1* and *Cf4* may be alleles and gave them the symbols $Cf1^1$ and $Cf1^2$. This certainly is not true for *Cf2* and *Cf3* because they are located on two chromosomes other than those carrying *Cf1* and *Cf4*. A *Cf5* gene has also been found, but later it was given the symbol $Cf1^3$ (6). If the *Cf4* is an allele of *Cf1*, this would mean that for each of two *Cf1* alleles a selective toxin is available in the pathogen. The avirulence genes A_1 and A_4 responsible for these toxins cannot be alleles of one gene because in certain races A_1 and A_4 are present simultaneously and *C. fulvum* is haploid. This means that the two avirulence genes of the fungus correspond to two alleles of *Cf1*.

So much for our model experiments. Let us now see what happens when an incompatible race penetrates a tomato plant, as in 'Vetomold' *(Cf2)* resistant to the races 0, 1 and 4. Bond (1) showed that an incompatible race lives intercellularly in immediate contact with the cell walls and membranes of a resistant plant. At these sites, production of even small amounts of toxin could damage membranes and cause leakage and hypersensitivity. Although hypersensitivity to *C. fulvum* may be explained along these lines the question remains as to why a hypersensitive reaction stops further development of *C. fulvum*. Of several possible explanations the most plausible one may be that the hypersensitive reaction desiccates host tissue and this prevents further development of the pathogen.

Our observations indicate strongly that the primary event which induces resistance is the reaction between toxin produced by an incompatible pathogen and a host receptor. Thus we have toxins which induce resistance which are in contrast to the host-specific toxins which induce diseases such as those caused by certain species of *Helminthosporium*.

The postulated reaction of fungal toxin A_1 with plant receptor R_1, of A_2 with R_2, A_3 with R_3 and A_4 with R_4 represents four different systems because the toxins and receptors are different in each case. This corresponds to the fact that in plants the nature of the hypersensitive reaction depends on the *Cf*-gene involved. Thus gene *Cf1*- and *Cf3*-resistance appear as more or less extended hypersensitive areas, whereas *Cf2*- and *Cf4*-resistance are localized as pinpoint lesions. When several resistance genes operate in one plant the pin-point lesions of *Cf2* and *Cf4* appear to be epistatic to the larger hypersensitive areas associated with *Cf1* and *Cf3*. This possibly depends on a more rapid reaction between toxin and host receptors in pin-point than in the larger lesions.

SPECULATIONS ON MECHANISMS OF HYPERSENSITIVITY

The resistance of new varieties often becomes ineffective after they have been grown for a few years in the field. For tomato and *C. fulvum* we suggest that this arises through spontaneous mutation of an avirulence allele so that the original toxin is not produced or is replaced by another of altered molecular composition which no longer reacts with the receptor site in the host membrane so as to cause leakage. If the fungal product is a protein, the change of one triplet in its genetic code may cause sufficient change to prevent recognition by the host membrane. In this respect it is interesting to note that by irradiation of spores of *C. fulvum* Day (3) has obtained a mutation from avirulence to virulence in 'Vetomold', from race 0 to race 2. Such material might be ideal for use in our model experiments to check whether irradiation causes loss of the ability to produce the selective toxin. In other words, allele A_2 should have mutated into a_2.

The proposed mechanism of resistance of tomato to *C. fulvum* may operate in other cases of hypersensitivity which are controlled either by gene-for-gene relations or more simply. However, the *C. fulvum* system has obvious advantages. Thus in contrast to rusts and mildews the pathogen can easily be cultivated *in vitro* so that culture filtrates are readily obtained. Also, many well-defined host cultivars and fungal races are available. The fact that *C. fulvum* is haploid simplifies its study in some respects but the absence of a sexual stage makes genetical studies difficult. In certain other diseases where hypersensitivity is controlled by gene-for-gene relations, physiologic races are distinguished by avirulence not by virulence genes because avirulence is dominant. It is tempting to assume that in these diseases, too, the set of avirulence genes may control production of selective fungal products which induce in appropriate host plants membrane leakage and hypersensitive resistance reactions. The resistance genes in the host plant would again control receptor sites for the fungal products. Rohringer's paper in this Institute suggests a similar system in stem rust of wheat.

Studies at our Institute by Raa (9, 10), and later work by Pellizari *et al.* (8) indicate that avirulence genes of *Venturia inaequalis* may control production of selective toxins which cause

membrane leakage in incompatible apple plants. Resistance in other
diseases such as those caused by *Phytophthora infestans, Bremia
lactucae* and rust fungi may be of a similar nature.

For bean and *Colletotrichum lindemuthianum* neither Skipp and
Deverall (11), nor Mercer *et al.* (7) discovered selectivity in the
toxicity of filtrates from cultures of the various fungal races in
compatible and non-compatible host-fungus combinations. Effects on
the host were non-specific but because the culture filtrates were
not purified, specific effects may have been masked by the non-
specific effects.

Hypersensitive resistance reactions are not limited to diseases
characterized by a gene-for-gene relation. They also occur in diseases for which only one resistance gene and only one race of the
pathogen are known. Here, too, hypersensitivity could be caused by
products of the fungus which react only with the resistant host.
But it would be rash to assume that hypersensitivity will always be
explained in this way.

I should now like to add some more data and speculate further
on the chemical nature of the toxins of *C. fulvum* and of their receptors in tomato plants. In view of their specificity and high
molecular weight the fungal products might well be immediate gene
products or closely related substances, possibly proteins or glycoproteins. The high molecular weight fractions of the filtrates from
cultures of 15 isolates of *C. fulvum* were submitted to gel electrophoresis and their protein patterns were compared (16). These isolates represented 9 physiological races and we wondered whether certain protein bands would correlate with certain avirulence alleles
in the fungus. Surprisingly, each of the 15 isolates including some
with the same avirulence genotype had a different pattern of protein
bands. None of the protein bands of the different isolates could be
correlated with avirulence genes. However, even if the toxins are
polypeptides they may be only a very small fraction of the total
protein excreted by *C. fulvum* and hence be difficult to detect by
electrophoresis.

What could be the chemical basis of toxin-receptor interaction ?
In nature many cases are known of highly selective reactions between
different organisms or products originating from them. This problem
is dealt with by other lecturers in this Institute. We know that
invading organisms, or their metabolites, are often selectively recognized by receptors of the invaded host, with recognition leading
in many cases to inhibition of growth or activity of the invader,
such as in immunological phenomena involving antibody response to
invading organisms or to foreign blood cells that carry antigens.
Recognition often depends on protein recognizing a specific carbohydrate structure that may be present as a glycoprotein. In plant
pathology such phenomena have hardly been studied so far but a fas-
cinating example of recognition seems to be the interaction of the

protein-like receptor in membranes of maize with the toxin excreted by *Helminthosporium sacchari* as proposed by Strobel (12). This toxin carries a carbohydrate chain which selectively combines with the receptor protein. It is of interest, too, that in recent years many plants proteins, the lectins, have been found to possess highly specific binding sites for sugars. The receptors in the tomato plant controlled by the *Cf*-genes may be lectins and the fungal toxins controlled by avirulence genes may be glycoproteins or carbohydrates located on the cell surface and excreted into the medium. This would not be surprising since it is known that many fungal cells carry antigenic cell wall compounds.

Future work will tell us in how far the mechanism of hypersensitivity encountered in tomato and *C. fulvum* also operates in other host-pathogen relations. It is surely a field of study which demands close cooperation between research workers belonging to various disciplines.

REFERENCES

1. BOND, T.E.T. (1938). Infection experiments with *Cladosporium fulvum* Cooke and related species. *Ann. appl. Biol.,* 35, 277 - 307.

2. DAY, P.R. (1956). Race names of *Cladosporium fulvum. Rep. Tomato Genet. Coop.,* 6, 13 - 14.

3. DAY, P.R. (1957). Mutations to virulence in *Cladosporium fulvum. Nature, Lond.,* 179, 1141 - 1142.

4. HUBBELING, N. (1966). Identificatie van nieuwe fysio's van *Cladosporium fulvum* en selectie op resistentie daartegen bij de tomaat. *Meded. Rijksfac. Landbouwetensch. Gent,* 31, 925 - 930.

5. KOOISTRA, E. (1964). Recent experience of breeding leaf mould resistant tomatoes. *Euphytica,* 13, 103 - 109.

6. LENHARDT, L.P. and KERR, E.A. (1972). New genes for resistance to tomato leaf mold, *Cladosporium fulvum. Rep. Tomato Genet. Coop.,* 22, 15 - 16.

7. MERCER, P.C., WOOD, R.K.S. and GREENWOOD, A.D. (1974). Resistance to anthracnose of French bean. *Physiol. Pl. Path.,* 4, 291 - 306.

8.   PELLIZZARI, E.D., KUČ, J. and WILLIAMS, E.B. (1970). The hyp-
     ersensitive reaction in Malus species : changes in the
     leakage of electrolytes from apple leaves after inoculat-
     ion with *Venturia inaequalis*. *Phytopathology,* 60, 373 -
     376.

9.   RAA, J. (1968). Natural resistance of apple plants to *Venturia*
     inaequalis. A biochemical study of its mechanism. Ph.D.
     Thesis, State University Utrecht, The Netherlands.

10. RAA, J. and KAARS SIJPESTEIJN, A. (1968). A biochemical mech-
     anism of natural resistance of apple to *Venturia inaequalis*.
     Neth. J. Pl. Path., 74, 229 - 231.

11. SKIPP, R.A. and DEVERALL, B.J. (1973). Studies on cross-prot-
     ection in the anthracnose disease of bean. *Physiol. Pl.*
     Path., 3, 299 - 313.

12. STROBEL, G.A. (1974). Phytotoxins produced by plant parasites.
     A. Rev. Pl. Physiol., 25, 541 - 566.

13. VAN DIJKMAN, A. (1972). Natural resistance of tomato plants to
     Cladosporium fulvum. A biochemical study. Ph.D. Thesis,
     State University Utrecht, The Netherlands.

14. VAN DIJKMAN, A. and KAARS SIJPESTEIJN, A. (1971). A biochemic-
     al mechanism for the gene-for-gene resistance of tomato
     to *Cladosporium fulvum*. *Neth. J. Pl. Path.,* 77, 14 - 24.

15. VAN DIJKMAN, A. and KAARS SIJPESTEIJN, A. (1973). Leakage of
     pre-absorbed ^{32}P from tomato leaf disks infiltrated with
     high molecular weight products of incompatible races of
     Cladosporium fulvum. *Physiol. Pl. Path.,* 3, 57 - 67.

16. VAN DIJKMAN, A., DIELEMAN, S.J. and KAARS SIJPESTEIJN, A.
     (1973). Differences in disc gel electrophorese pattern
     of soluble proteins excreted by different physiological
     races and isolates of *Cladosporium fulvum*. *Neth. J. Pl.*
     Path., 79, 70 - 80.

CONTRIBUTIONS

HIGGINS, V.J.   Ultrastructural studies of the tomato - *Cladosporium*
     fulvum interaction, in relation to the proposed involvement
     of toxin activity in specificity.

PAXTON, J.D.   Hypersensitivity in the *Phytophthora megasperma* var.
     sojae - soybean interaction.

TANI, T. The lack of a causal relationship between the initiation
of resistance and hypersensitive necrosis in oat crown rust
(*Puccinia coronata* var. *avenae*).

SUMMARY OF POINTS FROM CONTRIBUTIONS AND DISCUSSIONS
BY
D. D. CLARKE

Chairman and Discussion Leader

Discussion of the main lecture centred on two themes, firstly,
the high rates of non-specific leakage which occurred and their
probable relationship to the methods used to infiltrate leaf tissues,
and, secondly, the relationship between the specific toxins and the
hypersensitive response and the relationship of this response to
resistance. The account of this later discussion will include relevant material and discussion from the short papers given in this
session.

The high rates of non-specific leakage attracted many comments.
It was suggested that the method of infiltration probably caused
much of the leakage and that some cultivars, for example, 'Moneymaker' may be more susceptible to this kind of damage than are other
cultivars. A gentler method of infiltration might overcome this but
it was pointed out that since the active compounds had molecular
weights in excess of 1 000, rates of uptake by discs floating on
culture filtrates would be slow and bacterial contamination might be
a problem. One of the early slides shown in the lecture indicated
that infiltration at room temperature resulted in less leakage from
controls than did infiltration at $4^{O}C$ and Mussell made the interesting observation that phosphate uptake enzymes and transport enzymes
are among the most temperature sensitive in tomato plants. Thus,
low temperature infiltration may predispose tissues to leakage. However, Kaars Sijpesteijn, while acknowledging this, stated that infiltration at room temperature led to excess foaming and low temperatures were used to overcome this problem. The choice of buffer
was also questioned since it could be responsible for some leakage,
but it was stated that Tris-HCl was selected as the least toxic of
a number tested.

However, despite doubts about background leakage the impressive
outcome of the work was that the pattern of leakage of resistant
against susceptible combinations was correct in all cases.

In reply to further comments Kaars Sijpesteijn stated that
studies on the activity of the specific toxins had been confined to
leakage of phosphate. However, the accompanying chlorosis and other
changes in treated tissues indicated that general disorganization and
leakage of cell metabolites almost certainly did occur. Direct in-

volvement of the selective toxins in induction of the hypersensitive
response would require their production during the early stages of
infection prior to, or when the hyphae first made contact with epidermal cells. However, toxin production during the early stages
of growth in cultures has not been measured mainly because of the
large amounts of material required to obtain sufficient of the active components after purification.

Johnson made the interesting point that if a specific toxin is
specified by each avirulence gene, then more rapid leakage might be
expected to occur *in vitro* in interactions between hosts and parasites involving several avirulence and resistance gene pairs than
in interactions involving only one gene pair. Kaars Sijpesteijn
replied that the cultivars and fungal races used were not suitable
for this kind of investigation; a different set of host genotypes
and fungal races would be required.

The studies reported by Higgins were of considerable relevance
to this discussion. She reported that ultrastructural studies of
growth of hyphae of *Cladosporium fulvum* in tomato cultivar 'Vinequeen' *(Cf2 Cf4)* showed that changes in host cell structure occurred
in advance of the invading hyphae thus indicating involvement of a
toxin. Also, extra-cellular materials accumulated, probably as a
result of leakage from host cells. These results clearly supported
those detailed in the main lecture.

In the discussion on the relationship between hypersensitivity
and resistance it was pointed out that although loss of permeability
and necrotic reactions are components of the hypersensitive response
it is not yet clear which, or indeed if any, are involved in resistance. Possible relationships were explored in papers by Higgins,
Tani and Paxton. The main lecture had suggested tentatively that
desiccation, following tissue collapse was a factor in the cessation
of fungal growth. However, Durbin pointed out that if this were so,
then maintaining the tissues in a water saturated atmosphere should
enable fungal growth to continue. In reply, Kaars Sijpesteijn stated that in fact the hypersensitive response is best expressed in a
humid atmosphere and agreed that phytoalexins more probably explain
decrease in growth of the pathogen.

Day pointed out that while some resistance genes, e.g. *Cf2* and
Cf4, gave very restricted lesions, others, e.g. *Cf1* and *Cf3*, permitted a good deal of fungal growth and even some sporulation, yet the
results did not reveal differences in leakage from plants of these
genotypes. Kaars Sijpesteijn considered that there may be differences in amounts of toxin produced by different races, but it is
also possible that loss of permeability and cell leakage are not
directly involved in resistance, and in this connexion Higgins claimed that cessation of hyphal growth in the highly resistant cultivar
'Vinequeen' occurred before many cells were killed. She also re-

ported that hyphae became coated with osmiophilic materials and embedded in callose and suggested that this might result in an inadequate nutritional relationship between the fungus and its host. The early appearance of phenolic compounds in infection sites also suggests that a phytoalexin is involved.

Tani reported that the initial events leading to a resistance reaction in oat to crown rust *(Puccinia coronata* var. *avenae)*, occurred eight to twelve hours after inoculation, whereas collapse of host mesophyll cells did not occur until about 35 hours after inoculation. Leakage of electrolytes and metabolites was correlated with collapse of mesophyll cells and was also not related to resistance. Electron-micrographs revealed no changes in cell structure until about 20 hours after inoculation and thus at least 10 hours after resistance is initiated. Inhibitors such as blasticidin S, heat treatment and double inoculations with an incompatible and a compatible race also demonstrated that host cell collapse is not involved in inhibition of hyphal growth.

In a final paper Paxton reported that for the soybean - *Phytophthora megasperma* var. *sojae* system, the unidentified phytoalexin PA_k accumulated in living cells around the lesion. Thus resistance of a hypersensitive reaction is not necessarily confined to the necrotic cells.

In conclusion, the session covered many apsects of a range of components of hypersensitivity and their relation to resistance. It is clear that not all of the components are involved in resistance and not all the associated resistance resides in the necrotic tissue. While there are clearly common factors in the hypersensitive responses of the three host parasite systems discussed, care must be taken in attempting to extrapolate too freely from one system to another, because it must be remembered that in the final analysis each system is probably unique.

INDUCTION AND SUPPRESSION OF THE HYPERSENSITIVE REACTION CAUSED BY

PHYTOPATHOGENIC BACTERIA : SPECIFIC AND NON-SPECIFIC COMPONENTS [*]

LUIS SEQUEIRA

Department of Plant Pathology, University of Wisconsin

Madison, Wisconsin, U.S.A.

INTRODUCTION

The development of simple, yet effective techniques to introduce bacterial cells into the intercellular spaces of plant leaves
and the discovery that introduction of many incompatible phytopathogenic bacteria elicits the rapid cell collapse associated with a
hypersensitive reaction (HR) have stimulated numerous investigations
into the nature and specificity of this phenomenon (5, 12). Initial
results provided the basis for the attractive hypothesis that the
HR resulted from the interaction of a specific elicitor produced by
the bacterium and a specific receptor site on the host cell (4).
However, very little evidence has been presented for the existence
of either the elicitor or the receptor site, and the mechanism responsible for cellular collapse has remained largely unexplained.

Reports that the HR can be suppressed or interfered with by
various chemical or physical treatments have given hope that such
treatments may help to elucidate the nature of the HR. Of particular interest has been the finding that pre-treatment of tobacco
leaves with heat-killed bacterial cells would prevent the HR and
thus "protect" host cells from collapse (20). Unlike other host-
parasite systems (36), suppression of the HR by this method results
in a different type of resistance rather than in a shift to susceptibility to bacterial invasion.

* Certain of the author's studies included in this paper were sup-
  ported by the College of Agricultural and Life Sciences
  (Project 1474), and by the National Science Foundation (Grant
  No. BMS-74-17442).

The inherent complexities of the HR-induction and HR-suppression mechanisms have resulted in painfully slow progress in the study of this particular area of host-parasite interactions. Yet, sufficient data are available to allow some generalizations as to the nature of specific and non-specific components of the system. The purpose of this paper is to provide an overall view of the present status of the HR induced by phytopathogenic bacteria as it relates to the question of specificity. For convenience, I have divided the subject into two general areas : 1. the HR induction phenomenon, including the nature of the HR inducer and of the factors responsible for cellular collapse, and 2. the HR suppression phenomenon, including the nature of the suppression inducer and of the factors responsible for the protective response. It is hoped that analysis in this manner will allow a logical, stepwise examination of the various components of the system.

INDUCTION OF THE HYPERSENSITIVE REACTION

The physiological and anatomical changes that occur in tobacco cells following infiltration of the intercellular spaces with high populations (c. 5 x 10^6 cells/ml) of incompatible bacterial pathogens are well known. The changes in host cell permeability, as evidenced by electrolyte loss, that occur by six hours after infiltration (7) are accompanied by marked ultrastructural changes (9). The plasmalemma, tonoplast, and the bounding and internal membranes of chloroplasts and mitochondria are deranged. Water-soaking and desiccation of the affected tissues follow; within 18 hours, the process is complete, the host tissues have collapsed and a well delineated border separates the infiltrated area from the rest of the leaf tissues. The damage sustained by the affected host cells is : 1. very extensive in relation to the number of bacteria present per host cell, and 2. very rapid. These two facts are important; they point out several features of the possible mechanism of action of the inducer of the HR.

Number of Bacteria Necessary to Induce the HR

Opinions differ as to the ratio of bacteria to host cells that is necessary to result in visible collapse of the tissues. Stall and Cook (34) reported that in the tobacco-*Pseudomonas cichorii* system, this ratio must be near 100. When inoculum levels were below 10^6 cells/ml, the ratio of bacteria to host cells was less than 1.0 and macroscopic, visible collapse did not occur. These results are in agreement with Ercolani's conclusions (5) regarding the nature of heterologous host-pathogen combinations. From infectivity titration and mixed inoculation experiments, he determined that in heterologous combinations the production of symptoms was the result of joint, rather than independent, action of many inoculated bacteria.

This is consistent with the concept that the HR is a host reaction triggered by the concerted action of relatively large numbers of bacteria during a short incubation period.

A different point of view has been expressed by Turner and Novacky (35), who reported that at concentrations of *Pseudomonas pisi* below those necessary to cause a confluent lesion, dead tobacco leaf cells could be detected by staining with Evans blue. A 1:1 ratio between number of bacteria and dead host cells was obtained by six hours after inoculation. It was concluded that the HR occurs at the individual cell level, although the lesion is invisible to the unaided eye. The authors rationalized that massive cell death must be caused by bacteria acting independently of each other.

On the basis of the available evidence, it is difficult to reconcile these two points of view. The validity of the data presented by Turner and Novacky (35) is dependent on the specificity of the staining reaction. If, indeed, one single bacterium can cause death of a large mesophyll cell in a tobacco leaf, the mechanism involved must pertain to the action of products of this bacterium in a localized area of the host cell wall or membrane.

Rapidity of the HR Induction

The rapidity of the HR is the characteristic that separates this reaction from the compatible response. In the HR, symptoms first become apparent by six hours; in the compatible response, on the other hand, symptoms become apparent by 36 hours or later, after substantial multiplication of the pathogen has occurred. On the surface, the symptoms of the two reactions are not substantially different, but in the HR bacterial populations drop precipitously and no movement to adjoining tissues occurs; in the compatible reaction, bacteria move from the initial point of inoculation and may spread to other plant parts.

Early estimates considered that the reaction was irreversible after 20 minutes had elapsed from the time tobacco tissues were infiltrated with incompatible bacteria (14). Although this "induction" period varies with the species and population of the bacterium used, it appears to be longer than 20 minutes. In tobacco leaves infiltrated with avirulent (B1) isolates of *Pseudomonas solanacearum* at c. 10^8 cells/ml, irreversible induction occurred at approximately three hours after infiltration. This was determined by infiltrating either streptomycin or rifampicin at concentrations sufficient to cause a substantial reduction in bacterial populations (below the threshold level necessary for visual HR symptoms) at various intervals after infiltration (Fig. 1). The results with rifampicin are particularly interesting because this compound selectively inhibits bacterial RNA synthesis. Our results agree with revised estimates of the HR induction period published by Klement (13).

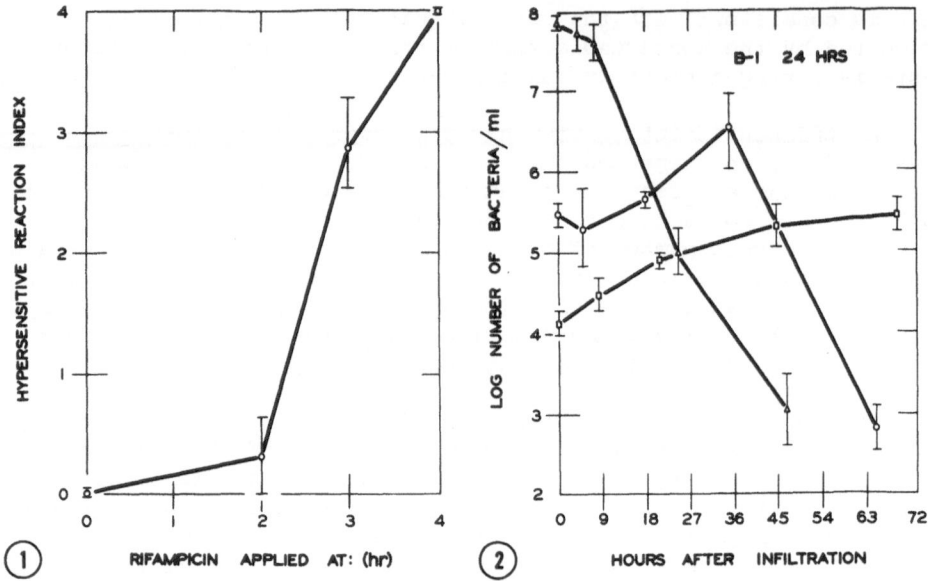

Figure 1. *Effect of rifampicin (50 μg/ml) on development of the
hypersensitive reaction (HR) in tobacco leaves inoculated
with* Pseudomonas solanacearum *B-1 (10⁸ cells/ml).   The
HR index ranged from 0 = no symptoms to 4 = complete collapse of infiltrated leaf area.*

Figure 2. *Populations of* Pseudomonas solanacearum *B-1 in tobacco
leaves following infiltration at 10^9 (Δ——Δ——),
10^7 (0——0——) and 10^5 (□——□——) cells/ml.*

There is no evidence that populations of the B1 strain of $P.$
solanacearum change during the three hour induction period.   If introduced at 10^8 cells/ml, populations remain steady for six hours
and then begin to decline after six hours;   if introduced at lower
levels, populations may increase after this initial lag until the
threshold level for induction of the HR is reached, or may decline
slowly in the absence of a visible HR (Fig. 2).

Specificity of the HR Induction

The question of specificity is related to the rapidity of induction of the HR.   In the case of $P.$ *solanacearum*, the change from
the compatible wild type (K60) to the avirulent, HR-inducing form
(B1) involves many physiological changes, including loss of the ability to form polysaccharide slime.   This is reflected in important
changes in the characteristics of the cell wall associated with its

ability to bind to lectins. Bl avirulent cells agglutinate when
treated with purified potato lectin, but most wild types of *P. solanacearum* do not agglutinate under similar treatment. The polysacha-
ride of the wild types apparently blocks the binding sites, because
agglutination of Bl cells is prevented if this polysaccharide is incorporated with the cells prior to addition of the lectin (Sequeira
and Donald, unpublished). It is possible, therefore, that differences in the cell wall characteristics allow B1, but not K60 cells
to attach to some infectible site on the host cell wall. Once attached, a metabolic product that affects the integrity of the host
plasmalemma or tonoplast appears to be released.

Regarding the infectible site hypothesis, there is evidence
from fine structure studies that attachment of bacteria to the host
cell does occur (1). At the electron microscope level, material
sectioned at three and six hours after infiltration indicated that
many B1 cells are attached to the cell wall and that the cell wall
is degraded at the point of attachment (Sequeira, de Zoeten and
Gaard, unpublished). Fibrillar material appears to attach the bacterium to the host cell wall. At the site of attachment, the host
cell wall presents a depression and the cell wall material surrounding this depression is eroded, the plasmalemma may be intact but
separated from the cell wall and is convoluted, and numerous membrane-bound vesicles accumulate in the space between the plasmalemma
and the cell wall. Thus, fine structure work provides some evidence
that there is an infectible site, as originally proposed by Ercolani
(4).

Nature of the HR Inducer (HRI)

Numerous attempts have been made to isolate an inducer from
several species of bacteria that cause the HR. To my knowledge, all
efforts have failed. Gardner and Kado (6) indicated that they had
isolated from cells of *Erwinia rubrifaciens*, by osmotic shock, a
compound that reproduced the symptoms of the HR. Similarly, Sequeira
and Ainslie (30) reported that a fraction from *P. solanacearum*, obtained by high speed grinding of the cells with glass beads, caused
a reaction similar to the HR on tobacco leaves. These reports have not
been confirmed. The problem is that it is difficult to distinguish
the typical HR from the non-specific toxicity of bacterial fractions
containing very high concentrations of protein. When these preparations are diluted down to protein concentrations similar to those
of cell populations that cause the HR, no necrosis is obtained. That
non-specific toxicity is involved may be surmised from the fact that
similar, active preparations are obtained from virulent strains of
P. solanacearum that do not induce the HR.

It is apparent that the HRI is : 1. produced only by live cells in close contact with the host cell, and 2. very unstable or readily degraded. The first point is supported by the fact that there is strong adhesion of the bacteria to the plant cell wall. At this site, induction of the HR probably is the direct result of the metabolic activity of the bacterium rather than the result of a physical change at the surfaces of the host or bacterial cell walls. This may be inferred from the fact that the induction process is time-dependent.

The available evidence on induction of the HR by phytopathogenic bacteria appears to support Ercolani's hypothesis that : 1. incompatible bacteria attach to an infectible site on the host cell wall, 2. that, as a result, bacteria become metabolically activated and release a highly unstable inducing factor (HRI), and 3. that this substance interacts with a "sensitivity locus" on the host cell, resulting in the release of the factor (HRF) that causes the HR. The HRF is ultimately responsible for the collapse of the host cell and also causes a rapid reduction in bacterial populations (Fig. 3).

Nature of the HR Factor (HRF)

That the HRF is distinct from the HRI is not clearly evident from the available information; the existence of the HRF is assumed from purely theoretical considerations. These involve : 1. the fact that the host cell literally "self-destructs" upon activation by the inducer released by relatively few or perhaps a single bacterium (35), and 2. bacterial populations begin to decrease at about the same time as electrolyte leakage first becomes evident, and drop precipitously before complete collapse and desiccation of the host cells occur. Evidently, materials that leak out of the affected cells have a bactericidal effect; it seems plausible that these materials are also responsible for death of the host cell.

A search for the HRF has been only partially successful. In one study, Lozano (19) extracted intercellular fluids at two hour intervals for 24 hours after infiltration of tobacco leaves with compatible or incompatible strains of P. solanacearum and then assayed these fluids for their effects on bacterial growth. An inhibitory effect was observed only with sterile fluid extracted from leaves 12 - 18 hour after infiltration with incompatible cells. Infiltration of tobacco leaves with the extracts that produced the maximum bacterial inhibition induced a reaction that was indistinguishable from the HR. The compounds responsible for these effects on bacterial and tobacco cells appeared to be heat-stable (95°C. for ten minutes), but it has been difficult to extract a sufficient amount to determine their nature.

That materials inhibitory to bacterial growth are released during the HR has been documented also by Stall and Cook (33) working

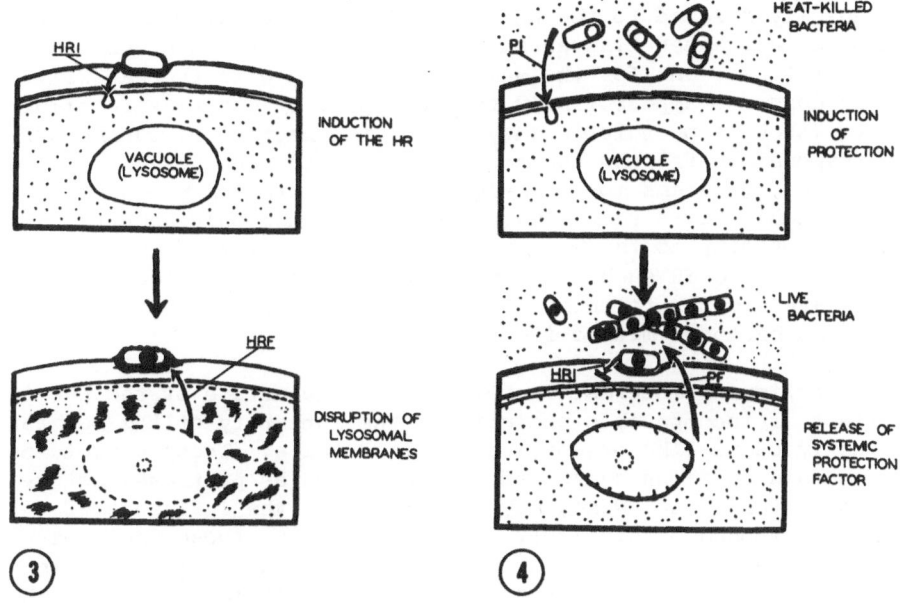

Figure 3. *Diagrammatic representation of the proposed mechanism of action of the hypersensitive reaction inducer (HRI) released by incompatible bacteria attached to the host cell wall. The inducer may be taken up by pinocytosis and cause disruption of lysosomal membranes. Compounds released from lysosomes (HRF) are thought to cause death of both host and pathogen cells. Dying bacterial cells contain numerous granules of poly-β-hydroxybutyrate. Adapted from Wilson (39).*

Figure 4. *Diagrammatic representation of the proposed mechanism of action of the protection inducer (PI) released from heat-killed bacterial cells. The inducer may be taken up by pinocytosis and cause slight injury to cell membranes. Substances involved in injury repair are released into the intercellular fluid and prevent either bacterial multiplication or the separation of bacterial cells after division.*

with resistant pepper leaves infiltrated with *Xanthomonas vesicatoria*.

These preliminary results support the contention that the HRI and the HRF are separable entities. A different point of view has been expressed by Lovrekovitch *et al.* (17) who reported that increases in pH and ammonia, the result of bacterial multiplication in the tissues, were sufficient to explain cellular collapse. However, it was found later that neither ammonia accumulation nor increased pH

were causally related to the HR (8). In bean leaves, measurement
of ammonia in treated tissues showed that concentrations were insufficient to account for the damage associated with the HR (25).
Thus, it does not appear likely that ammonia, or other extracellular
products (such as pectolytic enzymes) of the bacterium, can be directly responsible for the numerous physiological changes associated
with the HR. It seems more likely that the bacterium produces a
highly unstable toxin that damages host membranes only when released
in close proximity to the host cell wall. Many of the symptoms associated with the HR such as respiratory increase and electrolyte
leakage, are similar to those obtained as a result of membrane injury caused by toxic metabolites.

Of particular interest in relation to the nature of the HRF is
the effect of incompatible bacteria on the integrity of the membrane
bounding lysosome-like organelles in plant cells. These structures,
which contain numerous hydrolytic enzymes such as protease, ester-
ase, phosphatase, and RNase, may rupture under the influence of
toxins of pathogen origin (39). In animal cells, these "suicide
bags" are thought to be responsible for the rapid autolysis of injured tissues. By analogy, autolysis of plant cells undergoing the
HR may result from decompartmentalization of hydrolytic enzymes
contained in the vacuole or other portions of the lysosomal system
(26). The HRF may consist of hydrolytic enzymes released from ruptured lysosomes; upon release, these enzymes may interact with
phenolic glycosides or free phenols, leading to the formation of
low molecular weight compounds that affect multiplication of the
bacterium in the intercellular spaces.

The lysosome rupture hypothesis is attractive, but enzymologic-
al studies of tobacco tissues undergoing the HR indicate no significant increases in activities of proteases, polyphenoloxidase, peroxidase, and phosphatidases, and only minor increases in RNase in
supernatant fractions (11, 22).

SUPPRESSION OF THE HYPERSENSITIVE REACTION

Since the HR involves alterations in permeability, leading to
electrolyte leakage at very early stages of the reaction, it is not
surprising that substances that reduce permeability also affect development of the HR. For instance, infiltration of calcium into
leaves, prior to or simultaneous with inoculation with bacteria,
suppresses electrolyte loss from leaf tissues (3). Uranyl compounds,
which bind to the plasma membrane, also counteract the effects of
bacterial infiltration, but are not as effective as calcium. Calcium, on the other hand, was ineffective in suppressing electrolyte
loss resulting from treatment of leaf tissues with ammonia.

O'Brien and Wood (25) have reported that treatment of bean leaves with cycloheximide, at concentrations that did not affect bacterial growth, suppressed the leakage of electrolytes and tissue collapse at bacterial populations sufficient to induce the HR. As in the case of calcium, cycloheximide did not affect the permeability changes resulting from treatment with ammonia. The effects of cycloheximide are difficult to interpret; they could be related to protein turnover and depletion of a receptor for the HRI at the plasma membrane or elsewhere. Alternatively, cycloheximide may interfere with the accelerated senescence associated with the HR. It is well known that protein synthesis is a prerequisite for the changes in enzyme activity that characterize autolysis of plant cells (21). As might be expected, compounds that delay senescence, such as cytokinins, also suppress the development of the HR (23).

Suppression of the HR by high temperature (13) is a phenomenon that has not been thoroughly explored. No HR develops in tobacco plants kept at 37°C for at least five hours after infiltration with *P. phaseolicola*. The effect of high temperature might be on the bacteria themselves, but this has not been properly defined. Klement (13) concluded that there is a thermo-sensitive period in host tissues immediately following the HR induction period.

Novacky *et al.* (24) reported that tobacco cells pretreated with low concentrations of living bacteria do not undergo the HR when challenged 24 hours later with populations of the bacterium that would normally induce the HR. However, entirely opposite results were obtained by Cook (2) working with pepper.

Protection, in the sense that tissue collapse does not occur in certain pretreated tissues, may be envisioned as a progressive development of tolerance to the bacterium or its products. Activation of self-repair mechanisms in response to moderate injury may be involved in this response (10). However, we have been unable to obtain protective responses in tobacco cells injured by a variety of chemical or mechanical agents. In our system, protection is specifically associated with bacterial components that elicit specific changes in permeability and other characteristics of the host cell. These components are heat-stable; thus, they can be detected most conveniently in preparations from heat-killed bacterial cells.

Suppression of the HR by Heat-Killed Bacteria

That heat-killed bacterial cells can protect plants from subsequent invasion by plant pathogens has been known for decades, but was demonstrated most effectively by Lovrekovich and Farkas in 1965 (16). Tobacco leaves infiltrated with heat-killed cells of *Pseudomonas tabaci, P. syringae,* and *Corynebacterium flaccumfaciens* were protected against infection by compatible, live cells of *P. tabaci.*

The finding that heat-killed cells of *P. solanacearum* also suppressed the development of the HR in tobacco leaves (20) indicated that the protective response was also effective against incompatible bacteria. Because the HR can be adequately controlled and is highly reproducible, the development of this reaction provided a convenient assay to study the nature of the bacterial constituents that are responsible for the protection phenomenon.

The protection phenomenon depends on : 1. the time allowed between treatment and challenge, 2. the concentrations of the heat-killed cells used to induce protection and of the live cells used to challenge protected leaves, 3. temperature, and 4. light regime (20). If the concentration of bacteria used in the challenge inoculation is much higher than that in the pre-treatment, or if the plants are subjected to continuous high temperature (36°C) or low light intensity, only partial protection is obtained. The time- and light-dependency of the response constitutes presumptive evidence that the protective mechanism depends entirely on a host response.

Light is an absolute requirement for a protective response; action spectra indicate that the effective areas of the spectrum are at 450 - 500 and 600 - 660 nm; the minimum energy levels for full protection are in the order of 5 $\mu w/cm^2$ (Kraus, unpublished). It seems likely that the effective receptor is chlorophyll b (absorption peaks at 453 and 642 nm). On the surface, it would appear that the protective response is dependent on energy derived from photophosphorylation.

Protein synthesis appears to be essential for protection. In experiments in which actinomycin D was added at various intervals after infiltration of tobacco leaves with heat-killed bacteria, the protective effect was eliminated if this antibiotic was applied at 0 to two hours, and partially suppressed if applied at six hours after infiltration. The effect of actinomycin D was mainly on the host cells; bacterial populations were only slightly reduced by concentrations of actinomycin D that effectively suppressed the protective response (Sequeira and Donald, unpublished).

The protective response is systemic. If sufficient time is allowed after infiltration with heat-killed cells, tissues adjoining the pre-treated areas become fully protected and those of untreated leaves, below and above the treated leaf, will show a partial or delayed HR. The systemic nature of the response indicates that the active substance(s) (the protection inducer, or PI) is separable from the substance(s) that affords protection and is translocated (the protection factor, or PF).

Nature of the Protection Inducer (PI)

Suppression of the HR by heat-killed bacterial cells is associated with specific constituents of the bacterial cell wall. Frac-
tionation of bacterial constituents indicate that the effective components of the PI are glycoproteins, particularly those found in the
periplasmic space (31, 37). Other *P. solanacearum* constituents, such
as purified extracellular polysaccharide, DNA, or cell wall peptidoglycan (38) failed to give a protective response. In addition, it
has not been possible to obtain a protective response by infiltrating
leaves with a wide variety of proteins (casein, trypsin, peroxidase)
and injurious chemicals (mercuric chloride, trichloroacetic acid,
ammonium hydroxide) at a wide range of concentrations. Although
the PI is a specific constituent of bacterial cell walls, it is found
in a wide variety of phytopathogenic pseudomonads and xanthomonads.

Active protection-inducing fractions are obtained by grinding
bacterial cells at high speeds with glass beads, adding ethanol to
the supernatant in order to precipitate macromolecules, and then re-
suspending and eluting this fraction through Sephadex G-200. In
general, fractions from these Sephadex columns were active at protein
concentrations of 0.1 mg/ml and induced partial protection within two
hours after infiltration and complete protection within seven hours.
Most of the protective activity of these fractions could be removed
by treatment with hot trichloroacetic acid (5%) or after treatment
with proteolytic enzymes, such as trypsin and pronase (31).

The active material from glass bead-ruptured cells consistently
eluted from Sephadex G-200 columns at a ratio of elution volume to
void volume (Ve/Vo) of 2.37 (37). This corresponds to an approximate
molecular weight of 260 000. Separation by isoelectric focusing on
a sucrose gradient indicated that most of the protective activity
was associated with a band in the pH range 4.1 - 4.4. This band was
separated further by electrophoresis on acrylamide gels; the major
constituent was a band which showed positive staining for both protein (Coomassie blue) and carbohydrate (Schiff's reagent). The concentration of protein in this band was not sufficiently high to allow
assaying for biological activity after elution, even when bands from
several gels were pooled. Attempts to dissociate the active materials into component moieties by addition of sodium dodecyl sulfate
and mercaptoethanol failed. In fractions tested before separation
by gel electrophoresis, the ratio of protein to carbohydrate was approximately 6:4.

Substances capable of inducing a protective response have been
obtained also by osmotic shock of bacterial cells (37). This procedure extracts substances from the periplasmic space, i.e. from
outside the peptidoglycan layer. Purification procedures similar
to those described above yielded a biologically active glycoprotein

of physical properties different from those of the material obtained
by the cell rupture technique. The isoelectric point for the osmotic shock material was between 5.5 and 6.2; a Ve/Vo value of 3.20
indicated a molecular weight of approximately 160 000. These two
active glycoproteins may represent aggregates of smaller units, but,
so far, attempts to dissociate them have failed. Much remains to
be done to confirm the homogeneity and nature of the components of
these active fractions.

Growth of Compatible and Incompatible Bacteria in Protected Tissues

     The changes in bacterial populations following challenge in-
oculation of protected tissues provide evidence for the gradual development of an environment which is unfavourable for bacterial
growth. With compatible strains of *P. solanacearum,* initial populations introduced in protected leaves drop rapidly and then sta-
bilize at levels well below the threshold necessary for symptom
development. With incompatible strains, introduced at populations
above those necessary to induce the HR, numbers of bacteria decline
in both protected and unprotected tissues, but the decline is much
more pronounced in the protected tissues, even though there are no
outward symptoms of necrosis (32).

     The simplest interpretation of these results is that, under
the influence of the PI, materials which inhibit bacterial growth
are progressively released into the intercellular spaces. These
substances appear to constitute at least part of the protection
factor (PF).

                    Nature of the Protection Factor (PF)

     The fluid extracted from the intercellular spaces of protected
leaves contains substances that are inhibitory to the growth of *P.
solanacearum*. Because greater inhibition is obtained in fluid from
plants maintained in the light than in those in the dark, this component of the PF appears to be actively secreted only under conditions that result in a protective response (28).

     In experiments in which one-half of a tobacco leaf was infil-
trated with heat-killed bacterial cells and the other half was infiltrated with water, intercellular fluids obtained after incubation
for 24 hours supported bacterial growth to markedly different ex-
tents. Doubling times were always considerably longer in fluid
from the protected halves than in that from the unprotected halves.
In fluids from similar leaves that had been in the dark for 24 hours,
no differential growth response was obtained. Fluids from protected
leaves sometimes inhibited bacterial growth completely, or, when

limited growth occurred, the bacteria grew in chains of 30 or more
cells.

The inhibitory components in fluids from protected leaves are
heat stable and of low molecular weight. The crude fluids contain,
in addition to a specific bacterial growth inhibitor, substantial
amounts of salts, sugars, amino acids, and other compounds that affect bacterial growth. Separation of these various compounds has
been accomplished by column and thin-layer chromatography. Examination of the effect of the different components of fluids from
protected tissues on bacterial growth indicate that sugars and salts
contribute to the total inhibitory effect (Kraus, unpublished). The
osmotic pressure (approx. 6 atm) of protected fluids is within the
range where growth of *P. solanacearum* is inhibited and is substantially higher (approx. 2 x) than that from unprotected leaves.
After sugars and salts are separated from the inhibitory fraction
by thin-layer chromatography, the major inhibitory component is associated with a compound(s) that has some of the properties of a
terpenoid. It has not been possible to obtain the PF in pure form,
however, due to difficulties in obtaining reasonable amounts of
intercellular fluid.

Protection and Its Relation to Injury Repair Mechanisms

Infiltration of tobacco leaves with heat-killed cells of *P.
solanacearum* results in significant changes in protein content and
peroxidase activity of the intercellular fluid (28). These changes
in the intercellular fluid do not occur in leaves shaded immediately
after infiltration, indicating that there is a light-dependent transfer of proteins from the leaf cell to the intercellular fluid.
Because the amount of glucose-6-phosphate dehydrogenase, a cytoplasmic enzyme, also increases after infiltration, whether or not the
leaves are shaded, there appears to be some leakage of cellular contents into the intercellular spaces which may be due to injury.
The active transfer of proteins, on the other hand, indicates that
affected cells must be metabolically active.

Injury repair mechanisms, therefore, may be involved in the
increased tolerance to the presence of the pathogen and in the reduced necrotization that characterize the protective response. However, we do not know which component of this phenomenon is the important one. Increased peroxidase activity is characteristic of
early stages of injury to plant tissues (15). The active transfer
of peroxidase into the intercellular spaces occurs only in the light
and may be associated with the protective response in some indirect
fashion, i.e. in association with other injury responses. Lovrekovich *et al*. (18) suggested that this enzyme was involved in protection of tobacco leaves against *P. tabaci*. However, there are
substantial amounts of peroxidase in the intercellular fluid of

unprotected tobacco leaves (27) and we have been unable to obtain
protection in tobacco leaves by direct infiltration of horseradish
peroxidase.

CONCLUSIONS

The factors that provide protection are clearly complex. It
is unlikely that all of the different components, as they affect a
variety of different compatible and incompatible pathogens, can be
separated and their roles elucidated. The key protective response
that prevents necrotization associated with the HR appears to be
associated with changes in membrane structure that block the action
of the HRI or makes cells less permeable to water loss and, thus,
prevents the eventual collapse of the cells.

Necrotization does not appear to be a necessary component of
the resistant response to invasion by incompatible pathogens. In
protected leaves, populations of incompatible B1 cells of *P. solanacearum* decline earlier and at a faster rate than in unprotected
ones undergoing the HR, even though there is no tissue collapse or
any outward symptoms of injury. Saprophytes and certain incompatible bacterial strains such as race 3 of *P. solanacearum* do not induce the HR in tobacco leaves, yet their numbers decline steadily
and these bacteria are gradually eliminated. A defense reaction
appears to be involved, apparently due to the response of host cells
to proteinaceous constituents (the PI) of potential pathogens, and
is not accompanied by necrotization. A defense response that does
not involve extensive damage to the host cell would seem more likely
to become established through evolution than one that involves rapid
necrotization.

That the release of the PF is dependent on moderate injury
caused by the PI is supported by the systemic effects of wounding
described by Ryan (29). Wounding of many different types of plants,
including tobacco, releases a hormone that is translocated throughout the plant and induces the accumulation of proteinase inhibitors.
Ryan (29) has suggested that this is part of an immune response resulting in inhibition of the extracellular proteinases of invading
pathogens. The widespread occurrence of this hormone, and the
rapidity of its translocation following wounding, suggests that
similar but more specific responses to injury may be involved in
the protective response of tobacco leaves induced by bacterial
constituents.

REFERENCES

1. BOGERS, R.J. (1972). On the interaction of *Agrobacterium tumefaciens* with cells of *Kalanchoë daigremontiana*. In : *Proc. 3rd Int. Conf. Pl. Pathogenic Bacteria* (MAAS GEESTERANUS, H.P., Ed.). Wageningen, The Netherlands, 239 - 250.

2. COOK, A.A. (1975). Effect of low concentrations of *Xanthomonas vesicatoria* infiltrated into pepper leaves. *Phytopathology*, 65, 487 - 489.

3. COOK, A.A. and STALL, R.E. (1971). Calcium suppression of electrolyte loss from pepper leaves inoculated with *Xanthomonas vesicatoria*. *Phytopathology*, 61, 484 - 487.

4. ERCOLANI, G.L. (1970). Bacterial canker of tomato. IV. The interaction between virulent and avirulent strains of *Corynebacterium michiganense* (E.F. Sm.) Jens. *Phytopathol. Mediter.*, 9, 151 - 159.

5. ERCOLANI, G.L. (1973). Two hypotheses on the aetiology of response of plants to phytopathogenic bacteria. *J. gen. Microbiol.*, 75, 83 - 95.

6. GARDNER, J.M. and KADO, C.I. (1972). Induction of the hypersensitive reaction in tobacco with specific high molecular weight substances derived from the osmotic shock fluid of *Erwinia rubrifaciens*. *Phytopathology*, 62, 759 (Abstr.).

7. GOODMAN, R.N. (1968). The hypersensitive reaction in tobacco : a reflection of changes in host permeability. *Phytopathology*, 58, 872 - 873.

8. GOODMAN, R.N. (1971). Re-evaluation of the role of NH_3 as the cause of the hypersensitive reaction. *Phytopathology*, 61, 893 (Abstr.).

9. GOODMAN, R.N. and PLURAD, S.R. (1971). Ultrastructural changes in tobacco undergoing the hypersensitive reaction caused by plant pathogenic bacteria. *Physiol. Pl. Path.*, 1, 11 - 16.

10. HANCHEY, P., PASTALKA, T. and NOVACKY, A. (1974). Ultrastructure of tobacco protected against the hypersensitive response. *Proc. Am. Phytopath. Soc.*, 1, 74.

11. HUANG, J. and GOODMAN, R.N. (1970). The relationship of phos-
     phatidase activity to the hypersensitive reaction in
     tobacco induced by bacteria. *Phytopathology*, 60, 1020 -
     1021.

12. KLEMENT, Z. (1963). Methods for the rapid detection of the
     pathogenicity of phytopathogenic pseudomonads. *Nature,
     Lond.*, 199, 299 - 300.

13. KLEMENT, J. (1972). Development of the hypersensitive reaction
     induced by plant pathogenic bacteria. *In : Proc. 3rd Int.
     Conf. Pl. Pathogenic Bacteria*,(MAAS GEESTERANUS, H.P.,Ed.).
     Wageningen, The Netherlands, 157 - 164.

14. KLEMENT, Z. and GOODMAN, R.N. (1967). The role of the living
     bacterial cell and induction time in the hypersensitive
     reaction of the tobacco plant. *Phytopathology*, 57, 322 -
     323.

15. LIPETZ, J. (1970). Wound-healing in higher plants. *Int. Rev.
     Cytol.*, 27, 1 - 28.

16. LOVREKOVICH, L. and FARKAS, G.L. (1965). Induced protection
     against wildfire disease in tobacco leaves treated with
     heat-killed bacteria. *Nature, Lond.*, 205, 823 - 824.

17. LOVREKOVICH, L., LOVREKOVICH, H. and GOODMAN, R.N. (1970).
     Ammonia as a necrotoxin in the hypersensitive reaction
     caused by bacteria in tobacco leaves. *Can. J. Bot.*, 48,
     167 - 171.

18. LOVREKOVICH, L., LOVREKOVICH, H. and STAHMANN, M.A. (1968).
     The importance of peroxidase in the wildfire disease.
     Phytopathology, 58, 193 - 198.

19. LOZANO, J.C. (1969). Host responses to different isolates of
     Pseudomonas solanacearum. M.S. Thesis, University of
     Wisconsin, Madison, 68 pp.

20. LOZANO, J.C. and SEQUEIRA, L. (1970). Prevention of the hy-
     persensitive reaction in tobacco leaves by heat-killed
     cells of *Pseudomonas solanacearum*. *Phytopathology*, 60,
     875 - 879.

21. MATILE, P. and WINKENBACH, F. (1971). Function of lysosomes
     and lysosomal enzymes in the senescing corolla of the
     morning glory (*Ipomoea purpurea*). *J. exp. Bot.*, 23, 759 -
     771.

22. NEMETH, J., KLEMENT, Z. and FARKAS, G.L. (1969). An enzymol-
    ogical study of the hypersensitive reaction induced by
    Pseudomonas syringae in tobacco leaf tissues. *Phytopath.
    Z.*, 65, 267 - 278.

23. NOVACKY, A. (1972). Suppression of the bacterially induced
    hypersensitive reaction by cytokinins. *Physiol. Pl. Path.*,
    2, 101 - 104.

24. NOVACKY, A., ACEDO, G. and GOODMAN, R.N. (1973). Prevention
    of bacterially induced hypersensitive reaction by living
    bacteria. *Physiol. Pl. Path.*, 3, 133 - 136.

25. O'BRIEN, F. and WOOD, R.K.S. (1973). Role of ammonia in in-
    fection of *Phaseolus vulgaris* by *Pseudomonas* spp. *Physiol.
    Pl. Path.*, 3, 315 - 326.

26. PITT, D. and COOMBES, C. (1968). The disruption of lysosome-
    like particles of *Solanum tuberosum* cells during infection
    by *Phytophthora erythroseptica* Pethybr. *J. gen. Microbiol.*,
    53, 197 - 204.

27. RATHMELL, W.G. and SEQUEIRA, L. (1974). Soluble peroxidase in
    fluid from the intercellular spaces of tobacco leaves.
    Plant Physiol., Lancaster, 53, 317 - 318.

28. RATHMELL, W.G. and SEQUEIRA, L. (1975). Induced resistance
    in tobacco leaves : the role of inhibitors of bacterial
    growth in the intercellular fluid. *Physiol. Pl. Path.*,
    5, 65 - 73.

29. RYAN, C.A. (1974). Assay and biochemical properties of the
    proteinase inhibitor-inducing factor, a wound hormone.
    Plant Physiol., 54, 328 - 332.

30. SEQUEIRA, L. and AINSLIE, V. (1969). Bacterial cell-free
    preparations that induce or prevent the hypersensitive
    reaction in tobacco. *XI Int. bot. Congr.* (Abstr.), 195.

31. SEQUEIRA, L., AIST, S. and AINSLIE, V. (1972). Prevention of
    the hypersensitive reaction in tobacco by proteinaceous
    constituents of *Pseudomonas solanacearum*. *Phytopathology*,
    62, 536 - 542.

32. SEQUEIRA, L. and HILL, L.M. (1974). Induced resistance in
    tobacco leaves : the growth of *Pseudomonas solanacearum*.
    in protected tissues. *Physiol. Pl. Path.*, 4, 447 - 455.

33. STALL, R.E. and COOK, A.A. (1968). Inhibition of *Xanthomonas*
     vesicatoria in extracts from hypersensitive and susceptible
     pepper leaves. *Phytopathology*, 58, 1584 - 1587.

34. STALL, R.E. and COOK, S.A. (1973). Hypersensitivity as a
     defense mechanism against natural infection. *Proc. 2nd
     Int. Congr. Pl. Pathol.*, Abstr. No. 0586.

35. TURNER, J.G. and NOVACKY, A. (1974). The quantitative relation
     between plant and bacterial cells involved in the hyper-
     sensitive reaction. *Phytopathology*, 64, 885 - 890.

36. VARNS, J.L. and KUČ, J. (1971). Suppression of rishitin and
     phytuberin accumulation and hypersensitive response in
     potato by compatible races of *Phytophthora infestans*.
     Phytopathology, 61, 178 - 181.

37. WACEK, T.J. (1974). Isolation and characterization of bacter-
     ial fractions that induce resistance in tobacco leaves.
     Ph.D. Thesis, Univ. of Wisconsin, Madison, 97 pp.

38. WACEK, T.J. and SEQUEIRA, L. (1973). The peptidoglycan of
     Pseudomonas solanacearum : chemical composition and biol-
     ogical activity in relation to the hypersensitive reaction
     in tobacco. *Physiol. Pl. Path.*, 3, 363 - 369.

39. WILSON, C.L. (1973). A lysosomal concept for plant pathology.
     A. Rev. Phytopath., 11, 247 - 272.

CONTRIBUTIONS

DURBIN, R.   Factors affecting the "induction period" of bacterially
    induced hypersensitivity.

ERCOLANI, G.   Site heterogeneity within bean leaf tissues in rela-
    tion to multiplication and local lesion formation by *Pseudo-
    monas phaseolicola*.

LYON, F.M.   Hypersensitive responses of bean and tobacco leaves
    to *Pseudomonas* spp.

PATIL, S.S.   Induction of hypersensitivity in pepper by an endo-
    toxin isolated from *P. phaseolicola*.

SUMMARY OF POINTS FROM CONTRIBUTIONS AND DISCUSSIONS
BY
A. NOVACKY

Chairman and Discussion Leader

The discussion began with a series of questions from Wood concerning the involvement in the hypersensitive reaction (HR) of pectolytic and cellulolytic enzymes, of compounds such as capsidiol from Keen, and of lectins from Albersheim. Since *Pseudomonas solanacearum* can rapidly synthesize pectolytic and cellulolytic enzymes, the observed degradation of the host cell wall at the point of the attachment of bacteria is presumably due to these enzymes. However, their production is not related to the ability of different strains of *P. solanacearum* to induce HR. Sequeira suggested that attachment of bacteria to the plant cell wall may be a critical factor but to what extent cell wall degrading enzymes may be involved in the HR induction after attachment has occurred is not yet known. In response to the question of the involvement of capsidiol in protection against HR, Sequeira stated that the major component of the protective factor appears to be a terpenoid but its identity has not yet been determined. He did not know if capsidiol is active against *P. solanacearum*. To the possible involvement of lectins in protection against HR, Sequeira mentioned that he was unable to obtain a lectin from tobacco leaves. However, tobacco roots do contain a factor which agglutinates bateria and mimics the potato lectin but it is inactive on rabbit red blood cells. He is currently attempting to resolve this apparent inconsistency.

The discussion turned to the question of HR induction in cultured cells (Day) or in isolated protoplasts (Keen). Sequeira found tobacco cells in suspension remarkably insensitive to incompatible bacteria in an adjoining chamber separated by a filter membrane. Based on these preliminary experiments he concluded that it would be difficult to study HR in isolated tissue or isolated cells. He stated that Hoitnink was unable to induce HR in protoplasts. At this point Novacky mentioned that Huang and Goodman found a selective response of separated tobacco leaf cells to plant pathogenic and saprophytic bacteria. The oxygen consumption (monitored by an O_2 probe) immediately after mixing suspensions of bacteria and plant cells was greatest in HR combinations, less in compatible combinations and only slightly increased in combinations containing saprophytic bacteria.

In his system Sequeira did not obtain protection with bacterial constituents other than the specific glycoproteins he described. Others have described protective responses following infiltration with albumin, globulin and another protein or by mechanically wounding the tissue (Durbin). This discrepancy, as pointed out by Novacky, may be due to the fact that Sequeira used an avirulent form of a compatible bacterium (*P. solanacearum*) in his challenge experiments

whereas others used pathogens of other plant species such as *P. pisi*.
Sequeira agreed that it is tempting to suggest that there are basic
differences in the mechanisms of HR induction among the various
pseudomonad strains used as test organisms.

Sequeira then discussed Daly's question on the relation between bacterial populations in chains and bacterial populations in
compatible and incompatible interactions. He has observed growth
of *P. solanacearum* in chains only in the intercellular fluid from
protected leaves, that is, in leaves pre-treated with heat killed
bacteria. Because changes in populations observed in challenged,
protected leaves may reflect this type of growth, dilution plate
techniques may give erroneous results. For this reason he used an
haemocytometer to count individual bacteria.

Ercolani briefly discussed his study on site heterogeneity
within bean leaf tissues in relation to multiplication and formation
of local lesions by *Pseudomonas phaseolicola*. When inoculum contained a 1:1 mixture of two recognizable bacterial variants, single
local lesions from leaves inoculated with up to 4 x ED 50 yielded
virtually pure cultures of one variant independently of time of
formation whereas the proportion of those doing so from leaves inoculated with higher doses increased with time of formation after
inoculation. Local lesion counts in the same experiments did not
increase in proportion to the doses of inoculum. Recent evidence
combined with earlier data indicates that a host response taking
place early after inoculation is a factor in determing heterogeneity
of sites *in vivo* with respect to bacterial multiplication and local
lesion formation. DeVay asked if variability in the size of lesions
on both the leaf with many lesions and the leaf with few lesions was
due to the number of bacteria initially aggregated at the infection
site. Ercolani was not aware of evidence that this was the case.
In view of Ercolani's experiment, the question was posed by Paxton
that if only one serotype of *Rhizobium* is isolated from each legume
nodule on a root in soil containing several serotypes, would this
indicate that only one bacterium initiates a nodule ? Ercolani replied that this would be the case if the hypothesis of independent
action were applicable to the nodule system.

Lyon then discussed her work on HR of *Phaseolus vulgaris* and
Nicotiana tabacum to *Pseudomonas phaseolicola* and *P. mors-prunorum*
and halo blight caused by race 2 of *P. phaseolicola* on *P. vulgaris*.
She found antibacterial compounds in ethanol extracts of HR bean
leaf tissue one day after inoculation and in susceptible tissue
after five days. Two such compounds, phaseollin and coumestrol,
were identified at 100 μg/ml in liquid media. Phaseollin only inhibited saprophytic bacteria whereas coumestrol inhibited growth of
both HR inducing bacteria at 20 to 100 μg/ml. The antibacterial
compounds which accumulate in HR tissue of tobacco coincide with
terpenoid compounds on thin layer chromatograms but are at present

unidentified. Accumulation of these compounds in bean and tobacco
leaves can also be induced by injection of cell free extracts (after
French press treatment and 30 000 x g centrifugation) of *P. phaseolicola* and *P. mors-prunorum* but not of *P. fluorescens*.

Durbin then gave a short account of his recent study on the
induction period of HR. After establishing the length of induction
period (90 - 105 minutes) he exposed plants (*Nicotiana tabacum)* to
37°C before completion of the induction period and found that HR
was prevented. Since *P. solanacearum* strains insensitive to 37°C
induce normal HR he suggested that temperatures of 37°C suppress
the HR by temporarily destroying some labile factors of the bacteria
necessary for its induction.

Finally Patil described experiments with high molecular weight
endotoxins isolated from two strains of *Pseudomonas phaseolicola*.
When injected in the leaves of bell pepper (*Capsicum annuum*) the
endotoxins mimic the parent bacteria in inducing HR cell collapse.
One endotoxin (from the strain 650 Tox⁻), further purified by gel
filtration had a mol. wt. approximately 93 000. Confluent cell
collapse, without browning, senescence symptoms or leaf abscission
was induced with concentrations as low as 2 μg/ml and appeared at
a similar interval after infiltration as when intact bacterial cells
were used. The endotoxin is thermostable and is degraded by pronase.

SPECIFIC INTERACTIONS IN HIGHER PLANTS

HANS F. LINSKENS

Department of Botany, University of Nijmegen

Nijmegen, The Netherlands

INTRODUCTION

Interaction between cells is a characteristic of living systems (20). Unicellular systems interact in their environment in the liquid and/or gaseous phase. Multicellular living systems interact on two levels.

1. Between individuals, referred to as *external interaction*.

2. Between cells and tissues within a system, referred to as *internal interaction*.

Both types may display a high degree of specificity.

The special cases of an interrelation in which one partner of the interacting system harms the other, so that one speaks of a plant disease, has been emphasized in most lectures at this Institute. However, there are many other relations in higher plants in which the interactions are neither harmful nor neutral, but useful, at least in terms of biological performance.

In order to widen the scope of the problem of specificity in plant diseases (24), I shall deal with specific interactions between higher plants as now known. It will become evident that in most cases the basic mechanisms of specificity are still unknown and much work will have to be done to elucidate systems which involve specific reactions. As examples of interactions I shall report on allelopathy, epiphytism and parasitism in higher plants. For internal interactions I shall refer briefly to the systems

311

preceeding fertilization which guide the male motile cells to the
egg cells. But undoubtedly, the best known examples of specific
interaction in higher plants are the incompatibility barriers in
the fertilization of higher plants. The reactions involved here
represent a system of internal interactions of high specificity
and, from an evolutionary point of view, a very efficient way of
preventing inbreeding and one that favours recombination to maint-
ain the species characteristics.

EXTERNAL INTERACTIONS

Allelopathy

In an ecosystem higher plants interact with one another as do
micro-organisms which interact both with each other and with higher
plants. Mutual injury is caused mainly by competition for space,
water, light and other edaphic and climatic factors. However, competition of higher plants with micro-organisms and animals cannot be
disregarded. Species specific reactions to water supply, light and
temperature result in selective participation of organisms in a bio-
coenosis. The way in which individuals and species effect their
influence is called competition. However, mutual interaction includes positive and negative influences other than competition for
food and space.

Mutual interaction by means of chemical compounds is called
allelopathy (2, 11, 14, 17, 32, 33). The substances which mediate
allelopathy are synthesized by higher plants and may influence,
directly or indirectly, through the soil, other individuals of the
same species or individuals of other species. This influence can
be directed to the germination of diaspores, to the growth and to
the abundance of individuals in a plant association. These plant-
plant interactions are found in many species-combinations and are
believed to be highly specific in many cases. In some the chemical
mechanism has been elucidated.

Three examples will be mentioned. They are representative of
systems with regard to external interactions of higher plants.

Exclusion. A well known case is the allelopathic effect of the
walnut tree *(Juglans regia)*, which is widely distributed in Eurasia
and North America (37). The substance which inhibits growth of
other species has been identified as juglone (5-hydroxy-1, 4 naph-
thoquinone). In most tissues of the walnut tree this principle is
present in the non-oxidized form hydroxyjuglone which is non-toxic.
However, when it leaches out of the leaves or bark, as well as out
of litter of the walnut tree, it is oxidized to juglone which inhibits the undergrowth. The effect of juglone is strongly selective.

Some species, such as broomsedge, shrubs belonging to the *Ericaceae*, and many broad-leaved herbs are largely excluded whereas others, such as black raspberry and Kentucky bluegrass are tolerated or even favoured (3). Because of this exclusion mechanism, plant associations in walnut woods are highly species specific and the ground-cover under walnut trees is very poor. Similar mechanisms are known for *Eucalyptus globosus* where the specific substances are chlorogenic-*p*-coumarylquinone and gentisinic acid which are transferred by fog drip (30).

In the American desert *Encelia farinosa* has been found to excrete 3-acetyl 6-methoxybenzaldehyde (2).

Extension. Mutual interaction may result in a specific vegetation-pattern which is caused by chemical interaction. The classical example is the vegetation of the chaparral in southern California (31) where the vegetation consists of aromatic, dominant species such as mint (*Salvia leucophylla*) and a sagebrush (*Artemisia californica*). The (soft) chaparral vegetation invades the neighbouring grassland. It could be shown that invasion is preceded by a zone of grass-free belts which are not caused by shade, soil-drought, exhaustion of nutrients or competition between roots. A water-soluble mixture of compounds which is washed out from the leaves contains terpenes notably cineole and camphor. The gaseous substances are absorbed by soil particles and exclude annual plants by inhibition of seed germination. The terpenes also inhibit respiration and growth of seedlings which do germinate and in this manner the substances increase the sensitivity of the prairie plants to other environmental stresses during the dry season.

Similar mechanisms have been observed for the Mediterranean therophytes in the Rosmarino-Ericion association (4).

Succession. Specific allelopathic substances affect not only plant association and species diversity but also influence the succession (37). This can happen in various ways involving the dynamics and composition of plant communities. Allelopathic self-intoxication together with allelopathic repression favour distinct species. Retardation of colonization by certain species determines the species composition of a plant community; replacement of one species by another can be regulated by allelopathic substances which affect potential or dominant invaders.

All these systems of interaction in higher plants are included under the term "chemical ecology". In many cases specific mechanisms are involved or at least suspected.

Epiphytism

One of the most common, but generally overlooked systems involving specific interactions between higher plants is epiphytism,
a special interaction without direct withdrawal of nutrients from
either of the partners.

Epiphytism among algae is well known to be species-specific
(12, 21, 22). Unfortunately little is known about the mechanism.
However, the study of the epiphytic flora in tropical rain forests
has revealed a high degree of specificity (35). Certain epiphytic
orchids were found to grow usually or exclusively on distinct bas-
iphytic trees. Also, when trees of different species grow near
each other or are even intertwined, the attachment of many orchids
remained specifically associated with a particular basiphyt. Proximity and facilities for dispersal do not break down the specific
barrier. The specificity of this epiphytism in mountain-forests
of Java was found by Went (35, 36) to be so pronounced that trees
could be identified by their epiphytic orchids. It seems that the
physical consistency of branches and twigs is decisive for the
growth of epiphytes. This is similar to the finding that surface-
tension is decisive for specificity of attachment to the basiphyt
of marine algae (21, 22). There are also indications that specific
chemical substances, possibly as gases, are responsible for specificity in the epiphyt-basiphyt relationship. The problem of the
specific epiphytism cannot be explained in terms of a lack of inter-
relations. The mechanisms in these cases are far from being elucidated.

Parasitism

One of the most studied examples of interaction between species
of higher plants is the parasitism between *Cuscuta* spp. (*Convolvulaceae*), and its hosts. Host-specificity is not pronounced but the
various species have a certain preference for host species. For
example *Cuscuta epilinum* settles specifically on *Linum usitatissimum*,
and *Cuscuta gronovii* prefers *Salix* species. *Cuscata lupuliformis*
in the Netherlands grows specifically on blackberry bushes.

Cuscuta spp. are stem parasites and holoparasitic on flowering
plants, although they contain some chlorophyll and can photosynthes-
ize as shown by several authors during recent years (16). The need
for a host is thus not absolute. *In vitro* studies have shown that
when sugar was added to a mineral-nutrient medium, it was possible
to bring the parasite to the flowering stage (1). *Cuscuta* specifically inhibits flowering, fruit setting and growth of fruits in
severe infections. When the host was infected after fruit-setting,
Cuscuta seriously interfered with further growth of the fruit probably because of withdrawal of assimilates by the parasite, a process so efficient that only a negligible amount or nothing finds

its way into pods and seeds of parasitized hosts(38). The sink activity of developing fruits is thus reduced to a very low level.
The interaction takes place through the haustoria which are natural
grafts between host and parasite. After intrusive growth the haustorium attaches itself to the sieve elements of the host. The
"searching hyphae" from special contact-cells resemble a hand with
many fingers which grasp the sieve tubes (8, 34) increasing the contact surface between parasite- and host-tissue. Recent studies (39)
suggest a membrane-mediated transfer of solutes from sieve tubes
through the free-space to the adsorptive hyphae of *Cuscuta*.

     The better investigated examples of inter-relationships
between *Cuscuta* species and their hosts indicate that during evolution the specific interaction results in very efficient mechanisms
in which the participation of growth hormones and/or specific substances is likely but this is not yet proven.

     Another example of specificity of interaction between a higher
plant parasite and its host is that between mistletoes and their
host trees. Within the species *Viscum album* there are at least
three sub-species or biological varieties (13) which have an inborn,
genetically determined aptitude for distinct hosts (Fig. 1). *Viscum
album* f. sp. *mali* is found on a wide range of deciduous trees, whereas *V. album* f. sp. *abietis* is restricted to *Abies* species; *V. album*
f. sp. *pini* mainly uses *Pinus* species as host but is occasionally
found on *Picea, Larix* and *Salix* (Fig. 1). Each of the biological
forms of the mistletoe has one common host with two other special
forms. However, there is no common host for all three biological

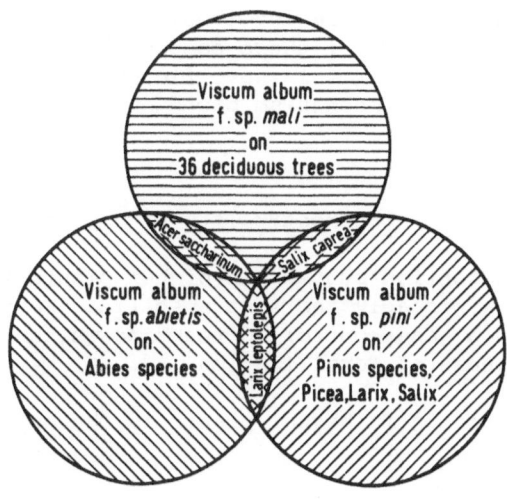

Figure 1

forms. By using the selective effect of differential hosts determination of the parasite species can be simplified. Normally the
seeds of the mistletoe are dispersed by birds, such as the mistle
thrush (*Turdus viscivorus*), and stick with viscin to the surface
of the bark of tree branches. However, after germination of seeds,
the host reacts with a defence reaction to the parasite which results in formation of cork. As we have seen with infection experiments this defence reaction is highly specific in some cases,
whereas in others such as in apple trees it is very weak. Therefore, on apple nearly all infections succeed and the mistletoe
extracts water and mineral nutrients from the xylem in the wood.

INTERNAL INTERACTIONS

Chemical Attraction in the Progamic Phase

In the Archegoniates chemotactic mechanisms control the attraction of male spermatozoids to the archegonia. Liverworts react
to protein solutions, mosses to sugar, *Lycopodium* to citric acid,
Salvinia, *Isoetes*, *Selaginella* and *Equisetum* to malic acid, and
fern spermatozoids are attracted to malic among other di-basic organic acids. But these attraction systems have a low degree of
specificity and some probably act through a pH gradient. A fair
degree of specificity is observed in Angiosperms especially in selective fertilization. Here the affinity is directed by the genome
of the ovules and may differ within one ovary. Different ovules
are able to attract pollen tubes of different genotypes. From
genetical analysis it is clear that the chemotropic substances are
complex mixtures and chemical analyses have shown that amino acids,
peptides, amines and sugars are involved. Specificity seems to
depend on specific mixtures of these components. The situation is
complicated by the fact that this specific system is accompanied by
a non-specific attraction mechanism based on the strong chemotropic
effect of the calcium ion.

Incompatibility

Among the interaction systems in higher plants incompatibility
at fertilization is probably the most specific. Fertilization-
incompatibility is a gene controlled reaction which takes place during the progamic phase of fertilization.

Sexual reproduction in higher plants is characterized by the
following.

1.   Transfer of the male gametophyte, i.e. dispersion of
pollen grains by various agents such as wind, water and insects

to the female organ, the pistil, called pollination.

2.   The progamic phase, which includes landing of the pollen grains on the receptive surface of the stigma, pollen germination, formation of the pollen tube and its growth through the transmitting tissue or canal of the style to the ovary, including all the regulatory processes such as carbohydrate and protein metabolism in the pollen tube, metabolic interactions between the diploid female tissue and the haploid pollen tube and mechanisms directing growth such as hydrotropism, electrotropism and chemotropism.

3.   Syngamy, the penetration of the tip of the pollen tube into one of the synergids of the embryo sac, rupture of the tip region, release of the contents of the pollen tube including cytoplasm and sperm cells, distribution of one sperm cell into the egg cell on one side and of the other into the central cell in the field of the fused or fusing polar nuclei.

The incompatibility reaction of higher plants takes place during the progamic phase and is controlled with a high degree of specificity by incompatibility genes called S-genes (25, 27).

Genetic specificity. In phanerogams two forms of incompatibility can be distinguished with regard to fertilization.

1.   Interspecific incompatibility, characterized by an apparent barrier during crossing of plants of different species.

2.   Intraspecific incompatibility, manifested within a species or variety and which is the most important specific mechanism in preventing reproduction in closely related individuals.

The reaction that prevents self fertilization is determined by genetic factors which control production of certain substances which affect pollen germination and growth of pollen tubes within female transmitting tissue. If pollen contains the same self-sterility factors as tissues of the stigma and the style in which the pollen tube must develop to complete fertilization, then growth of the pollen tube will be blocked by the combination of similar hereditary factors *(homogenic incompatibility)*.   Only if pollen and the style contain different self-incompatibility factors will the progamic stage proceed so that fertilization becomes possible.

Genetic factors controlling prevention of fertilization are a series of multiple alleles which represent forms of the same gene. The alleles may control phenotypic differences but occupy the same sites of homologous chromosomes. Alleles can be transformed from one into another by mutations and reverse-mutations. Multiple alleles represent such a series of alleles arising from one gene by

mutation.   These alleles function at the same locus of a chromosome
and affect the same processes, although they differ among themselves.
They are the basis of the specificity.

The symbol S is used to designate the genetic factor of the
specific inhibition reaction.   Each diploid parent plant carries in
its chromosomes two alleles of a single gene which are usually designated S_1, S_2...S_x.   The alleles are divided into two equal parts
during meiosis : one appears in the pollen grain, the other in the
embryo sac.   In a plant with the S_1S_2 genotype, half of the pollen
contains the S_1-allele, the other half the S_2-allele.   The number
of alleles may vary from two to a huge number.

As mentioned before successful cross-pollination is, therefore,
possible only between individuals of different genotypes.   On the
other hand when pollen with a certain S-allele complex arrives on a
stigma with an identical S-allele composition, an inhibition reaction is induced which prevents fertilization.   Specificity of the
pollen tube growth is, therefore, determined by the combination of
the S-alleles.

Control of pollen tube behaviour may be sporophytic or gametophytic.   The sporophytic system is characterized by the fact that
it is the genotype of the diploid sporophyte that determines the
reaction of the pollen and not the one S-allele carried by pollen.
The S-alleles in pollen and style can have a dominance relation and
different possibilities of interaction.   In contrast, gametophytic
determination of incompatibility is directed by the genotype of the
haploid nucleus of the pollen grain (the male gametophyte);   there
is no interaction in the diploid style tissue, that is, the alleles
act independently of each other.

Gametophytic and sporophytic determination also manifest themselves at the site of action.   Generally in gametophytic systems the
growth of pollen tubes stops after being in contact with the transmitting tissue of the style whereas in sporophytic systems, in general, the contact between pollen grain wall and stigma papillae leads
to a reaction which either inhibits pollen germination or prevents
penetration of the tip of the pollen tube into the stigma.

There is also an interesting relation with the number of nuclei
in pollen grains.   Gametophytic systems generally have binucleate
pollen so that the second mitotic division of the generative nucleus
results in the formation of two sperm cells in the growing pollen
tube, whereas sporophytic systems are characterized by trinucleate
pollen grains so that the second mitotic division occurs before pollen germination, mostly during pollen development in the anthers.

Biochemical mechanisms. The biochemical mechanisms of the specific incompatibility reaction seems to be different in the two genetic systems (23) as follows.

1.   *Sporophytic systems inhibit germination or penetration respectively.* This reaction is strictly localized on the surface of the stigma and is characteristic of families such as *Cruciferae, Papilionaceae, Gramineae, Compositae* and others. By-passing of the stigma surface by introducing pollen grains into the stigma or directly into the style, or direct pollination of the ovule result in normal fertilization. On the other hand, compatible pollen tubes are able to penetrate the cuticular layer which covers the stigma papillae. The conclusion is that a special enzyme, cutinase, able to break down the cutin of the stigmatic papillae, is involved in the incompatibility reaction (18, 28). It resembles the enzyme involved in infection of host plants by many micro-organisms (10). Transfer experiments with *Arabis* or *Brassica* (18) showed that after a short period on the stigmatic surface of a compatible plant, the pollen tube becomes able to penetrate its own, originally "inhibiting" stigma. The conclusion is that during the short contact phase in which the pollen grain lies on the "foreign" stigma, cutinase is activated and then breaks down the cutin layer on its "own" stigma surface. In contrast, after self-pollination the cutinase system of the pollen is not activated and the pollen is unable to act on the barrier by hydrolysis of the cuticle (Fig. 2).

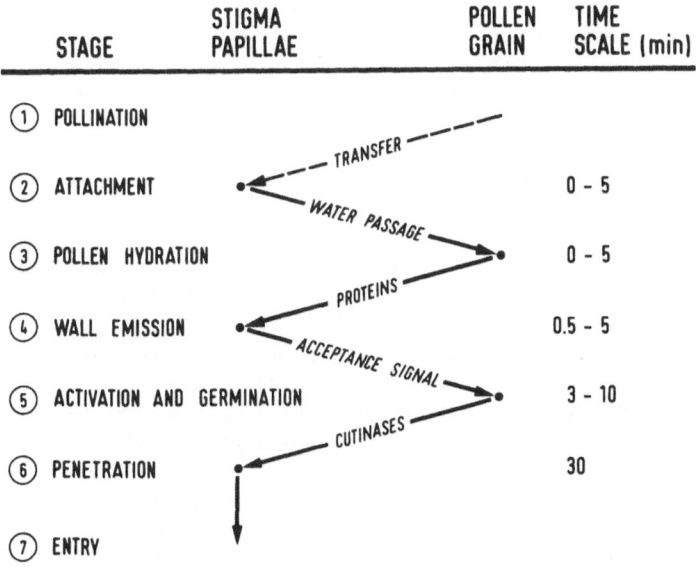

Figure 2

The assumption has been made (15) that the stigma papillae
do not autonomously in their defensive role passively relay signals from some other tissue of the style which is itself the
determinant of the specific response. This means that all the
information for the specific reaction must be present in the
papillae and in the proteinous coating of the pollen grain wall.
Thus Heslop-Harrison (15) speaks about a "dialogue" between a
pollen grain and the stigma papillae. This suggests that there
are several critical stages of recognition. However, whether
or not pollen is accepted is decided within the first ten minutes or so after the first contact of the pollen grains with
the stigma surface. The incompatibility reaction depends on
the combination of complementary sites involving specific molecules on the outer surface of the pollen grain and on the
stigma. There has to be a molecular code by which a compatible
pollen tube and a stigma cell recognize each other. For more
than 15 years we have known that the pollen grain wall is a physiologically active structure which contains protein and enzymes. Recent work (15) has shown that both the stigma and the
pollen grains are pre-programmed so that the signals of "acceptance" or   "rejection" do not need *de-novo* DNA transcription
and RNA translation. The very fast reaction suggests a preformed "trigger" system.

2. *Gametophytic systems with pollen tube inhibition.* The
second type of the incompatibility reaction occurs in representatives of the *Solanaceae* including many cultivated plants.
The pollen germinates normally, the pollen tube penetrate the
stigma and enters the style but its growth rate decreases a
few hours after pollination. In most cases growth of the incompatible pollen tubes stops at some distance from the stigma.
In contrast compatible pollen tubes germinate, penetrate the
style and grow at a constant rate (0.1 to 1.0 mm per hour) with
no sign of inhibition. The extent of the incompatible reaction
becomes evident from the fact that compatible tubes behave in
the same uninhibited way when incompatible pollen is present.

In styles of plants with a transmitting tissue the pollen
tubes make their path within this tissue by dissolving enzymatically the intercellular matrix consisting mainly of pectic
substances, and create in this way a tube-like space through
which they grow. In plants with an open stylar channel the
pollen tubes grow through a layer of slime which is secreted
into the free space of the style through an internal epithelium.
The abnormal behaviour of incompatible pollen tubes is striking
(27). In about 50% of all tubes the generative nucleus does
not divide. The normal carbohydrate metabolism is disturbed
and results in higher density of the cellulose microfibrils in
in the thickened walls of branched tips of tubes and in an increased number of callose plugs. Inhibited pollen tubes have

a higher respiration rate, increased absorption rate, increased enzyme activity and an altered protein synthesis. The altered protein pattern reflects the changed protein synthesis which has been demonstrated with stylar polysome preparations, injected into *Xenopus* eggs (7). The plant proteins synthesized in this animal system can be extracted and tested in styles. Certain fractions from incompatible-pollinated styles exhibited in these *in vivo* experiments an inhibitory effect, whereas other fractions from polysomes of compatible-pollinated styles induced synthesis of proteins which, *in situ*, promoted growth of pollen tubes. Furthermore, the specific activity of RNA-synthesis showed a distinct peak about 10 to 12 hours after self-pollination, whereas at the same time cross-pollinated styles showed no RNA-synthesis. Thus it can be concluded that for normal, compatible growth of pollen tubes all necessary information is present in the pollen tube-style system. Incompatible systems need a new messenger which is responsible for the specific protein included in the incompatibility-reaction. The conclusion could be that as a consequence of the interaction after incompatible pollination a special switch is turned on to induce incompatibility. In contrast, it has been found that in cross-pollinated styles, in which growth of tubes is not inhibited, certain enzymes have a higher specific activity than have those in self-pollinated styles. They include glucanhydrolases, which split complex proteins, such as glycoproteins which are produced as a consequence of the incompatible interaction.

Analogies between incompatibility and immunology. It is tempting to speculate on the analogy of the incompatibility reaction and certain similar phenomena in plant diseases. Apart from the high degree of specificity it is striking that the inhibition reaction does not occur, or is not so effective, when the flower is in the bud stage, the so-called bud fertility. The same is true when the plant is at the end of its life when the inhibition reaction becomes weaker, the so-called end-season fertility. East (9) suggested an analogy between incompatibility and immunity. Pollen tubes grow into intercellular spaces of stylar tissue in a parasitic way. They can indeed be considered as an organism with presumably many characteristics of parasites, especially their specificity. The pollen tube, bearing a certain S-gene, acts as an antigen against the stylar-tissue in which the same S-gene is present, and forms antibodies against the pollen tube; the complex subsequently formed could be responsible for inhibition of growth of the tube.

However, I want to make clear that we do not need an immunological analogy to explain incompatibility specificity in terms of molecular regulation. Recent experiments have shown (5) that RNA synthesized in incompatible- and compatible-pollinated styles are

different quantitatively, and qualitatively. The main differences
are found in the molecular masses of the fractions ranging between
600 000 and 40 000, i.e. the messenger-like fractions. These have
messenger activity as shown in *in vitro* protein-synthesis experiments. The protein- and RNA-synthesis during the progamic phase show
distinct differences under compatible and incompatible conditions
(6). There are different waves of synthetic activity in the poll-
inated styles; these involve protein- as well as RNA-synthesis and
also the nucleotide pool. There is evidence that the RNA synthesized in cross-pollinated, compatible styles is more stable than the
RNA synthesized in self-pollinated, incompatible styles. There is
also some evidence that synthetic processes during the incompatibility reaction occur during the first three hours after contact between the male pollen and the female stigma. So I have come to the
view (26), based on the fact that RNA- and protein-synthesis is
retarded in cross-pollinated compared with self-pollinated styles,
that "rejection" caused by self-recognition is the normal reaction.
The style has to switch on or activate another mechanism in the
pollen tube to produce an adequate enzyme-system to prepare a way
through the intercellular material in the stylar transmitting-tis-
sue. Finally, I think that a distinction has to be made between
the "recognition"-reaction which is the gene specific one, and the
"rejection"-reaction which builds the incompatibility barrier. Each
of the *S*-genes which control the specificity of the barrier produces
a highly specific protein in the pollen and a molecule with the same
specificity in the style (19). The interaction during fertilization
is a special case of interaction in and between organisms (24) and
may serve as a model for investigating specificity in plant diseases.

     Analysis of the basic molecular mechanism of interactions among
and between higher organisms is one of the most exciting problems
that confront us. Whereas control processes within single cells and
between cells are being studied intensively, interactions at high
levels also involve control of protein and RNA synthesis. Our main
interest is directed at the extent to which control is exerted by the
modification of proteins. This can be separated into the question
of post-translational control and the possible influence of protein
modifications on the control of translation itself. We can postulate analogies between the infection processes in plant diseases
and the incompatibility reaction in the fertilization of higher
plants.

                              REFERENCES

1.   BALDEV, B. (1962). *In vitro* studies of floral induction on
           stem apices of *Cuscuta reflexa* Roxb - a short-day plant.
           Ann. Bot., N.S., 26, 173 - 180.

2.   BONNER, J. (1950). The role of toxic substances in the inter-
         action of higher plants. *Bot. Rev.,* 16, 51 - 65.

3.   BROOKS, M.G. (1951). Effect of black walnut trees and their
         products on other vegetation. *Bull. W. Va. Univ. agric.
         Exp. Stn.,* 347, 1 - 31.

4.   DELEUIL, G. (1951). Origine des substances toxiques du sol
         des associations sans thérophytes du Rosmarino-Ericion.
         C. r. hebd. Séanc. Acad. Sci., Paris, 232, 2038 - 2039.

5.   DONK, J.A.W.M. VAN DER   (1974). Differential synthesis of RNA
         in self- and cross-pollinated styles of *Petunia hybrida.*
         Molec. gen. Genet., 131, 1 - 8.

6.   DONK, J.A.W.M. VAN DER   (1974a). Synthesis of RNA and protein
         as a function of time and type of pollen tube-style in-
         teraction in *Petunia hybrida* L. *Molec. gen. Genet.,* 134,
         93 - 98.

7.·   DONK, J.A.W.M. VAN DER   (1975). Translation of plant messen-
         ger in egg-cells of *Xenopus laevis. Nature,* Lond., 256,
         674 - 675.

8.   DÖRR, I. (1972). Der Anschluss der Cuscuta-Hyphen an die
         Siebröhren ihrer Wirtspflanzen. *Protoplasma,* 75, 167 -
         184.

9.   EAST, E.M. (1929). Self-sterility. *Biblphia genet.,* 5, 331 -
         370.

10.   ENDE, G. VAN DEN, and LINSKENS, H.F. (1974). Cutinolytic en-
         zymes and phytopathogenesis. *A. Rev. Pl. Pathol.,* 12,
         247 - 258.

11.   EVENARI, M. (1961). Chemical influences of other plants
         (allelopathy). *Hand. PflPhysiol.,* 16, 691 - 736.

12.   FUNK, G. (1955). Beiträge zur Kenntnis der Meeresalgen von
         Neapel. *Pubbl. Staz. zool. Napoli,* 25, Suppl., 178 pp.

13.   GÄUMANN, E. (1951). *Pflanzliche Infektionslehre.* 2nd ed.
         Birkhäuser, Basel, 611 pp.

14.   GRÜMMER, G. (1961). The role of toxic substances in the in-
         terrelationships between higher plants. *In : Mechanisms
         of biological competition* (MILTHORPE, F.L., Ed.). *Symp.
         Soc. exp. Biol.,* 15, 219 - 228.

15. HESLOP-HARRISON, J., HESLOP-HARRISON, Y. and BARNER, J. (1975).
     The stigma surface in incompatibility response. *Proc. R.
     Soc.*, B, 188, 287 - 298.

16. KERSTEKTER, R.E. and HULL, R.J. (1970). Autotrophic incor-
     poration of $^{14}CO_2$ in *Cuscuta pentagona* in relation to its
     parasitism. *Adv. Front. Pl. Sci.*, 25, 83 - 91.

17. KNAPP, R. (1967). *Experimentelle Soziologie und gegenseitige
     Beeinflussung der Pflanzen*. Ulmer, Stuttgart. 202 pp.

18. KROH, M. (1966). Reaction of pollen after transfer from one
     stigma to another (contribution to the character of the
     incompatibility mechanism in Cruciferae). *Züchter*, 36,
     185 - 189.

19. LEWIS, D. (1960). Genetic control of specificity and activity
     of the S antigen in plants. *Proc. R. Soc.*, B, 151, 468 -
     477.

20. LINSKENS, H.F. (1957). *Die Abwehrreaktionen der Pflanzen*.
     Dekker and van der Vegt , Utrecht-Nijmegen. 20 pp.

21. LINSKENS, H.F. (1963). Oberflächenspannung an marinen Algen.
     Proc. K. ned. Akad. Wet., C, 66, 205 - 217.

22. LINSKENS, H.F. (1963a). Beitrag zur Frage der Beziehungen
     zwischen Epiphyten und Basiphyten bei marinen Algen. *Pubbl.
     Staz. zool. Napoli*, 33, 274 - 293.

23. LINSKENS, H.F. (1964). Biochemistry of incompatibility. *In :
     Genetics to-day*. (GEERTS, S.J., Ed.). Vol. 3, 631 -
     636. Pergamon Press, Oxford.

24. LINSKENS, H.F. (1968). Host-pathogen interaction as a special
     case of interrelations between organisms. *Neth. J. Pl.
     Pathol.*, 74, Suppl. 1, 1 - 8.

25. LINSKENS, H.F. (1973). The reaction of inhibition during in-
     compatible pollination and its elimination. *Fiziologiya
     Rast.*, 29, 192 - 203; also *Soviet Pl. Physiol.*, 20,
     156 - 166.

26. LINSKENS, H.F. (1975). Incompatibility in Petunia. *Proc. R.
     Soc.*, B, 188, 299 - 312.

27. LINSKENS, H.F. (1975a). The physiological basis of incompat-
     ibility in angiosperms. *Biol. J. Linn. Soc.*, 7 Suppl.
     1, 143 - 152.

28. LINSKENS, H.F. and Heinen, W. (1962). Cutinase-Nachweis in
         Pollen. *Z. Bot.*, 50, 338 - 347.

29. MOLISCH, H. (1937). *Der Einfluss einer Pflanze auf die andere.
         Allelopathie.* Fischer, Jena, 106 pp.

30. MORAL, R. DEL and MULLER, C.H. (1969). Fog drip : a mechanism
         of toxin transport from *Eucalyptus globulus*. *Bull. Torrey
         bot. Club*, 96, 467 - 475.

31. MULLER, C.H. (1969). Allelopathy as a factor in ecological
         process. *Vegetation*, 18, 348 - 357.

32. RADEMACHER, B. (1959). Gegenseitige Beeinflussung höherer
         Pflanzen. *Hand. PflPhysiol.*, 11, 655 - 706.

33. RICE, E.L. (1974). *Allelopathy.* Academic Press, London,
         New York, 353 pp.

34. SCHUMACHER, W. (1934). Die Absorptionsorgane von *Cuscuta
         odorata* und der Stoffübertritt aus den Siebröhren der
         Wirtspflanzen. *Jb. wiss. Bot.*, 80, 74 - 91.

35. WENT, F.W. (1940). Soziologie der Epiphyten eines tropischen
         Urwaldes. *Ann. Jardin Botan. Buitenzorg*, 50, 1 - 98.

36. WENT, F.W. (1970). Plants and the chemical environment. *In :
         Chemical Ecology.* (SONDHEIMER, E. and SIMEONE, J.B. Eds.),
         71 - 82. Academic Press, New York and London.

37. WHITTAKER, R.H. (1970). The biochemical ecology of higher
         plants. *In : Chemical Ecology.* (SONDHEIMER, E. and
         SIMEONE, J.B. Eds.), 43 - 70. Academic Press, New York,
         London.

38. WOLSWINKEL, P. (1974). Complete inhibition of setting and
         growth of fruits of *Vicia faba* L. resulting from the
         draining of the phloem system by *Cuscuta* species. *Acta
         bot. neerl.*, 23, 48 - 60.

39. WOLSWINKEL, P. (1974a). Enhanced rate of ^{14}C-solute release
         to the free space by the phloem of *Vicia faba* stems para-
         sitized by *Cuscuta*. *Acta bot. neerl.*, 23, 177 - 188.

CONTRIBUTIONS

WOLSWINKEL, P.   Specific interactions between *Cuscuta* and host
    plants.

SUMMARY OF POINTS FROM CONTRIBUTIONS AND DISCUSSIONS
BY
H. KERN

Chairman and Discussion Leader

Wolswinkel discussed the specific interactions between *Cuscuta*
and its host plants. When broad bean plants are infected with
Cuscuta at the time of flowering, fruit setting is generally completely inhibited. When the host is infected in a later stage, *Cuscuta* seriously interferes with further growth of pods. Experiments
with ^{14}C-labelled assimilates showed that *Cuscuta* is able to with-
draw all assimilates which normally move from the host leaves to
the growing pods. The haustorial organ of the parasite is very well
suited for withdrawing assimilates from the host. It forms a contact cell resembling a hand with a conspicuous wall labyrinth, en-
larging the absorbing surface, and with transfer cell characteristics.

Transfer cells absorb assimilates from the free space between
plasmalemmae of host and parasite. An enhanced unloading rate of
the host phloem seems essential for the transfer of assimilates from
host to parasite. Higher amounts of labelled solutes are washed
out from the free space in parasitized stems than in healthy ones.
An active metabolism seems necessary for the normal movement of assimilates from the sieve tubes into the free space. Non-parasitized
stem parts, situated adjacent to parasitized parts, may be influenced by hormonal effects caused by *Cuscuta*. More potassium and mag-
nesium ions are released into the free space in parasitized stems
than in healthy ones. The transfer of solutes to the parasite seems
to be regulated by a metabolically controlled efflux from the phloem
at the infection site. The parenchyma cells of the host stem at the
infection site may take advantage of the supply of solutes into the
free space.

Parallels exist between the influence of *Cuscuta* on translocation patterns and that of rusts and mildews. Infections by these
fungi may cause an enhanced efflux from the phloem at the infection
site. Host cells at the infection site may absorb solutes released
into the free space and become suited for nourishing the parasite;
accumulation of nutrients, increased respiration, and green islands
will result at the infection site.

SPECIFICITY - AN ASSESSMENT [a]

RONALD K. S. WOOD

Department of Botany, Imperial College of Science and Technology
London, U.K.

In the context of plant pathology the study of specificity deals with why a pathogen grows and causes disease in some cells and tissues of plants but not in others. When these cells and tissues are in the same plant we have the tissue and organ specificity described by Graniti of which there are many striking examples. There are also the marked differences in reaction to a pathogen that occur in what seem to be similar cells sometimes close together in the same tissue. Mesothetic reactions to rust fungi are special examples of such differences. It is surprising that they have not been more studied in view of their implications for hypotheses on mechanisms of interactions believed to be controlled by gene-for-gene relations. They and other cell-from-cell differences would be difficult to analyse because of their unpredictability but this is not true of tissue and organ specificity with predictable responses and where differences in reaction can be as striking as in resistant and susceptible plants. Here, presumably, factors related to the position of cells over-ride those that control reactions in plants of different genotype. Similar considerations apply to factors such as temperature that alter the response of plants of a given genotype. Closer study of such systems might be profitable particularly when the genetics of the basic resistant and susceptible reactions of closely related plants is known. However, most plant pathologists have been and will continue to be more interested in the reactions based on different

[a] An extended and elaborated version of a summing-up paper given by the author at the end of the Institute.

Table 1   *Scheme of interactions between specialized*
             pathogens, saprophytes and higher plants

Pathogen A	Plant X			Plant Y		
	Cv1	Cv2	(Cvn)	Cv1	Cv2	(Cvn)
Race 0	S	S	(S)	Ry	Ry	(Ry)
Race 1	S	Rx	(S or Rx)	Ry	Ry	(Ry)
Race 2	Rx	S	(S or Rx)	Ry	Ry	(Ry)
(Race n)	(S or Rx)	(S or Rx)	(S or Rx)	Ry	Ry	(Ry)
Saprophyte B	Rb	Rb	(Rb)	Rb	Rb	(Rb)

1. A - example of specialized pathogens.

2. X - example of a group of plants some members of which are
        hosts of A. Cv1, Cv2, (Cvn) are cultivars of X; Cv could
        be replaced in the scheme by a group of related species.

3. Y - example of all species of plant other than those covered
        by 2.

4. B - example of species of micro-organisms not known to cause
        disease in plants.

5. S - susceptible, R - resistant reaction.

genotypes with which this Institute has been mainly concerned particularly when the pathogens are more or less highly specialized. Table 1 summarizes reactions between plants and such pathogens, and, also to saprophytic micro-organisms. To enlarge on the notes to the Table, Plant X is representative of the small group of plants some members of which are susceptible whereas the rest are resistant to Pathogen A. X is often a species with cultivars that differ in their reaction to Pathogen A or its Races. Plant Y is representative of all plants other than the group represented by Plant X. Because Pathogen A is specialized, the number of X plants is very small indeed compared with the number of Y plants. There are, of course, a good number of pathogens such as *Botrytis cinerea* in which the group of Plants X would be large in respect of Pathogen A though still small compared with Y. They are not our immediate concern. Pathogen A exemplifies the many thousands of pathogenic bacteria and fungi more or less specialized in their parasitism. Very probably, most occur as races with different capacities to cause disease in different types of Plant X though the majority have not been investigated from this point of view. Saprophyte B, in any of its strains, by definition cannot cause disease in X or Y plants.

Now to comment on the R reactions in Table 1. Rb is very high resistance. We know little about its nature though perhaps in the present context we should be emphasizing deficiencies of the saprophytic rather than attributes of the host. Ry is also high resistance but here we know that in some combinations pathogen A does infect plant Y and grow in it though to a very limited extent. The extremes of Rx may resemble Ry, at least visually; this is not to assert that the causes of both are the same or similar. But Rx is not confined to these extremes; for different combinations of pathogen A and plant X in their different forms there may be a continuous series of reactions between the extremes of Rx that appear similar to Ry to the extreme of susceptibility S. There are no such intermediate reactions between pathogen A and plant Y. The intermediate reactions between A and X tend to be ignored particularly when gene-for-gene systems are postulated, when we select model systems for research, and when we propose hypotheses to accommodate our data.

Consider now some implications of the mechanisms proposed to explain susceptibility S and resistance Rx and Ry (I shall return later to Rb). The one with most secure evidence is based on host selective toxins (*sensu* Wheeler). In systems so far studied pathogen A occurs essentially as Race 1 producing a toxin which damages susceptible Cv1 of plant A at concentrations much lower than those which similarly damage Cv2 of plant A and Cv1-n of plant Y (neglecting strains of pathogen A which do not produce toxins and so are avirulent). Note, however, the possibility of a system in which Race 2 of pathogen A produces a second toxin with the reverse of the above effects on Cv1 and Cv2. It was proposed many years ago that damage arises when the toxin binds to specific receptors probably in plasma

membranes. Recently Strobel has claimed to have identified the receptor for helminthosporoside, the host-selective toxin of *Helminthosporium sacchari*. This remains to be fully established but let
us assume it will be for this and for other host-selective toxins.
Does this provide a satisfactory explanation of specificity in such
diseases ? Some difficulties remain. Resistant plants of Cv2 are
infected by pathogen A and cells are killed but the pathogen grows
little if at all beyond them. What kills these cells ? Presumably
not the toxin because if it did it would hardly be selective. And
why does the pathogen not grow beyond the locus of the dead cells
because if it has killed some cells why does it not continue to kill
adjacent cells ? Pathogen A does not at first seem so active in
killing cells of Cv1 of plant X but later it kills many more than
it does in resistant plants, presumably after secreting the host-
selective toxin though, in fact, one would like to see more evidence
on this point. But because a resistant reaction is not invoked in
the early stages of infection, the toxin need not be selective in
the later stages; the toxins of *Pseudomonas phaseolicola* or the
pectic enzymes of soft-rot pathogens are not selective. Although
the genetic evidence relating to host-selective toxins is most per-
suasive, is it possible that the selectivity of these toxins is
fortuitous and may specificity depend on quite other factors ?

Another difficulty for some host-selective toxins arises from
their potency. Thus, victorin is some 100 000 times more active on
susceptible than on resistant plants but this figure has to be considered in relation to the value of 10^{-4} µg/ml which prevents growth
of susceptible roots. This means that toxicity to resistant plants
is still impressive and that not much toxin would be needed to kill
cells after infection. Also to be considered is that in susceptible
plants the hypersensitive response may be suppressed by the toxin, or
that susceptible but not resistant plants induce its synthesis. The
difficulty here would be that host-selective toxins are produced
readily *in vitro*. It would press this line of conjecture too far
to suggest that resistant plants inhibit synthesis !

The essential feature of specificity basis on host-selective
toxins is that disease occurs only when a toxin combines with a receptor in the host. This is the reverse of what is supposed to
happen in systems such as those described for facultative parasites
by Kaars Sijpesteijn and presumably, for obligate parasites by
Rohringer. Here Rx depends on reaction between a substance produced
by avirulent Race 1 of pathogen A and a receptor in Cv2 of Plant X,
or between a different substance and a different receptor for Race
2 and Cv1 and so on. These reactions lead to conditions in the host
that limit growth of the pathogen. In its highly expressed form
damage to the host and growth of the pathogen are severely limited,
the reaction is called hypersensitive and is characteristic of gen-
for-gene systems in which synthesis of the substance by the pathogen

is controlled by one gene matched by another gene that controls
synthesis of the receptor in the host. In susceptible responses S
either the pathogen does not produce the substance, or the host
does not contain the receptor, or both. The damage does not occur,
neither do consequences of that damage so that the pathogen is able
to grow in the host. Note that the scheme requires selectivity of
the substance produced by the pathogen, and toxicity, at least in
most of the systems so far studied. Such properties would seem to
make them host-selective toxins ! Though, of course, the consequences of their toxicity are quite different.

     This scheme is simple and satisfying but presents a number of
difficulties on close examination particularly when extended to
resistance of type Ry. First there is the awkward fact, stressed
by Daly in this Institute but long recognized, that interactions
apparently controlled by single genes are not always nearly so well
defined as would seem to be required by the hypothesis. In many
systems intermediate reactions are as common or more so than the
extremes which nicely fit the hypothesis. It may, however, be possible to explain a wide range of reactions in terms of different
rates of production of inducing substances by races of pathogen A
and differences in the affinities of the substances for the receptors.
The hypothesis in its simplest form also assumes in the susceptible
interaction that the pathogen does not invoke the response that kills
cells and leads to resistance. But certainly for facultative parasites this honeymoon period does not last long. The pathogen, the
host, or both change their behaviour and then the plant is damaged
on the much more intensive scale which we call disease. Is this
damage similar in cause and type to that which occurred in the resistant reaction ? If so, or indeed even if not, why does it not
now induce resistance as it did earlier ? And because it does not,
what are the differences in damage to cells, or in conditions in
which it occurs, that sometimes lead to resistance and sometimes do
not ? A possible explanation is that resistance develops after
damage only when damaged cells are backed up by a sufficiently large
number of undamaged cells. If, as in susceptible plants, many cells
were killed over a short period, this condition would not prevail
and resistance could not develop. Clearly, for facultative parasites,
there are many questions to be answered here. But perhaps not so
many for the obligate parasite where death of infected cells may be
sufficient to explain resistance though, of course, we must not
assume that this *is* the mechanism of resistance until we have evidence
better than that available at present. A final point on this subject;
although it is convenient to study systems in which Rx appears as
death of a few cells, we should not assume that Rx is always readily
visible. There are many stages between robust health and death and
resistance could well be associated with one or more of the intermediate stages as, of course, it seems to be in many diseases.

It is now time to extend the above, rather simple hypothesis
to resistance of type Ry. This can be visually similar to extreme
forms of Rx. Is it based on the similar mechanism of reaction between a specific product of the pathogen and a specific receptor in
the host ? If so, this may imply in the first place that each of
the thousands of species represented by plant Y has the same receptors for the series of substances produced by races of pathogen
A. And there would be different sets of receptors for each of the
many thousands of pathogens of type A which, presumably, produce
different substances that induce resistance. This seems improbable
but then so might be the capacity of each mammal to produce specific
antibodies to a seemingly endless number of antigens. But we know
that it does.

Alternatively we can postulate that Ry is different in nature
from Rx and that it is non-specific in the sense that all plants
have the capacity to react to some product or component, or class of
such substances, common to all pathogens to which the plants are resistant. If so, then here is a system about which we know virtually
nothing both as to the substances from pathogens and their targets
in plants. However, in this connexion we should remember the similarity of the responses of tissues of higher plants to wounds caused
in different ways. Some years ago I suggested that we might get
useful leads for the study of resistance mechanisms from information
on wound responses but until recently few seem to have been interested in these reactions. We should keep an eye on progress now
that physiologists are taking a renewed interest in this field.

To continue with the idea of a non-specific induction of resistance by micro-organisms. If this does occur generally then we
must ask whether it also occurs when Race 1 of pathogen A infects
Cv2 of plant X, the single gene controlled, specific response of
type Rx. Because if it does, then Rx would seem to be superfluous.
But perhaps Race 1 has lost the capacity to invoke the non-specific
response in Cv2 and then there is the dilemma that it has not lost
this capacity for the thousands of plants of type Y. There is con-
stantly in these comparisons the contrast between the generality of
the resistance of plants to almost all pathogens and the specificity
of the resistance in particular host-parasite combinations. This
contradiction may arise because we emphasize too much the positive
role of resistance in specificity. Perhaps we should make more of
susceptibility, the much rarer phenomenon, and its induction by
virulent pathogens as the key to specificity, a theme developed by
Daly but mainly for obligate parasites.

In the susceptible response infection and early growth of a
pathogen is not accompanied by the severe damage caused by an avirulent pathogen. On hypotheses of the type proposed by Kaars Sijpesteijn this is because there is no inducing substance from the pathogen,

no receptor in the host, or both. What are the implications of this hypothesis for a breeding programme involving pathogen A and plant X ? A stage may be reached in which each of the Races has lost any capacity it may have had to produce substances that induce Rx in one or other of the cultivars. A cultivar (Cvn+1) is introduced which contains a new gene for a receptor for a substance controlled by a gene present in each of Races 1-n. The new gene will be effective until one or more of the races changes so as not to produce the substance that can react with the receptor. This implies that populations of pathogens such as A contain genes with functions that do not become apparent until revealed by the introduction of a new genotype in the host. Unless, of course, such genes also have other functions. Otherwise why did such genes evolve in the first place and why do they persist especially against a background of non-specific resistance. Another point is that whereas pathogen A can change so as not to produce substances that induce Rx, apparently it cannot change so that it does not produce substances that induce Ry. But, presumably, it must have done so in the past in respect of plant X. Does this explain the emergence of genes that control Rx ?

If, in fact, Rx and Ry are based on reactions between products of pathogens and receptors of hosts then we know next to nothing about either apart from the results obtained by Kaars Sijpesteijn, and Rohringer and their colleagues for Rx and promising though the results may be I am sure that these workers will agree that they are not much more than this at present. It is surprising that so little progress has been made in this direction and perhaps a little ominous. In most of the systems so far studied preparations from Races of pathogen A do induce responses that resemble Rx but do so quite non-specifically. One can understand the lack of progress with fungal pathogens because the active substances could be closely bound to hyphae and be very unstable on separation but we had reason to be more hopeful for bacteria because here, presumably, a cell of Plant X responds to a substance produced by the pathogen in an intercellular space and after passage through the cell wall. Technically, it should not have been difficult to reproduce this system *in vitro*, or to extract the active substances from intercellular spaces but to my knowledge no success along these or other lines has been reported. This has suggested that synthesis of the active substance is induced only after the host has acted upon the bacterium. If so then for bacteria as for fungi we may still be some way from identifying the substances that induce Rx and, probably, still further from identifying the corresponding receptors in plants though we can always hope that studies of ultrastructure of reacting cells will provide clues as to their nature. In seeking these receptors it might be advisable to follow the lead of Wheeler, Hanchey and others and to consider the cell wall as the site of the earliest specific reactions between pathogen and host though in this connexion we should also

remember that the apparent discontinuity between cell wall and plasma
memberane may only reflect the crudeness of our techniques for
examining this interface. And although one recognizes the attraction
of identifying the substances and receptors that establish Rx for
what this will tell about specificity, we should also concern our-
selves with those that cause Ry because, possibly, Rx is needed only
when Ry has failed.

Having dealt at some length with Rx and Ry, I shall now comment
briefly on Rb, the reaction, or more likely, the lack of reaction of
all plants to saprophyte B. We know as little about Rb as we do
about Ry because, understandably, plant pathologists are not much
given to studying saprophytes. Nevertheless, it is not improbable
that we could learn something useful about specificity by trying to
find out why saprophytes do not cause disease although many of them
can synthesize all or most of the enzymes that will degrade all
important components of plant cells and not a few of them will produce, at least *in vitro,* substances toxic to plant protoplasts. What
is it, therefore, that they still require to become pathogens ? Do
they even lack the capacity of pathogens to evoke responses in non-
host plants of type Y ? On the other hand, it is possible that Rb
is essentially the same as Ry. What we call a saprophyte may be
just another "non-pathogenic" pathogen so far as the plant is concerned. In this connexion, it might be worth studying what makes a
few species of *Penicillium* pathogenic, albeit to storage tissue,
whereas the great majority of species of this large genus are saprophytes. Equally provoking is the co-existence of saprophytic forms
of *Fusarium oxysporum* and the range of *formae speciales* which are so
highly specialized in their parasitism on different species of higher
plants. But I must not speculate too much on this theme because
worthy though the objective, I think it might be difficult to get
financial support for research aimed at turning saprophytes into
pathogens !

A characteristic of Rx is its stability, hence the term "vertical
resistance". But in the occasional system it is possible to change
Rx to S or S to Rx by factors such as temperature. The mechanisms
that control such changes deserve more attention although the view
could be taken that interesting as are such effects, they are not
likely to tell us much about specificity until we know more about
how they control normal metabolism. This comment may apply particularly to growth-regulating substances. But it signifies a defeatist
attitude. I think that there is a good chance that something significant will be learned about specificity by trying to alter it
especially when the results are as exciting as those described by
Kuć and his colleagues for resistance induced in upper leaves of
susceptible plants by causing severe disease in lower leaves; and
the work along similar lines by Sequeira and his colleagues, and by
others. The next few years will reveal the potential of this line

of attack on the problem of specificity.

It is now time to consider further the expression of resistance particularly of type Rx about which we do have some information. In its most obvious form it appears as death of a few cells following infection. For the obligate parasite we could then assume that death of cells is sufficient to explain why its growth stops and leave the matter there. But it would be rash to do so. The simple nutritional hypothesis possibly applicable to obligate parasites, almost always, is not available for the non-obligate parasite. Here we must usually assume that the plant reacts so as to prevent growth by acting direct- ly on the parasite, by inactivating the agents by which it causes damage, or by nullifying the effects of these agents. In the last two categories we have, *inter alia*, the inactivation of enzymes and toxins, and lignification or other changes in cell walls. It is, however, the direct effects on pathogens that are currently attrac- ting most attention in the form of phytoalexins, to some plant patho- logists the key to the problem of specificity, to others a series of irrelevancies, but with the truth probably somewhere in between. No matter what our views about the significance of phytoalexins in specificity, I think that we must now accept that they can accumulate sufficiently at the locus of infection in some examples of Rx to explain why the pathogen is confined to the locus. Quantitative data for the very early stages of infection in which growth may also be curtailed are much less satisfactory. The accumulation of phyto- alexins in lesions in susceptible plants also has to be explained quantitatively. Here it must be emphasized that in resistant and in susceptible reactions the only concentrations of phytoalexins that matter are those to which the pathogens are exposed. It is, there- fore, necessary to compare concentrations at intervals after infection at a locus involving only one or a few cells as in a hypersensitive response, with concentrations in a large lesion of many thousands of cells. But this comparison must be established and give the ap- propriate results to substantiate a definitive role for a phytoalexin in resistance. To do so is technically a formidable task even for a simple lesion caused by a facultative parasite in a mass of paren- chyma. The task is still more daunting for a vascular wilt or an obligate parasite. However, assume now that it has been established that concentrations of phytoalexins can explain the differences of growth of a pathogen in resistant and susceptible reactions. This is only the end of the story and not the key to specificity. Now must be considered the factors that induce new metabolic pathways that end in phytoalexins and how these factors originate in resistant and in susceptible reactions. We face the troublesome anomaly that whereas Rx is induced specifically, many different types of agents will induce synthesis of at least certain phytoalexins, for example, pisatin and phaseollin, and do so in ways unlikely to be related to specificity as when inoculation with a virus leads to accumulation of a phytoalexin. This implies that synthesis of a phytoalexin may

be induced by substances produced when cells are damaged in one of
a number of ways and that damage caused by a pathogen only is one such
way. If this be true, then phytoalexins would be relegated to a
secondary role in specificity though, of course, their essential role
in resistance would remain. The critical events would be those that
cause the right type of damage. Here we face the problem as to what
is the right type of damage. It is often assumed to be associated
with death of cells but I would be surprised if this were always so.
Also, are phytoalexins synthesized in cells that later will die or
in adjacent cells that respond to stimuli that originate in moribund
or dead cells ? We have a little evidence on these points but we
need much more. How do "elicitors", in the sense used by Keen fit
into this picture ? They are substances produced by pathogens (only?)
that in low concentrations elicit synthesis of phytoalexins. If
elicitors from Races of pathogen A cause Rx in cultivars of plant X
in the same pattern as do the Races, then we can begin to see a role
for them in specificity though, of course, a lot of quantitative
data will be needed to establish this role. Even then there will
remain the question as to whether an elicitor causes the damage
which initiates a series of steps leading to a triggering of the
synthesis of a phytoalexin, or whether it independently triggers the
synthesis with the damage coming later and, possibly, caused by the
phytoalexin itself. Elicitors of types described by Albersheim which
are non-specific in their action are nevertheless of interest because of their activity at very low concentrations, and the possibility that they may be produced by all pathogens. Although it seems
that they have no role in resistance of type Rx there is the intriguing possibility that they are involved in Ry. If so, we must ask
again what happens to them in susceptible and in Rx type reactions.

Now I must make some points about susceptibility by far the
rarest of the reactions given in Table 1. I shall not refer again
to diseases associated with host-selective toxins. For other diseases there is first the simple concept that all plants can recognize
"non-self" organisms, tissues or cells and then react to reject them;
Rx is one form of these phenomena. The corresponding phenomena in
mammals are, of course, well known and intensively studied. In susceptible plants, recognition of the pathogen does not occur and neither does the chain of events that lead to rejection (resistance).
The pathogen grows in the plant and causes disease. Susceptibility,
therefore, would depend on something not happening. This is an en-
ticing approach but at present we have little or no information about
the substances involved in this hypothetical recognition though I
have no doubt that many of us will now be looking for them with an
eye on glycoproteins in plasma membranes and cell walls of host plants
and on elicitors of the type described by Albersheim from pathogens.
There is, however, the slightly discouraging point made by Bushnell
that there has not been much evidence of recognition and rejection
between isolated protoplasts from widely separated species of higher
plants.

Other views of susceptibility are more positive. Thus it is
known that inoculation of part of a plant with a virulent pathogen
may make an adjacent part of the plant susceptible to a pathogen
avirulent on parts of the plant not affected by the first inoculation. This could mean that the capacity for recognition was altered by the first inoculation. Alternatively, susceptibility may
not depend on non-recognition so much as on suppression of the
events that follow recognition in resistant plants. Still another
view, particularly for obligate parasites, is that the virulent
pathogen induces changes that allow it to parasitize susceptible
plants. If these are not induced then the pathogen cannot grow and
sooner or later dies. The reactions of host cells, usually death,
associated with the failure to induce susceptibility would, presumably, be regarded as of no consequence. But they should be explained.

One can understand the attraction of the hypothesis of induced
susceptibility for diseases caused by obligate parasites which are
based on what may be envisaged as delicately balanced reactions between host and pathogen. The hypothesis is less attractive for
facultative parasites which cause disease not by allowing cells to
remain alive but by killing them; in this context induced susceptibility is, for me, difficult to conceive.

Now,briefly, I shall refer to the less specialized pathogens
which, inevitably, would not attract much attention in a series of
meetings on specificity. *Botrytis cinerea* is a good example of
such pathogens. In a wide variety of plants it is an active necr-
otroph because soon after infection it rapidly secretes pectic and
other cell wall degrading enzymes which cause cell separation and
kill protoplasts and then continues to do so in susceptible tissues.
In this respect it differs from a more specialized pathogen such
as *Colletotrichum lindemuthianum* which does not kill protoplasts
for some days after infection though it does so later as a lesion
develops. Although this pathogen is able to secrete the same
enzymes as *B. cinerea* there is, in fact, not much evidence of their
activity in lesions. This also applies to diseases, other than soft
rots, caused by most facultative parasites. Can one argue from this
that an important characteristic of the less specialized pathogen
is that they produce cell wall degrading enzymes much more rapidly,
under a wider range of conditions and much more destructively than
do more specialized pathogens and in so doing more readily and frequently overwhelm defense reactions of host tissues? Against this
possible argument is the fact that certain soft rot pathogens are
quite specialized and that pathogens such as *Corticium solani* which
in many respects are less specialized, usually cause lesions similar
to those caused by pathogens such as *C. lindemuthianum*.

What may be another characteristic of the less specialized
pathogen is that for plants that it does parasitize there is little

evidence of gene controlled differences in reactions between races
and host cultivars.   On the other hand, pathogens such as *C. solani*
are known to contain strains with different host ranges, some relatively specialized, others much less so.   In some respects this
parallels the wide host range of *B. cinerea* and the highly specialized
parasitism of closely related species such as *B. fabae* and *B. tulipae*.
Is it a coincidence that it is for diseases caused by these two
pathogens in plants resistant to *B. cinerea* that we have the best
evidence for specificity based on inactivation of phytoalexins or
of pre-formed inhibitors ?   To my knowledge any similar evidence for
more specialized pathogens is much less satisfactory.

It is tempting to speculate still further on what may be the
significant characteristics of less specialized pathogens in the
context of specificity but instead I shall return to specialized
pathogens and end by referring, as Brian has done, to one of the most
puzzling problems in specificity, heteroecism in diseases caused by
certain rust fungi.   The classic example is, of course, *Puccinia
graminis* where hyphae with dikaryotic cells parasitize species of
Gramineae whereas hyphae with one or other of the nuclei of the dikaryon are parasitic on species of *Berberis* though there is the further intriguing fact that later in parasitism tissues of barberry
do contain dikaryotic cells.   The problem becomes still more provoking
when we consider that other species of rust fungi have wide host
ranges for both monokaryotic and dikaryotic stages, or wide ranges
for one stage and restricted ranges for the other.   Much ingenuity
will be needed to accommodate these facts to any of the hypotheses
about specificity proposed at this Institute or elsewhere.

In ending this assessment I am well aware that I may have spent
too much of it in asking questions for which we have no answers and
raising doubts and difficulties about our present hypotheses on
specificity and on the data that have been used to support them.
This somewhat negative approach is partly justified because for so
important a series of problems as those presented by specificity we
can hardly be too critical in assessments of our experimental results
and in ensuring that specific substances are present at suitable concentrations at the right places before they are assigned important
roles.   But, justifying the critical approach does not alter the
fact that progress in our understanding of specificity will depend
not on criticisms but on new ideas and enterprising research.   I
hope that our proceedings at this Institute will be productive of
both.

PARTICIPANTS

ALBERSHEIM, Prof. Peter. Department of Chemistry, University of
    Colorado, Boulder, Colorado 80302, U.S.A.

BERVILLÉ, Dr. André. Université de Paris-Sud, Laboratoire
    Amélioration des Plantes. Bat. 360. Orsay, France 91405.

BOTTALICO, Dr. Antonio. Istituto di Patologia vegetale dell'
    Università, Via Amendola 165/A, 70126 Bari, Italy.

BOUCHER, Mr. Christian. Station Centrale de Pathologie Végétale,
    INRA, Route de Saint-Cyr, 78000 Versailles, France.

BRAMBL, Dr. Robert M. Department of Plant Pathology, University of
    Minnesota, St. Paul, Minnesota 55108, U.S.A.

BRETHAUER, Mr. Todd S. Department of Plant Pathology, University of
    Illinois, Urbana, Illinois 61801, U.S.A.

BRIAN, Prof. Percy W. Department of Botany, University of Cambridge,
    Botany School, Downing Street, Cambridge, CB2 3EA, U.K.

BUSHNELL, Dr. William R. U.S.D.A., Agricultural Research Service,
    Cereal Rust Laboratory, University of Minnesota, St. Paul,
    Minnesota 55101, U.S.A.

CALLOW, Dr. James A. Department of Plant Sciences, University of
    Leeds, Baines Wing, Leeds, LS2 9JT, U.K.

CERVONE, Dr. Felice. Istituto di Patologia vegetale dell' Università
    di Napoli, 80055 Portici, Italy.

CICCARONE, Prof. Antonio. Istituto di Patologia vegetale dell'
    Università, Vía G. Amendola 165/A, 70126 Bari, Italy.

CIRULLI, Dr. Matteo. Istituto di Patologia vegetale dell' Università,
    Via G. Amendola 165/A, 70126 Bari, Italy.

CLARKE, Dr. Donald D. Department of Botany, University of Glasgow,
    Glasgow, G12 8QQ, Scotland, U.K.

DANIELS, Dr. Michael J. Department of Genetics, John Innes Institute,
    Colney Lane, Norwich, Norfolk, NR4 7UH, U.K.

DALY, Prof. Joseph M. Laboratory of Agricultural Biochemistry,
    University of Nebraska, Lincoln, Nebraska 68503, U.S.A.

DAY, Dr. Peter R. Connecticut Agricultural Experiment Station,
    P.O. Box 1106, New Haven, Connecticut 06504, U.S.A.

DÉFAGO, Miss Dr. Geneviève. Institut für Spezielle Botanik, Eidg. Technische Hochschule, Universitätstrasse 2, 8006 Zürich, Switzerland.

DEKHUIJZEN, Dr. Harold M. Stichting Centrum voor Plantenfysiologisch Onderzoek, Bornsesteeg 47, Postbus 52, Wageningen, The Netherlands.

DEVAY, Prof. James E. Department of Plant Pathology, University of California, Davis, California 95616, U.S.A.

DEVERALL, Prof. Brian J. Department of Plant Pathology and Agricultural Entomology, University of Sydney, Sydney, 2006 New South Wales, Australia.

DURBIN, Prof. Richard D. U.S.D.A. Pioneering Research Laboratory, Department of Plant Pathology, University of Wisconsin, Madison, Wisconsin 53706, U.S.A.

ELLINGBOE, Prof. Albert H. Department of Botany and Plant Pathology, Michigan State University, 166 Plant Biology Building, East Lansing, Michigan 48824, U.S.A.

ERCOLANI, Dr. Gian Luigi. Istituto di Patologia vegetale dell' Università, Via G. Amendola 165/A, 70126 Bari, Italy.

FUCHS, Dr. Adriaan. Laboratorium voor Fitopathologie, Agricultural University, Binnenhaven 9, Wageningen, The Netherlands.

GAY, Dr. John L. Department of Botany, Imperial College of Science and Technology, Prince Consort Road, London SW7 2BB, U.K.

GIL, Mr. Francisco. Department of Botany, Imperial College of Science and Technology, Prince Consort Road, London SW7 2BB, U.K.

GRANITI, Prof. Antonio. Istituto di Patologia vegetale dell' Università, Via Amendola 165/A, 70126 Bari, Italy.

HADWIGER, Prof. Lee A. Department of Plant Pathology, Washington State University, Pullman, Washington 99163, U.S.A.

HANCHEY, Miss Dr. Penelope J. Department of Botany and Plant Pathology, Colorado State University, Fort Collins, Colorado 80521, U.S.A.

HEATH, Mrs. Dr. Michèle C. Botany Department, University of Toronto, Toronto, Ontario, M5S 1A1, Canada.

HEUVEL, Dr. Joop van den. Phytopathologisch Laboratorium "Willie Commelin Scholten", Javalaan 20, Baarn, The Netherlands.

HIGGINS, Miss Dr. Verna J. Department of Botany, University of
    Toronto, St. George Campus, Toronto, Ontario M5S 1A1, Canada.

HUGHES, Dr. Colin R. National Institute for Medical Research,
    The Ridgeway, Mill Hill, London NW7 1AA, U.K.

JOHNSON, Dr. Roy. Plant Breeding Institute, Maris Lane, Trumpington,
    Cambridge, CB2 2LQ, U.K.

KAARS SIJPESTEIJN, Miss Dr. Antje。 Department of Biochemistry and
    Microbiology, Institute for Organic Chemistry TNO, P.O. Box
    5009, Utrecht, The Netherlands.

KEEN, Dr. Noel T. Department of Plant Pathology, University of
    California, Riverside, California 92502, U.S.A.

KERN, Prof. Heinz. Institut für Spezielle Botanik, Eidg. Technische
    Hochschule, Universitätstrasse 2, 8006 Zürich, Switzerland.

KUĆ, Prof. Joseph. Department of Plant Pathology, College of
    Agriculture, University of Kentucky, Lexington, Kentucky
    40506, U.S.A.

LINSKENS, Prof. Hans F. Botanisch Laboratorium, Katholieke
    Universiteit, Toernooiveld, Nijmegen, The Netherlands.

LYON, Mrs. Dr. Frances M. Department of Biological Sciences,
    University of Dundee, Dundee, DD1 4HN, Scotland, U.K.

MANNERS, Dr. John G. Department of Biology, University of
    Southampton, Southampton, S09 5NH, U.K.

MANSFIELD, Dr. John W. Biology Department, University of Stirling,
    Stirling, FK9 4LA, Scotland, U.K.

MARRAS, Prof. Franco. Istituto di Patologia vegetale dell'
    Università, Via E. De Nicola, 07100 Sassari, Italy.

MARTELLI, Prof. Giovanni. Istituto di Patologia vegetale dell'
    Università, Via G. Amendola 165/A, 70126 Bari, Italy.

MOREAU, Mrs. Prof. Mireille. Laboratoires de Biologie Végétale,
    Faculté des Sciences, 6 Avenue Victor Le Gorgeu, 28283 Brest
    Cedex, France.

MUSSELL, Dr. Harry. Boyce Thompson Institute for Plant Research,
    1086 North Broadway, Yonkers, New York 10701, U.S.A.

MYERS, Dr. Alan. Department of Biology, University of Southampton,
    Building 44, Southampton, S09 5NH, U.K.

NAEF-ROTH, Mrs. Dr. Stephi. Institut für Spezielle Botanik, Eidg.
    Technische Hochschule, Universitätstrasse 2, 8006 Zürich,
    Switzerland.

NOVACKY, Dr. Anton. Department of Plant Pathology, University of
    Missouri, 108 Waters Hall, Columbia, Missouri 65201, U.S.A.

OUCHI, Dr. Seiji. Laboratory of Plant Pathology, College of
    Agriculture, Okayama University, Tsushima, Okayama 700, Japan.

PATIL, Prof. Suresh S. Department of Plant Pathology, University
    of Hawaii, 3190 Maile Way, Honolulu, Hawaii 96822, U.S.A.

PAXTON, Dr. Jack D. Department of Plant Pathology, University of
    Illinois, Urbana, Illinois 61801, U.S.A.

PERSON, Prof. Clayton. Department of Botany, University of British
    Columbia, 2075 Wesbrook Place, Vancouver, British Columbia,
    V6T 1W5, Canada.

PROTA, Prof. Ulisse. Istituto di Patologia vegetale dell'
    Università, Via E. De Nicola, 07100 Sassari, Italy.

RAGGI, Dr. Vittorio. Istituto di Patologia vegetale dell' Università,
    S. Pietro, 06100 Perugia, Italy.

RAHE, Dr. James E. Department of Biological Sciences, Simon
    Fraser University, Burnaby, British Columbia, V5A 1S6, Canada.

RICCI, Dr. Pierre. Station de Botanique et Pathologie Végétale,
    Centre de Recherche d'Antibes, I.N.R.A., Villa Thuret,
    Boulevard du Cap, Boîte Postale 78, 06602 Antibes, France.

RICH, Dr. Daniel H. Center for Health Sciences, School of Pharmacy,
    University of Wisconsin, 425 North Charter Street, Madison,
    Wisconsin 53706, U.S.A.

ROHRINGER, Dr. Roland. Canada Agriculture, Research Branch,
    Research Station, Cereal Rust Section, 25 Dafoe Road, Winnipeg,
    Manitoba, R3T 2M9, Canada.

RUDOLPH, Dr. Klaus W.E. Institut für Pflanzenpathologie und
    Pflanzenschutz, Georg-August-Universität, Grisebachstrasse 6,
    34 Göttingen, W. Germany.

SCHEFFER, Prof. Robert P. Department of Botany and Plant Pathology,
    Michigan State University, East Lansing, Michigan 48824, U.S.A.

SCHLÖSSER, Prof. Eckart. Institut für Phytopathologie, Justus
    Liebig-Universität, Ludwigstrasse 23, 6300 Giessen, W. Germany.

SCHÖNBECK, Prof. Fritz. Institut für Pflanzenkrankheiten und
    Pflanzenschutz der Technischen Universitat Hannover, W. Germany.

SEQUEIRA, Prof. Luis. Department of Plant Pathology, University of
    Wisconsin, Madison, Wisconsin 53706, U.S.A.

SISTO, Dr. Daniele. Istituto di Patologia vegetale dell' Università,
    Via G. Amendola 165/A, Bari, Italy.

SIVAK, Dr. Bela. Department of Botany, University of British
    Columbia, Vancouver, British Columbia, V6T 1W5, Canada.

SMITH, Dr. Ian M. Department of Botany, Imperial College of Science
    and Technology, Prince Consort Road, London SW7 2BB, U.K.

SNYDER, Dr. William C. Prof. Em. University of California, 1926
    Napa Avenue, Berkeley, California 94707, U.S.A.

SOLHEIM, Dr. Bjørn. Institute of Biology and Geology, University of
    Trømso, N-9001 Trømso, Norway.

SPARAPANO, Dr. Lorenzo. Istituto di Patologia vegetale dell'
    Università, Via Amendola 165/A, 70126 Bari, Italy.

STAPLES, Dr. Richard C. Boyce Thompson Institute, 1086 North
    Broadway, Yonkers, New York 10701, U.S.A.

STAUB, Dr. Theodor. CIBA-GEIGY AG., Division of Agricultural
    Chemicals, 4002 Basel, Switzerland.

TANI, Dr. Toshikazu. Faculty of Agriculture, Kagawa University,
    Miki-Tyo, Kagawa - Ken, 761-07, Japan.

TJAMOS, Dr. Eleftherios C. Benaki Phytopathological Institute,
    Kifissia, Athens, Greece.

TOUZÉ, Prof. André. Laboratoire de Physiologie végétale et
    pathologique, Université Paul Sabatier, 118 Route de Narbonne,
    31077 Toulouse Cedex, France.

WARD, Dr. Edmund W.B. Canada Department of Agriculture, Research
    Institute, University Sub Post Office, London, Ontario N6A 3KO,
    Canada.

WELVAERT, Prof. Willy. Laboratoria voor Fytovirologie,
    Rijksuniversiteit, Coupure Links 533, B-9000 Gent, Belgium.

WHEELER, Prof. Harry E.   Department of Plant Pathology, S-305,
    Agricultural Science Center-N, University of Kentucky,
    Lexington, Kentucky 40506, U.S.A.

WOLF, Dr. Gerhard.   Institut für Pflanzenpathologie und
    Pflanzenschutz, Georg-August-Universität, Grisebachstrasse 6,
    34 Göttingen, W. Germany.

WOLSWINKEL, Dr. Pieter.   Botanisch Laboratorium van de Rijksuniver-
    siteite Utrecht, Lange Nieuwstraat 106, Utrecht 2501, The
    Netherlands.

WOOD, Prof. Ronald K.S.   Department of Botany, Imperial College of
    Science and Technology, Prince Consort Road, London, SW7 2BB,
    U.K.

WOODWARD, Mr. Richard.   Department of Plant Pathology, Pennsylvania
    State University, 211 Buckout Laboratory, University Park,
    Pennsylvania 16802, U.S.A.

YODER, Dr. Olen C.   Department of Plant Pathology, Cornell Univers-
    ity, 334 Plant Science Building, Ithaca, New York 14853, U.S.A.

ZENTMYER, Prof. George A.   Department of Plant Pathology, University
    of California, Riverside, California 92502, U.S.A.

Index

345